FUNDAMENTALS OF EXTRAPOLATION MODELING OF INHALED TOXICANTS:
Ozone and Nitrogen Dioxide

EDITORS

Fred J. Miller

U.S. Environmental Protection Agency
Health Effects Research Laboratory
Research Triangle Park, North Carolina

Daniel B. Menzel

Duke Medical Center
Durham, North Carolina

HEMISPHERE PUBLISHING CORPORATION

Washington New York London

DISTRIBUTION OUTSIDE THE UNITED STATES

McGRAW-HILL INTERNATIONAL BOOK COMPANY

Auckland Bogotá Guatemala Hamburg Johannesburg
Lisbon London Madrid Mexico Montreal New Delhi
Panama Paris San Juan São Paulo Singapore
Sydney Tokyo Toronto

NOTICE

Because this volume is appearing as an issue of the *Journal of Toxicology and Environmental Health* as well as a hardbound book, it was necessary to use a numbering system that would incorporate both the continuous journal page numbers and the book page numbers. The pages are double-numbered as follows: page 1 of the book is equivalent to page 181 of the journal. The numbers appear on the pages as journal page number followed by book page number in brackets.

FUNDAMENTALS OF EXTRAPOLATION MODELING OF INHALED TOXICANTS: Ozone and Nitrogen Dioxide

1 2 3 4 5 6 7 8 9 0   S P S P   8 9 8 7 6 5 4

This book was set in Theme by Hemisphere Publishing Corporation. The editors were Elizabeth Dugger, Brenda Brienza, and Roberta Robey; the production supervisor was Miriam Gonzalez and the typesetter was Sandra F. Watts. The Sheridan Press was printer and binder.

Library of Congress Cataloging in Publication Data

Main entry under title:

Fundamentals of extrapolation modeling of inhaled toxicants.

Bibliography: p.
Includes index.
1. Ozone—Toxicology. 2. Nitrogen dioxide—Toxicology.
3. Toxicity testing. I. Miller, Fred J. II. Menzel, Daniel B.
RA1247.096F86 1984   615.9'1   84-10820
ISBN 0-89116-373-5

CONTENTS

PREFACE

The goals of experimental studies of toxicology in animals are to estimate the toxicity in man and to define the mechanisms of toxicity. Extrapolation of data from animal to man has always been implied. However, at present, only a qualitative extrapolation is usually possible. If a pollutant causes an effect in several animal species, most toxicologists would agree that it is likely to cause the same type of effect in man, albeit at an unknown exposure level. Such qualitative extrapolations have been used indirectly by the Environmental Protection Agency (EPA) in the risk assessment and risk management processes in setting national ambient air quality standards. Studies with animals are the only research approach that can provide both cause-effect data under controlled and defined conditions with virtually unlimited exposure conditions (including chronic), and can provide methods enabling a relatively complete description of a very broad array of reversible or irreversible toxic effects.

For regulatory and scientific purposes, it is essential to go beyond the qualitative extrapolation. In risk assessment we must know, with a reasonable degree of certainty, at what concentration humans might experience a toxic effect observed in animals. Such quantitative extrapolation is essential before an informed judgment can be made. The essentiality of such quantitative extrapolations is well recognized.

To perform quantitative extrapolation, three major categories have been identified: dosimetry, species sensitivity, and health effects. Dosimetry refers to determining the amount of pollutants that reach specific sites in animals and humans after exposure to a given concentration of pollutant (e.g., tissue dose resulting from exposure to a given concentration of pollutant). However, even when equivalent doses are delivered to these target sites, there is a possibility that the responses may differ among species. Thus, species sensitivity data are needed, and dosimetry and sensitivity results need to be linked to achieve the quantitative extrapolation of effective pollutant concentrations between animals and humans. Even when extrapolation becomes possible, there must be an adequate understanding of health effects from which to extrapolate. There will always remain some degree of scientific uncertainty, which will require judgments at various stages of the standard-setting process.

In view of the importance of this topic to the mission of EPA, the EPA Health Effects Research Laboratory, Office of Research and Development sponsored the workshop, "Fundamental Approaches to Extrapolation Modeling of Inhaled Toxicants: Ozone and Nitrogen Dioxide" in Washington, D.C. on June 27–29, 1983. The papers in this volume are from the workshop.

Key issues that are discussed include qualitative and quantitative extrapolation of animal toxicological data to humans. Emphasis is on integrating existing knowledge and on identifying areas that need additional research. To provide a focus, information concerning ozone and nitrogen dioxide is used for an overview of relevant topics. Part 1 of this volume provides an introduction to the health effects of both O_3 and NO_2. Part 2 provides a background on respiratory tract structure and function in relationship to mass transport of gases in the lung and their integration into the formulation of dosimetry models. Parts 3, 4, and 5 explore the extrapolations from animals to humans of structure-function changes, metabolic responses to oxidant exposure, and host defense responses against infection.

These papers demonstrate the feasibility of dosimetry modeling and the integration of species sensitivity data based on physical, physiological, anatomical, and toxicological information. Novel methods are presented to estimate regional lung doses from inhaled gases and aerosols. Current concepts of analogy between animals and humans are discussed. While these papers deal with pulmonary responses, the general approach and outline point to new fundamental directions in toxicology, applicable to any other specialization in toxicology. Extrapolation modeling is a new aspect of toxicology, especially relevant to regulatory decisions and a growing methodology in fundamental toxicological research.

Fred J. Miller
Daniel B. Menzel

Part One

INTRODUCTION

OZONE: AN OVERVIEW OF ITS TOXICITY IN MAN AND ANIMALS

Daniel B. Menzel

Departments of Pharmacology and Medicine, and Comprehensive Cancer Center, Duke Medical Center, Durham, North Carolina

Ozone is one of the most toxic and ubiquitous air pollutants. This review focuses on the toxic effects of ozone in animals and on the similarities and disimilarities between the toxic effects in animals and humans. The molecular basis for the toxicity of ozone is discussed, based on the vigorous oxidizing properties of ozone. Despite the existence of anatomical differences between human, subhuman primate, and dog lungs versus common experimental rodent lungs, the anatomical lesion of ozone inhalation occurs at the functionally equivalent site of the junction between the conducting airway and the respiratory region. Ciliated cells of the upper airways and the type 1 cell of the centriacinar region are most affected. Type 2 cell proliferation is a hallmark of ozone toxicity. A wide variety of biochemical and physiological changes have been noted in several animal species and in humans. Considerable evidence for a free-radical-mediated or lipid peroxide-mediated toxicity is evident, especially in the induction of the glutathione peroxidase system and the protective effects of vitamins C and E. Ozone appears to be a weak mutagen and to produce chromosomal abnormalities. Defects in defense against airborne infection are present in animals, which are more susceptible to airborne infection after ozone exposure. Epidemiological studies, however, fail to detect increased respiratory infections in humans due to ozone. Despite the variety of toxic effects, few qualitative differences between species are apparent; rather, quantitative differences do occur. Ozone may thus be an ideal compound for quantitative extrapolation of toxicity from animals to humans.

INTRODUCTION

Ozone is recognized as one of the most ubiquitous urban air pollutants. As a consequence, the biological effects of ozone have been studied extensively in plants, animals, and humans. This review will center principally on the effects of ozone on animals and the correlation between

Much of the work reported from this laboratory appearing in this review was supported by U.S. Environmental Protection Agency (EPA) grant R-809283 and EPA Cooperative Agreement CR-809715 and by National Institutes of Health grants ES-01859 and ES-02916.

I wish to thank my colleagues, Dr. Judith Graham, Dr. Frederick Miller, and Dr. Donald Gardner, and Ms. Elaine Smolko, for their help in developing the concepts brought out in this review. I especially wish to thank Dr. Jose Valentini, Dr. John Overton, Dr. Ted Martonin, and Dr. Robert Wolpert for their help in quantitating some of the ideas presented here; their mathematical modeling efforts have done much to help me differentiate between alternatives.

effects observed in animals and those found in humans. While much has been accomplished to demonstrate the similarities in effects between animals and humans, much of the present data was not accumulated with the idea that quantitative comparisons would be drawn. In many cases only qualitative comparisons can be made. Since recognition and quantitation of the health risk of exposure to ozone is the ultimate objective of both this paper and the research on the toxicity of ozone, an added objective of this review will be to emphasize those areas where appropriate quantitative experiments have been reported or where quantitative extrapolation is theoretically possible.

CHEMISTRY OF OZONE

It is generally agreed that the toxic effects of ozone can be ascribed to its oxidative capacity (Menzel, 1970, 1980). As a powerful oxidant, ozone can react with virtually every class of biological substance, making identification of the critical or essential toxic lesion almost impossible.

Two general hypotheses have been put forth to explain the toxic effects of ozone on a molecular basis: first, that the oxidation of thiols or amino acids in tissue proteins or low-molecular-weight peptides is responsible (Mudd and Freeman, 1977); second, that the oxidation of unsaturated fatty acids (UFA) to fatty acid peroxides (FAP) results in toxicity (Menzel, 1970). A careful study of the biochemical, physiological, and morphological effects of ozone exposure points to cellular membranes as the site of toxicity. Both molecular mechanisms could occur at the membrane. Because membranes are composed of both proteins and lipids, damage to one is difficult to separate from damage to the other or both. Since the metabolism of FAP is linked to the glutathione peroxidase system (Chow and Tappel, 1972), depletion of thiols such as glutathione (GSH) by ozone could enhance the toxicity of FAPs. The two theories are thus not separate, but complementary, and it is most likely that oxidation of UFAs and proteins occurs simultaneously.

While ozone is a general reactant, the relative rates of reaction with different biological substances are remarkably different. Ozone oxidizes ethylene groups 1000 times faster than other compounds. Consequently, attention has been focused on UFAs as targets for ozone reaction. The Criegee mechanism for addition of ozone across unsaturations still remains in favor. In the presence of protonic solvents such as water, peroxides can result directly from the decomposition of the initial ozone adduct. Pryor et al. (1983) have also suggested that free radicals are generated directly by other mechanisms during the ozonation of UFAs. Regardless of the chemical mechanism, once free radicals are generated in membranes the toxic effects are similar. These are a loss of regulation of electrolytes and water (changes in permeability), leakage of essential intracellular enzymes from the cell, and inhibition of metabolic chains imbedded in mitochondrial or microsomal matrices (Menzel, 1976). If damage is sufficiently severe, the cell cannot recover and dies. Necrosis is commonly reported in the lungs of animals exposed to ozone.

Perhaps the most compelling evidence for UFA peroxidation *in vivo* is the protective effect of dietary vitamin E on ozone toxicity (Roehm et al., 1971a,b, 1972; Chow and Tappel, 1972; Fletcher and Tappel, 1973; Donovan et al., 1977; Sato et al., 1976a; Chow and Kaneko, 1979; Plopper et al., 1979; Chow et al., 1981; Chow, 1983; Mustafa et al., 1983). Vitamin E as a phenolic antioxidant easily inhibits the peroxidation of UFAs initiated *in vitro* by ozone (Roehm et al., 1971b). Indirect evidence that vitamin E protects against ozone-initiated peroxidation of UFAs *in vivo* is presented by the studies of Dumelin et al. (1978a, b), who found that the exhaled ethane and pentane in the breath of macaque monkeys, rats, and mice was decreased when vitamin E was provided in their diets. Ethane and pentane are thought to be produced by peroxidation of UFAs initiated by ozone inhalation. Changes in lung fatty acid composition due to peroxidation are not easily intrepreted because of the complexity of lung UFA composition and difficulties in quantitative analysis. Direct evidence of peroxidation of UFAs in human subjects has, however, been lacking.

Amino acids are oxidized by ozone (Mudd et al., 1969; Mudd and Freeman, 1977; Previero et al., 1964). Cysteine, methionine, and tryptophan are the three most easily oxidized amino acids. Human alpha-1-proteinase inhibitor is destroyed by ozonolysis *in vitro*, presumably by oxidation of methionine, tyrosine, and tryptophan amino acid residues in the molecule (Johnson, 1980). Johnson has proposed that the oxidation of alpha-1-proteinase inhibitor could lead to emphysema and explain much of the long-term toxicity of ozone. A large number of enzymes are inhibited immediately after ozone exposure to 1–4 ppm for 1 h or more (DeLucia et al., 1972; Mountain, 1963; Kyei-Aboagye et al., 1973; Holtzman et al., 1979; Menzel et al., 1976). Lung mitochondria lose their respiratory control following ozone exposure (Mustafa et al., 1973). DeLucia et al. (1975) also reported the depletion of GSH and the formation of mixed disulfides in the lungs of ozone-exposed rats.

In contrast to the inhibition of enzymes by ozone exposure, the glutathione peroxidase system consisting of glucose-6-phosphate dehydrogenase, glutathione reductase, and glutathione peroxidase is increased in the lungs of rats exposed to ozone (Chow and Tappel, 1972, 1973). These authors favor the concept that ozone increases UFA peroxidation, which in turn induces the glutathione peroxidase system as a protective mechanism.

MORPHOLOGICAL EFFECTS OF OZONE

Morphological studies of the effects of ozone have been confined to the respiratory tract. Morphological studies following controlled human exposure to ozone have not been reported and are not likely to be reported due to ethical considerations. Damage to all parts of the respiratory tract occurs in animals, with location and intensity dependent on the ozone concentration. At low concentrations of ozone (below 1 ppm), damage is principally confined to the junction between the alveoli and the conducting airways. Dogs, monkeys, and humans have respiratory and nonrespiratory bronchioles, while the common experimental animals—mice, rats, and guinea pigs—have only nonrespiratory bronchioles. The location of the ozone lesion is thus different depending on the species examined (Plopper et al., 1979; Castleman et al., 1977, 1980; Dungworth et al., 1975a; Eustis et al., 1981). In both types of lungs, the effects of ozone have been found at concentrations as low as 0.5 ppm for times as short as 2 h of exposure (Stephens et al., 1974). The type 1 pneumocytes are the most affected cell types of this region (Stephens et al., 1974; Evans et al., 1976a; Castleman et al., 1980; Eustis et al., 1981). Type 1 cell death reaches a maximum at 12 h after exposure of 0.8 ppm, for example (Castleman et al., 1980), paralleling the time delay observed in measurements of respiratory function (see the section on lung function). In a number of species, ozone damage is clearly limited to the centriacinar region and does not extend to peripheral alveoli (Stephens et al., 1973a, b, 1974; Schwartz et al., 1976; Mellick et al., 1977). Clara cell damage is also observed even on prolonged exposure for 90 or 180 d to 0.5 or 0.8 ppm for 8 h/d (Boorman et al., 1980; Moore and Schwartz, 1981).

In the upper airways, ciliated cells appear to be the most sensitive cell type, being damaged by 0.2–0.5 ppm for 8 or 24 h for 7 d in rats (Schwartz et al., 1976), bonnet monkeys (Castleman et al., 1977), rhesus monkeys (Dungworth et al., 1975b; Mellick et al., 1977), and mice (Ibrahim et al., 1980). Uniform damage is not always noted; most often, reports are of patchy damage. Damage is present in both trachea and bronchi.

Following and during continued exposures, type 2 cells begin to proliferate, and their proliferation is a hallmark of ozone injury regardless of species. In rats, DNA synthesis in type 2 cells was reported as early as 4 h after exposure to 0.2 ppm (Stephens et al., 1974) and reached a maximum at 2 d following continuous exposure to either 0.35 or 0.5 ppm (Evans et al., 1976a, b) or 0.8 ppm (Boorman et al., 1980). While type 2 cells proliferated in the ozone-exposed lung, complete maturation of type 2 cells to type 1 cells did not occur, even after exposures as long as 180 d (Moore and Schwartz, 1981). In the normal progression, type 2 cells would have matured into type 1 cells. Continued ozone exposure inhibits both ciliagenesis and type 2 cell maturation.

Inflammation occurs in all species so far examined. The inflammatory response is seen as early as 4 h after exposure to 0.8 ppm in monkeys (Castleman et al., 1980). In both rats and monkeys, inflammation persists on continued exposure, although at reduced levels. Quantitative estimates of the degree of inflammation are, however, lacking at present. The contribution of the inflammatory response to subsequent long-term features of ozone toxicity has not been studied in detail, even though techniques are available.

While some delay has been observed between the exposure to ozone and maximum manifestation of morphological changes, the time required for ozone exposure to result in morphological changes is 2 h on exposure to 0.2 or 0.5 ppm in rats (Stephens et al., 1973b) or on exposure by endotracheal tube in cats to 0.26, 0.5, or 1 ppm for 4.7 or 6.6 h (Boatman et al., 1974). These data are in agreement with the molecular theories above that ozone acts rapidly to oxidize some critical tissue component. Little difference in severity was noted by Schwartz et al. (1976) between exposure to 8 h/d or 24 h/d when the exposure was continued for 7 d. Quantitative studies of different exposure regimens are lacking, but are technically feasible.

Morphological studies of vitamin E deficient or supplemented rats have been undertaken to correlate the biochemical findings with morphological alterations (Plopper et al., 1979; Chow et al., 1981; Schwartz et al., 1976; Sato et al., 1976a, b, 1978, 1980). Scanning and transmission electron microscopy were clearly shown to be better than light microscopy for visualization of the ozone lesions. Vitamin E deficient or marginally supplemented rats were damaged more than highly supplemented rats, supporting mortality (Donovan et al., 1977) and biochemical studies that vitamin E is protective. Despite the presence of vitamin E in the diets of these animals, the morphological lesion of ozone exposure was unchanged. Vitamin E thus alters the rate and extent of toxicity, but not the lesion itself.

When comparisons are made at the analogous anatomical site, the morphological effects of ozone on the lungs of a number of species of animals are remarkably similar. Despite the inherent differences in anatomy between most experimental animals and humans, the junction between the conducting airways and the gas exchange region is most affected by ozone exposure in animals and probably will be the principal site affected in humans.

BIOCHEMICALLY DETECTED EFFECTS OF OZONE

Induction of the glutathione peroxidase system is one of the most sensitive indices of ozone exposure measured biochemically. Increases in the glutathione peroxidase system have been reported on exposure of rats to 0.1 ppm for 7 d (Chow et al., 1981). The dietary vitamin E fed to the rats had major effects on the induction of the glutathione peroxidase system (Chow and Tappel, 1972, 1973). A dietary level of about 66–110 mg vitamin E/kg diet was necessary to prevent induction of this system with an exposure of 0.7 ppm for 8 h/d for 5 d (Chow et al., 1981). Mustafa et al. (1983) summarized a number of experiments to note that oxygen consumption, monoamine oxidation, glucose utilization, NADPH formation, GSH reductase, disulfide reductase, nonprotein sulfhydryl content, GSH peroxidase, and superoxide dismutase all increased, by as much as 50%, but mostly by 30–40% of control lung values on ozone exposure. A concentration-related increase was found for the succinate oxidase, succinate-cytochrome c reductase, NADPH-cytochrome c reductase, glucose-6-phosphate dehydrogenase, and glutathione peroxidase content following continuous exposure to ozone at concentrations between 0.2 and 0.8 ppm for 7 d. Chow et al. (1974) noted similar increases with continuous or intermittent exposures to concentrations from 0.2 to 0.8 ppm for 7 d. Induction of the glutathione peroxidase system has also been reported for monkeys exposed to 0.2 ppm for 8 h/d for 7 d (Mustafa and Lee, 1976). Other species have not been investigated.

The increases in the glutathione peroxidase system are thought to be protective. The vitamin E content of the diet may not regulate the glutathione peroxidase system directly, but indirectly, by reducing the amount of FAPs formed on exposure to ozone. Since the cellular localization of the glutathione peroxidase system is not known, major changes in cell populations following ozone exposure, such as increase type 2 cell number, could also be responsible for the increases in glutathione peroxidase system content.

Ozone exposure inhibits the phenol red active transport mechanism of the lung in a concentration-related manner (Williams et al., 1980). The non-carrier-mediated diffusion of phenol red from the lung increased simultaneously with the inhibition of the carrier-mediated transport. Diffusion of plasma proteins into the airways following ozone exposure has been reported in dogs (Reasor et al., 1979), rats (Alpert et al., 1971), and guinea pigs (Hu et al., 1982). The protein content of lung lavage fluid from vitamin C deficient guinea pigs was much greater than from those receiving adequate vitamin C, and was proportional to the ozone concentration to which the guinea pigs had been exposed in the concentration range of 0.1 to 0.75 ppm for 3 h. Since only guinea pigs, primates, and the Dalmatian dog require vitamin C, analogous experiments in rats are not possible. Vitamin C (ascorbic acid) functions as an antioxidant in the aqueous phase, contrasted with the lipid-phase antioxidant function of vitamin E. Through this antioxidant function, vitamin C might protect lung cells from oxidative injury by ozone.

Leakage of lactate dehydrogenase into the plasma has been reported as another index of increased lung permeability following ozone exposure (Chow et al., 1977). Lysosomal enzymes were increased in rats (Dillard et al., 1972; Castleman et al., 1973) and probably represent part of the inflammatory process seen throughout periods of ozone exposure.

Inhibition of several enzymes in the lung has also been reported.

Gordon et al. (1981) studied the inhibition of guinea pig lung, brain, and diaphragm acetylcholinesterase in an attempt to explain the bronchial hyperreactivity observed in ozone-exposed guinea pigs and humans. Acetylcholinesterase activity decreased by 17% on exposure to 0.1 ppm and by 16% on exposure to 0.8 ppm. The lack of a clear concentration-response relationship weakens the argument that ozone inhibition of acetylcholinesterase is directly responsible for the much greater increase in reactivity of the airways following ozone exposure. A number of other enzymes and enzyme systems have been investigated, but few concentration-response or species comparisons have been done.

Based on the limited comparisons between rats and monkeys, ozone exposure is likely to cause major alterations in intermediary metabolism and the availability of reducing equivalents in the form of NADPH within lung cells. Because the glutathione peroxidase system has a high turnover, the increased availability of NADPH may still not be adequate to meet all of the demands of the other enzyme systems of the lung. In the liver, for example, the glutathione peroxidase activity may drain away all of the available NADPH when driven to its maximum velocity. NADPH and NADH are also compartmentalized in cells, and although the total NADPH may be increased, the amount available for vital functions may not be. The quantitative comparison of intermediary and energy metabolism in several species has no technological limitations. Since these functions are fundamental and are present in all mammalian species, direct comparisons between species should be possible.

A scattering of extrapulmonary effects, which are difficult to integrate into a single concept, has been reported. An important series of experiments has been conducted by Gardner et al. (1974), Graham (1979), and Graham et al. (1981, 1982) investigating the effects of ozone inhalation of xenobiotic metabolism. Pentobarbital narcosis is terminated by hepatic metabolism. The duration of pentobarbital sleeping time serves as an indirect bioassay of pentobarbital hepatic metabolism. In rats, hamsters, and three strains of mice, females had longer sleeping times following exposure to 1 ppm ozone for 5 h. A lesser response was observed in male hamsters, with no response observed in male rats or mice. When the ozone concentration was varied from 0.1 to 5 ppm, the number of daily 3-h exposures required to elongate the sleeping time was greater for lower concentrations. Subsequent experiments demonstrated that the metabolism of pentobarbital to its hydroxylated metabolites was decreased by ozone exposure. These experiments are the first to indicate a sex difference in ozone effect. While sex differences in xenobiotic metabolism are known, the ozone effect on a xenobiotic metabolic pathway is also unique. The nature of the interaction with the xenobiotic metabolizing system is unknown. While ozone itself is unlikely to reach the liver through the blood, ozone-produced peroxides and other oxygenated products may have a sufficient lifetime to reach the liver.

LUNG FUNCTION

Changes in lung function following ozone exposure have been studied in mice, rats, guinea pigs, sheep, rabbits, cats, bonnet monkeys, dogs, and humans. Only short-term exposures of up to 6 h have been conducted in both humans and animals. In all species, rapid shallow breathing develops on exposure to concentrations as low as 0.26 ppm ozone for durations as short as 2 h (Amdur et al., 1978; Yokoyama, 1969; Watanabe et al., 1973; Folinsbee et al., 1978) in guinea pigs, cats, rabbits, and humans, for example. The onset of this pattern of breathing is rapid, and the abnormal breathing pattern usually disappears within 30 min of cessation of exposure. These effects are thought to be due to stimulation of sensory receptors in the airways that, through the vagal reflex, produce a reflex bronchoconstriction. In humans, exercise produces substantial increases in the pulmonary effects of ozone (Hazucha et al., 1973; Kerr et al., 1975; Folinsbee et al., 1978). With increased work load, effects at concentrations as low as 0.3 ppm for 15-min periods of exercise have been observed (Folinsbee et al., 1978). While the effect of ozone on the breathing pattern is immediate, other changes in pulmonary function from a single short-term exposure to ozone reach a maximum at 1 d after exposure. These delayed effects are associated with premature airway closures (Inoue et al., 1979).

The effects of short-term ozone exposure may be mediated through damage to the respiratory epithelium, which sensitizes the underlying nerves to mechanical and chemical stimulation through increased permeability. Molecules as large as ovalbumin can reach immunologic receptors in the lungs of mice exposed to 0.5 or 0.8 ppm ozone continuously for 3–5 d (Osebold et al., 1980). Increased sensitivity to histamine or cholinomimetic drugs by aerosol or injection has been noted in several species. Easton and Murphy (1967) showed that the lethal dose of histamine in guinea pigs was much lower in those animals exposed to concentrations as low as 0.5 or 1 ppm ozone for 2 h. The pulmonary resistance due to subcutaneous injection of histamine also increased in guinea pigs exposed to 0.8 ppm ozone for 1 h (Gordon and Amdur, 1980). Similarly, dogs exposed to 0.7 ppm ozone for 2 h had greater changes in pulmonary resistance following histamine aerosol inhalation (Lee et al., 1977). Sheep exposed to 0.5 ppm ozone for 2 h experienced increased pulmonary resistance from the cholinomimetic drug carbachol (Abraham et al., 1980). Humans experienced ozone-related increases in pulmonary resistance following histamine (Holtzman et al., 1979; Dime et al., 1981), methacholine (Holtzman et al., 1979), or acetylcholine exposures (Koning et al., 1980). Since the exposure protocol and subjects varied considerably among the investigators, one can only estimate the minimum effective concentration of ozone in humans as between 0.2 to 0.4 ppm ozone for less than 30 min and generally 15 min. The similarities in pharmacological effects of ozone among these species is qualitatively the same, but the experiments are so different that direct scaling between them is not possible.

The effects of long-term exposure to ozone, of course, cannot be directly evaluated in humans. Age has a pronounced effect on susceptibility to ozone in rats. Neonatal rats exposed to 0.12 or 0.25 ppm ozone for 12 h/d for 7 d/wk for 7 wk had decreased peak inspiratory flows and increased lung volumes (Raub et al., 1983). Similar increases in lung volume were found in adult rats exposed to 0.8 ppm ozone 6 h/d for 5 d/wk for 62 exposure d (Costa et al., 1983).

OZONE EFFECTS ON THE SUSCEPTIBILITY TO INFECTION

To protect humans and other mammals from inhaled infectious agents, the mammalian lung possesses an array of defense mechanisms. One or more of these defense mechanisms could be impaired by ozone or other inhaled toxicants. Epidemiological studies suggest higher incidences of respiratory infections in areas of urban pollution. Considerable effort has been expended on studies of the potential effects of ozone on such defense mechanisms, in hopes of understanding the underlying mechanisms and predicting concentration-response relationships.

Inhaled bacteria and viruses are removed from the lung by two mechanisms. Ciliated cells of the bronchi and tracheae propel mucus carrying entrapped particles and infectious agents upward from the distal regions of the lung to be swallowed and excreted. Alveolar macrophages (AM) engulf, kill, and remove infectious agents from the respiratory regions of the lung by migrating with the particle to either the mucus lining layer to be propelled upwards or the lymphatic system to be drained from the lung.

Ciliated cells are damaged by ozone inhalation, as demonstrated by major morphological changes in these cells, including necrosis and sloughing, or by the shortening of the cilia in cells remaining attached to the bronchi. Ciliated cell damage should result in decreased transport of viable and nonviable particles from the lung. Rats exposed to 0.4 or 0.8 ppm for times as short as 4 h have decreased short-term clearance of particles from the lung (Phalen et al., 1980; Frager et al., 1979; Kenoyer et al., 1981). Short-term clearance is mostly due to mucus removal of particles, and the decreased short-term clearance is an anticipated functional result predicted from morphological observations. The mucous glycoprotein production of the bronchi is also altered by ozone exposure, but not in a simple manner. Mucus glycoprotein biosynthesis, as measured *ex vivo* in cultured tracheal explants from exposed rats, was inhibited by short-term exposure to ozone (0.8 ppm continuously for 3–5 d) (Last and Cross, 1978; Last and Kaizu, 1980; Last et al., 1977). Glycoprotein synthesis and secretion recovered to control values after 5–10 d of exposure and increased to greater than control values after 10 d of exposure. Glycoprotein synthesis by tracheal explants from bonnet monkeys exposed to 0.5 or 0.8 ppm 8 h/d for 7 d was not affected. Since no other reports have appeared, it is not clear whether the lack of response by monkeys is specific to that species or the ozone inhibition is specific to the rat. There are no technical barriers to the use of other species or concentration-response studies.

The long-term clearance of particles from the lung is a function carried out by the alveolar macrophage (AM). AMs are damaged by ozone inhalation. Most studies have been with rabbits where a 3-h exposure to 0.5 ppm reduces phagocytosis (Coffin and Gardner, 1972; Coffin et al., 1968a), disrupts AM membranes (Dowell et al., 1970; Goldstein et al., 1977; Hadley et al., 1977), reduces AM lysosomal hydroxylase activity (Hurst et al., 1970; Hurst and Coffin, 1971), and reduces AM numbers (Coffin et al., 1968b; Alpert et al., 1971). The bacteriostatic capacity of AMs may also be impaired, since the superoxide production capacity is also inhibited by ozone exposure (Witz et al., 1983; Amoruso et al., 1981).

An integrated measure of both of these mechanisms of viable organism removal and the more general measures of host defense against airborne infections can be determined by the carefully controlled airborne infection of animals following ozone exposure. This method, known as the infectivity model, has recently been described in detail by Gardner (1982). A number of different animals and infectious agents, both bacterial and viral, have been used to illustrate the increased susceptibility of animals to airborne infections following very modest concentrations and time of exposure. To date it is one of the most sensitive measures of ozone toxicity. Mice exposed to 0.08–0.1 ppm for 3 h are significantly more susceptible to infection than mice exposed to clean air (Coffin et al., 1968a; Ehrlich et al., 1977; Miller et al., 1978; Aranyi et al., 1983).

Immunological effects of ozone have only been scantily studied in mice and guinea pigs. Depression of the cell-mediated immune response has been measured at high ozone concentrations in the guinea pig (Thomas et al., 1981). The systemic immune response is depressed in mice, as measured by the blastogenic response of splenic lymphocyte to mitogens and allomitogens (Aranyi et al., 1983). The mice were exposed to 0.1 ppm for 5 h/d, 5 d/wk for 90 d.

CHROMOSOMAL AND MUTATIONAL EFFECTS

The putative chromosomal and mutagenic effects of ozone are most controversial. In both cases, very little comparative data is available, and the concentrations of ozone used were generally outside of ranges considered relevant for other physiological or biochemical parameters.

Ozone reacts *in vitro* with both DNA (Christensen and Giese, 1954; Prat et al., 1968; Shinriki et al., 1981) and RNA (Shinriki et al., 1981).

Thymidine and guanine are the two bases most easily destroyed by ozone in DNA and RNA, respectively. Destruction of the other bases also occurred, but to a lesser extent. Single-strand breaks in DNA appear responsible for the genotoxicity of ozone in *E. coli* (Hamelin and Chung, 1974), and polymerase I deficient strains are more sensitive (Hamelin and Chung, 1975a). Strains with other defects in DNA repair are not more sensitive to ozone (Hamelin et al., 1977). These studies were undertaken at high concentrations of ozone, where cytotoxicity through other mechanisms may have influenced interpretation of the results.

Aberrations in the peripheral blood lymphocytes from Chinese hamsters exposed to 0.2 ppm for 5 h were found by Zelac et al. (1971a,b). Tice et al. (1978) attempted to repeat these experiments, but introduced sufficient variables in the protocol to make the two studies independent. Tice et al. found no differences between clean air-exposed and ozone-exposed (0.43 ppm for 5 h) hamsters or mice in either sister-chromatid exchange rates or chromosomal abnormalities. Because Zelac et al. used polkweed mitogen, while Tice et al. used phytohemagglutinin (PHA) mitogen, to induce DNA replication, the two groups measured the response of different populations of lymphocytes. Zelac et al. measured principally B-lymphocytes, while Tice et al. measured T-lymphocytes. B-Lymphocytes may be more sensitive to ozone, since Savino et al. (1978) found rosette formation depressed in the B- but not the T-lymphocytes from human subjects exposed to 0.4 ppm for 4 h. Hadley et al. (1977) found perturbations in surface receptors of AM from ozone-exposed animals, suggesting that surface membrane effects occur with ozone. Ozone may inhibit or inactivate the PHA receptors (Peterson et al., 1978; Gooch et al., 1976). Without a means to induce a cycling state in ozone-damaged lymphocytes, ozone damage cannot be observed. In an extensive study of the cytogenetic effects of ozone on human subjects, McKinzie et al. (1977) and McKinzie (1982) studied chromosomal and chromatid aberrations and sister chromatid exchange rates in subjects exposed to 0.4 ppm for 4 h. No statistically significant differences were found in any measurement of cytogenetic effect. No changes were found in sister-chromatid exchange rates in the circulating lymphocytes of male and female subjects exposed to 0.5 ppm for 2 h (Guerrero et al., 1979). A dose-related increase was found, however, with human diploid fetal lung cells (WI-38) exposed *in vitro* to ozone at concentrations up to 1 ppm for 1 h.

Mutational effects of ozone have been studied in *E. coli* (Hamelin and Chung, 1975a,b), where mutations to a number of growth-factor auxotrophic states were found. Dominant lethal mutations were also induced in *Drosophila virilis* (Erdman and Hernandez, 1982). Both base-substitution and frame-shift mutations were found in *Saccharomyces cerevisae* (Dubeau and Chung, 1979, 1982). The mutagenic properties of ozone have therefore been demonstrated in prokaryotic and eukaryotic cells. The potency appears very weak. Studies with human and animal cells are needed before the mutational properties of ozone can be properly assessed.

CARDIOVASCULAR EFFECTS

Few cardiovascular effects have been ascribed to ozone inhalation (Brinkman et al., 1964; Bloch et al., 1971; Revis et al., 1981). Any effects may be secondary to pulmonary damage.

HEMATOLOGICAL AND SERUM CHEMISTRY EFFECTS

Examination of blood is a method easily adapted to studies in several species, including humans. Most ozone studies have been with rodents, even though differences may exist between rodent red blood cells (RBCs) and human RBCs. Calabrese and Moore (1980) and Moore and Schwartz (1981) point to a difference in the lack of ascorbic acid synthesis and the relatively low level of glucose-6-phosphate dehydrogenase (G-6-PD) in humans compared to active ascorbic acid synthesis and high levels of G-6-PD in mice and rats. G-6-PD is essential for the function of the GSH peroxidase system in the RBC to decompose FAP or hydrogen peroxide formed by ozone-initiated UFA peroxidation. This group has suggested the use of the C57L/J strain of mice and the Dorset sheep as better animal models for hematological studies, since these species have levels of G-6-PD closer to those of humans, especially the G-6-PD deficient patient. However, the RBCs of Dorset sheep are no more sensitive to ozone than normal human RBCs, although the G-6-PD levels are very low.

Ozone exposure does produce a number of hematological and serum chemistry changes both in rodents and in humans. Most of the hematological changes appear to be linked to a decrease in RBC GSH content (Menzel et al., 1975; Buckley et al., 1975; Posin et al., 1979; Linn et al., 1978) at concentrations of 0.2 ppm for 30–60 min in humans, 0.5 ppm for 2.75 h in sheep, or 0.5 ppm continuously for 5 d in mice and rats. Heinz bodies, disulfide cross-linked methemoglobin complexes attached to the inner RBC membrane, were detected in mice exposed to ozone (Menzel et al., 1975). RBC acetylcholinesterase inhibition was found in mouse (Goldstein et al., 1968), human (Buckley et al., 1975), and squirrel monkey RBCs (Clark et al., 1978) at concentrations of 0.4–0.75 ppm and times as short as 2.75 h in humans or 4 h/d for 4 d in monkeys. Loss of RBC acetylcholinesterase could be mediated either by membrane peroxidation or by loss of acetylcholinesterase thiol groups at the active site. Dorsey et al. (1983) observed that the deformability of CD-1 mouse RBCs decreased on exposure to 0.7 and 1 ppm ozone for 4 h. Deformability also decreased at 0.3 ppm, but was not statistically significant. These data also support the concept of membrane damage to circulating RBCs. While it is theoretically possible that differences in ozone sensitivity may exist between human and rodent RBCs because of differences in G-6-PD, experiments so far reported do not support this contention. Rather, rodent and human RBCs appear closely similar in their response to ozone exposure.

A number of serum changes have been reported in animals and humans, but none seems specific to lung injury or extrapulmonary effects of ozone.

REPRODUCTIVE AND TERATOLOGICAL EFFECTS

Only experiments with mice and rats have been reported examining the reproductive and teratological effects of ozone. Perinatal exposure of mice to 0.2 ppm reduced infant survival and increased the incidence of unlimited incisor growth (Brinkman et al., 1964). Midgestational exposure of rats to 1.49 ppm resulted in a higher embryo resorption (Kavlock et al., 1979), while late gestational exposure to 1 ppm slowed neonatal growth rates and delayed early reflex development and the onset of grooming and rearing behavior (Kavlock et al., 1980). Comparative studies with other mammals have not been reported. The implications for human reproduction and development are difficult to judge at present.

CONCLUSION

While the precise molecular mechanism of ozone toxicity remains obscure, the strong oxidative properties of ozone point to oxidation with or without the production of free radicals as the underlying molecular mechanism. Support for free-radical mechanisms in ozone toxicity are found in the impressive protective effects of the lipid antioxidant vitamin E in rats, mice, and monkeys, and of the aqueous antioxidant, vitamin C, in guinea pigs. The reaction of ozone with unsaturated fatty acids is thought to be more rapid than with other tissue constituents. The glutathione peroxidase system, a protective enzymatic mechanism for destroying organic peroxides, is induced in mice, rats, and monkeys by ozone inhalation. Induction of the glutathione peroxidase system may be a measure of the fatty acid peroxides produced by ozone. Evidence for protection by vitamins E and C in humans has not, however, been obtained. The lack of an effect in humans may be due to the chronic, rather than the acute, exposures needed to demonstrate protection by either vitamin C or E.

Ozone chemical reactions are rapid, compared to the time required for breathing and demonstration of the pathophysiological effects of ozone. In both animals and humans, shallow rapid breathing occurs on ozone exposure and, in short-term exposures, disappears within 30 min on cessation of exposure. In both animals and humans, a delay of 12–24 h occurs between a single short-term exposure and maximum expression of toxicity, as measured by more complex biochemical or physiological indices. In short-term continuous exposures, maximum anatomical lesions in animals occur after 3 d of exposure. The time sequence beyond 7 d of continuous exposure is not well documented, but after intervals of 14–30 d, and up to 180 d, of exposure, the anatomical lesions persist and become more pronounced. Inflammation persists, although at a reduced intensity, after about 7 d and throughout continued ozone exposure.

A characteristic lesion is produced by ozone in the centriacinar region of the lung. Because common experimental rodents have only nonrespiratory bronchioles, while humans, subhuman primates, and dogs have both respiratory and nonrespiratory bronchioles, the ozone lesion may be located at different anatomical levels. However, it is always at the junction between the conducting airway and the respiratory region. This junction is also the junction between mucus and surfactant. The protection provided by either mucus or surfactant is minimal and may be one reason for the susceptibility of this region. Increasing the concentration of inhaled ozone causes a progression of the lesion to more distal sites. No clear-cut concentration-response relationship has been reported between the intensity of the centriacinar lesion and the time or concentration of ozone to which an animal is exposed. Periods of exposure of as short as 2 h to 0.2 ppm produce observable changes. Intermittent and continuous exposures produce anatomical alterations of comparable intensity, but only limited concentrations and times of exposure have been investigated. The subjective nature of conventional anatomical observation also tends to obscure quantitative concentration-response relations. The use of quantitative morphometry should resolve this issue.

A characteristic response of the centriacinar region to ozone exposure in all species so far examined is a proliferation of type 2 cells to replace dead or injured type 1 cells. Proliferation of type 2 cells is rapid, beginning in rats after as short a time of exposure as 4 h to 0.35 or 0.5 ppm.

In the upper airways, the ciliated cells are the most susceptible to ozone damage and are sloughed off in patches. Regions injured are apparently no different from those that are not. Ciliagensis also is impaired, with ciliated cells having flatter profiles with shorter cilia after ozone exposure. Clara cells at the junction of the airways and alveoli are affected in a similar manner. On continued exposure to ozone, type 2 cells, which are proliferating to replace type 1 cells, fail to mature completely to normally appearing type 1 cells. Mucus glycoprotein synthesis measured *ex vivo* in rats is decreased on short-term exposure (3–5 d), returns to normal (5–10 d), and is stimulated on longer term exposure. Mucus glycoprotein synthesis is unaffected in monkeys exposed to the same ozone concentration (0.8 ppm). Species differences may exist.

These changes in mucus secretion and cilia function are reflected in decreased short-term clearance of particles from the lung.

Ozone inhalation also impairs mechanisms of defense against inhaled infectious agents aside from mucociliary clearance. Alveolar macrophages isolated from exposed animals have major defects in systems thought to be important in their bacteriostatic and phagocytic capacity. Increased susceptibility to airborne infections is one of the most sensitive indices of ozone toxicity. Mice exposed to 0.08–0.1 ppm for 3 h are more sensitive than mice exposed to clean air. Since these ozone concentrations are within the ambient levels found in many parts of the country, experiments in this range are difficult to conduct in animal facilities not equipped with sophisticated exposure systems. Other species are also more susceptible to airborne infections, but generally at greater concentrations. A variety of species and infectious agents have been used. Epidemiological studies do not, however, support an increased susceptibility to airborne infections in humans from ambient ozone concentrations.

Ozone also increases the permeability of the lung to both large and small molecules. The active transport of phenol red is inhibited, and diffusional transport predominates. Proteins diffuse into the airways from the blood, especially in guinea pigs made vitamin C deficient prior to ozone exposure. Inhaled antigens, such as ovalbumin, penetrate to receptors more easily. Histamine and cholinomimetic drugs acting on bronchial smooth muscles in humans and animals are more potent, suggesting greater penetration to receptors. The potentiation of histamine and cholinomimetic drugs is quite similar in all species so far studied, making potential extrapolation attractive.

Chromosomal and mutational effects of ozone are controversial. In hamsters and rats, ozone produces chromosomal alterations in circulating lymphocytes. B-Lymphocytes appear most affected. Human lymphocytes may be affected also, but the concentrations used in single exposures of human subjects may have been insufficient to detect alterations. Evidence of weak mutagenic effects of ozone has been found in microorganisms, but the applicability of these effects to human or animal cells remains to be elucidated. Clearly, this is an important area of comparative study needing work.

The issue of extrapulmonary effects of ozone is now in need of investigation. Because of the reactivity of ozone, it was not intuitively obvious that ozone would ever reach the circulation. More recent studies in mice, rats, and hamsters have shown that hepatic metabolism of xenobiotic compounds is impaired by ozone inhalation. The inhibition of xenobiotic metabolism is sex specific, being evident in females only. Alterations in red blood cell morphology, deformability, and enzymatic activity, as well as the chromosomal effects in circulating lymphocytes, now indicate either a direct or indirect effect of ozone exposure that can ultimately involve all of the organ systems of the body. Central nervous system effects have also been demonstrated in humans, where decreases in visual acuity and night vision have been found. Auto accident frequencies have been associated with increased ozone concentrations. Decreased reproduction has also been found in rats and mice. Extrapulmonary effects of ozone may be due to the release of oxygenated lipids mimicking prostanoid activities. Hepatic cytochrome P-450 enzymes, for example, degrade fatty acid peroxides to fatty acid alcohols, but are inhibited in the process. Ozonation of unsaturated fatty acids produces peroxides having activity on both fundic and uterine smooth muscle preparations.

A remarkable array of toxic effects has been ascribed to ozone inhalation. Most likely, this array of effects is due to the vigorous oxidation of a number of compounds. Since ozone does oxidize some compounds more rapidly than others, a quantitative balance between different sites may influence the toxic outcome. In the lung, immediate membrane effects are the dominant feature, while in the RBC, membrane and intracellular effects are measurable. In extrapulmonary organs, metabolic or hormonal effects are more likely. When corrections are made for inherent anatomical differences between the lungs of different species, the pulmonary lesion of ozone toxicity is remarkably similar between species. Because of the vigorous nature of its chemical reactivity, ozone may be an ideal chemical to study in the quantitative extrapolation of toxicity from animals to humans.

REFERENCES

Abraham, W. M., Januszkiewicz, A. J., Mingle, M., Welker, M., Wanner, A., and Sackner, M. A. 1980. Sensitivity of bronchoprovocation and tracheal mucous velocity in detecting airway responses to O_3. *J. Appl. Physiol. Respirat. Environ. Exercise Physiol.* 48:789–793.

Alpert, S. M., Schwartz, B. B., Lee, S. D., and Lewis, T. R. 1971. Alveolar protein accumulation. A sensitive indicator of low level oxidant toxicity. *Arch. Intern. Med.* 128:69–73.

Amdur, M. O., Ugro, V., and Underhill, D. W. 1978. Respiratory response of guinea pigs to ozone alone and with sulfur dioxide. *Am. Ind. Hyg. Assoc. J.* 39:958–961.

Amoruso, M. A., Witz, G., and Goldstein, B. D. 1981. Decreased superoxide anion radical production of rat alveolar macrophages following inhalation of ozone or nitrogen dioxide. *Life Sci.* 28:2215–2221.

Aranyi, C., Vana, S. C., Thomas, P. T., Bradof, J. N., Fenters, J. D., Graham, J. A., and Miller, F. J. 1983. Effects of subchronic exposure to a mixture of O_3, SO_2, and $(NH_4)_2SO_4$ on host defenses of mice. *J. Toxicol. Environ. Health* 12:55–71.

Bloch, W. N., Miller, F., and Lewis, T. R. 1971. Pulmonary hypertension in dogs exposed to ozone. Biological Research Branch, Air Pollution Control Office, Cincinnati, Ohio, p. 10.

Boatman, E. S., Sato, S., and Frank, R. 1974. Acute effects of ozone on cat lungs. II. *Am. Rev. Respir. Dis.* 110:157–169.

Boorman, G. A., Schwartz, L. W., and Dungworth, D. L. 1980. Pulmonary effects of prolonged ozone insult in rats. Morphometric evaluation of the central acinus. *Lab. Invest.* 43:108–115.

Brinkman, R., Lamberts, H., and Veninga, T. 1964. Radiomimetic toxicity of ozonized air. *Lancet* 1:133–136.

Buckley, R. D., Hackney, J. D., Clark, K., and Posin, C. 1975. Ozone and human blood. *Arch. Environ. Health* 30:40–43.

Calabrese, E. J., and Moore, G. S. 1980. Does the rodent model adequately predict the effects of ozone induced changes to human erythrocytes? *Med. Hypotheses.* 6:505–507.

Castleman, W. L., Dungworth, D. L., and Tyler, W. S. 1973. Cytochemically detected alterations of lung acid phosphatase reactivity following ozone exposure. *Lab. Invest.* 29:310–319.

Castleman, W. L., Tyler, W. S., and Dungworth, D. L. 1977. Lesions in respiratory bronchioles and conducting airways of monkeys exposed to ambient levels of ozone. *Exp. Mol. Pathol.* 26:384–400.

Castleman, W. L., Dungworth, D. L., Schwartz, L. W., and Tyler, W. S. 1980. Acute respiratory bronchiolitis: An ultrastructural and autoradiographic study of epithelial cell injury and renewal in rhesus monkeys exposed to ozone. *Am. J. Pathol.* 98:811–840.

Chow, C. K. 1983. Influence of dietary vitamin E on susceptibility to ozone exposure. In *Biomedical Effects of O_3 and Photochemical Oxidants,* pp. 75–93. Princeton, N.J.: Princeton Scientific.

Chow, C. K., and Kaneko, J. J. 1979. Influence of dietary vitamin E on the red cells of ozone-exposed rats. *Environ. Res.* 19:49–55.

Chow, C. K., and Tappel, A. L. 1972. An enzymatic protective mechanism against lipid peroxidation damage to lungs of ozone-exposed rats. *Lipids* 7:518–524.

Chow, C. K., and Tappel, A. L. 1973. Activities of pentose shunt and glycolytic enzymes in lungs of ozone-exposed rats. *Arch. Environ. Health* 26:205–208.

Chow, C. K., Dillard, C. J., and Tappel, A. L. 1974. Glutathione peroxidase system and lysozyme in rats exposed to ozone or nitrogen dioxide. *Environ. Res.* 7:311–319.

Chow, C. K., Cross, C. E., and Kaneko, J. J. 1977. Lactate dehydrogenase activity and isoenzyme pattern in lung, erythrocytes, and plasma of ozone-exposed rats and monkeys. *J. Toxicol. Environ. Health* 3:877–884.

Chow, C. K., Plopper, C. G., Chiu, M., and Dungworth, D. L. 1981. Dietary vitamin E and pulmonary biochemical and morphological alterations of rats exposed to 0.1 ppm ozone. *Environ. Res.* 24:315–324.

Christensen, E., and Giese, A. C. 1954. Changes in absorption spectra of nucleic acids and their derivatives following exposure to ozone and ultraviolet radiations. *Arch. Biochem. Biophys.* 51:208–216.

Clark, K. W., Posin, C. I., and Buckley, R. D. 1978. Biochemical Response of Squirrel Monkeys to Ozone. *J. Tox. Environ. Health* 4:741–753.

Coffin, D. L., and Gardner, D. E. 1972. Interaction of biological agents and chemical air pollutants. *Ann. Occup. Hyg.* 15:219–235.

Coffin, D. L., Gardner, D. E., Holzman, R. S., and Wolock, F. J. 1968a. Influence of ozone on pulmonary cells. *Arch. Environ. Health* 16:633–636.

Coffin, D. L., Blommer, E. J., Gardner, D. E., and Holzman, R. 1968b. Effect of air pollution on alteration of susceptibility to pulmonary infections. In *Proc. 3rd Ann. Conf. Atmospheric Contaminants in Confined Spaces.* AMRL-TR-67-200, Aerospace Medical Research Laboratories, Wright-Patterson AFB, Dayton, Ohio, pp. 71–80.

Costa, D. L., Kutzman, R. S., Lehmann, J. R., Popenoe, E. A., and Drew, R. T. 1983. A subchronic multi-dose ozone study in rats. In *Advances in Modern Environmental Toxicology,* Vol. V, eds. S. D. Lee, M. G. Mustafa, and M. A. Mehlman, in press.

DeLucia, A. J., Hogue, P. M., Mustafa, M. G., and Cross, C. E. 1972. Ozone interaction with rodent lung, effect on sulfhydryls and sulfhydryl-containing enzyme activities. *J. Lab. Clin. Med.* 80:559–566.

DeLucia, A. J., Mustafa, M. G., Hussain, M. Z., and Cross, C. E. 1975b. Ozone interaction with rodent lung. III. Oxidation of reduced glutathione and formation of mixed disulfides between protein and nonprotein sulfhydryls. *J. Glin. Invest.* 55:794–802.

Dillard, C. J., Urribarri, N., Reddy, K., Fletcher, B., Taylor, S., de Lumen, B., Langberg, S., and Tappel, A. L. 1972. Increased lysosomal enzymes in lungs of ozone-exposed rats. *Arch. Environ. Health* 25:426–431.

Dimeo, M. J., Glenn, M. G., Holtzman, M. J., Sheller, J. R., Nadel, J. A., and Boushey, H. A. 1981.

Threshold concentration of ozone causing an increase in bronchial reactivity in humans and adaptation with repeated exposures. *Am. Rev. Respir. Dis.* 124:245–248.

Donovan, D. H., Williams, S. J., Charles, J. M., and Menzel, D. B. 1977. Ozone toxicity: Effect of dietary vitamin E and polyunsaturated fatty acids. *Toxicol. Lett.* 1:135–139.

Dorsey, A. P., Morgan, D. L., and Menzel, D. B. 1983. Filterability of erythrocytes from ozone-exposed mice. *Third Int. Congress on Toxicol.*, San Diego, Calif., in press.

Dowell, A. R., Lohrbauer, L. A., Hurst, D., and Lee, S. D. 1970. Rabbit alveolar macrophage damage caused by *in vivo* ozone inhalation. *Arch. Environ. Health* 21:121–127.

Dubeau, H., and Chung, Y. S. 1979. Ozone response in wild type and radiation-sensitive mutants of *Saccharomyces* cerevisiae. *Mol. Gen. Genet.* 176:393–398.

Dubeau, H., and Chung, Y. S. 1982. Genetic effects of ozone induction of point mutations and genetic recombination in *Saccharomyces* cerevisiae. *Mut. Res.* 102:249–259.

Dumelin, E. E., Dillard, C. J., and Tappel, A. L. 1978a. Effect of vitamin E and ozone on pentane and ethane expired by rats. *Arch. Environ. Health* 33:129–134.

Dumelin, E. E., Dillard, C. J., and Tappel, A. L. 1978b. Breath ethane and pentane as measures of vitamin E protection of *Macaca radiata* against 90 days of exposure to ozone. *Environ. Res.* 15:38–43.

Dungworth, D. L., Cross, C. E., Gillespie, J. R., and Plopper, C. G. 1975a. The effects of ozone on animals. In *Ozone Chemistry and Technology. A Review of the Literature: 1961–1974*, eds. J. S. Murphy and J. R. Orr, pp. 29–54. Philadelphia: Franklin Institute.

Dungworth, D. L., Castleman, W. L., Chow, C. J., Mellick, P. W., Mustafa, M. G., Tarkington, B., and Tyler, W. S. 1975b. Effects of ambient levels of ozone on monkeys. *Fed. Proc.* 34:1670–1674.

Easton, R. E., and Murphy, S. D. 1967. Experimental ozone preexposure and histamine. *Arch. Environ. Health* 15:160–166.

Ehrlich, R., Findlay, J. C., Fenters, J. D., and Gardner, D. E. 1977. Health effects of short-term inhalation of nitrogen dioxide and ozone mixtures. *Environ. Res.* 14:223–231.

Erdman, H. E., and Hernandez, T. 1982. Adult toxicity and dominant lethals induced by ozone at specific stages in spermatogenesis in *Drosophila virilis. Env. Mutagen.* 4:657–666.

Eustis, S. L., Schwartz, L. W., Kosch, P. C., and Dungworth, D. L. 1981. Chronic bronchiolitis in nonhuman primates after prolonged ozone exposure. *Am. J. Pathol.* 105:121–137.

Evans, M. J., Johnson, L. V., Stephens, R. J., and Freeman, G. 1976a. Renewal of the terminal bronchiolar epithelium in the rat following exposure of NO_2 or O_3. *Lab. Invest.* 35:246–257.

Evans, M. J., Johnson, L. V., Stephens, R. J., and Freeman, G. 1976b. Cell renewal in the lungs of rats exposed to low levels of ozone. *Exp. Mol. Pathol.* 24:70–83.

Fletcher, B. L., and Tappel, A. L. 1973. Protective effects of dietary alphatocopherol in rats exposed to toxic levels of ozone and nitrogen dioxide. *Environ. Res.* 6:166–175.

Folinsbee, L. J., Drinkwater, B. L., Bedi, J. F., and Horvath, S. M. 1978. The influence of exercise on the pulmonary changes due to exposure to low concentrations of ozone. In *Environmental Stress: Individual Human Adaptations*, eds. L. J. Folinsbee, J. A. Wagner, J. F. Borgia, B. L. Drinkwater, J. A. Gliner, and J. F. Bedi, pp. 125–145. New York: Academic.

Frager, N. B., Phalen, R. F., and Kenoyer, J. L. 1979. Adaptations to ozone in reference to mucociliary clearance. *Arch. Environ. Health* 34:51–57.

Gardner, D. E. 1982. Use of experimental airborne infections for monitoring altered host defenses. *Environ. Health Persp.* 43:99–107.

Gardner, D. E., Illing, J. W., Miller, F. J., and Coffin, D. L. 1974. The effect of ozone on pentobarbital sleeping time in mice. *Res. Commun. Chem. Pathol. Pharmacol.* 9:689–700.

Gooch, P. C., Creasia, D. A., and Brewen, J. G. 1976. The cytogenetic effect of ozone: Inhalation and *in vitro* exposures. *Environ. Res.* 12:188–195.

Goldstein, B. D., Pearson, B., Lodi, C., Buckley, R. D., and Balchum, O. J. 1968. The effect of ozone on mouse blood *in vivo. Arch. Environ. Health* 16:648–650.

Goldstein, B. D., Hamburger, S. J., Falk, G. W., and Amoruso, M. A. 1977. Effect of ozone and nitrogen dioxide on the agglutination of rat alveolar macrophages by concanavalin A. *Life Sci.* 21:1637–1644.

Gordon, T., and Amdur, M. O. 1980. Effects of ozone on respiratory response of guinea pigs to histamine. *J. Toxicol. Environ. Health* 6:185–195.

Gordon, T., Taylor, B. F., and Amdur, M. O. 1981. Ozone inhibition of tissue cholinesterase in guinea pigs. *Arch. Environ. Health* 36:284–288.

Graham, J. A. 1979. Alteration of hepatic xenobiotic metabolism by ozone. Ph.D. dissertation, Duke University, Durham, North Carolina.

Graham, J. A., Menzel, D. B., Miller, F. J., Illing, J. W., and Gardner, D. E. 1981. Influence of ozone on pentobarbital-induced sleeping time in mice, rats, and hamsters. *Toxicol. Appl. Pharmacol.* 61:64–73.

Graham, J. A., Miller, F. J., Gardner, D. E., Ward, R., and Menzel, D. B. 1982. Influence of ozone and nitrogen dioxide on hepatic microsomal enzymes in mice. *J. Toxicol. Environ. Health* 9:849–856.

Guerrero, R. R., Rounds, D. E., Olson, R. S., and Hackney, J. D. 1979. Mutagenic effects of ozone on human cells exposed *in vivo* and *in vitro* based on sister chromatid exchange analysis. *Environ. Res.* 18:336–346.

Hadley, J. G., Gardner, D. E., Coffin, D. L., and Menzel, D. B. 1977. Enhanced binding of autologous cells to the macrophage plasma membrane as a sensitive indicator of pollutant damage. In *Pulmonary macrophage and epithelial cells*, eds. C. L. Sanders, R. P. Schneider, G. E. Dagle, and H. A. Ragan. *ERDA* Symposium Series, 43, pp. 66–77.

Hamelin, C., and Chung, Y. S. 1974. Resistance a l'ozone chez *Escherichia coli*. I. Etude comparative de differanismes de veparation de l'ADN. *Mol. Gen. Genet.* 129:177–184.

Hamelin, C., and Chung, Y. S. 1975a. The effect of low concentrations of ozone on *Escherichia coli* chromosome. *Mut. Res.* 28:131–132.

Hamelin, C., and Chung, Y. S. 1975b. Characterization of mucoid mutants of *Escherichia coli* K-12 isolated after exposure to ozone. *J. Bacteriol.* 122:19–24.

Hamelin, C., Sarhan, F., and Chung, Y. S. 1977. DNA degradation caused by ozone in mucoid mutants of *Escherichia coli* K-12. *FEMS Microbiol. Lett.* 2:149–151.

Hazucha, M., Silverman, F., Parent, C., Field, S., and Bates, D. V. 1973. Pulmonary function in man after short-term exposure to ozone. *Arch. Environ. Health* 27:183–188.

Holtzman, M. J., Cunningham, J. H., Sheller, J. R., Irsigler, G. B., Nadel, J. A., and Boushey, H. A. 1979. Effect of ozone on bronchial reactivity in atopic and nontopic subjects. *Am. Rev. Respir. Dis.* 120:1059–1067.

Hu, P. C., Miller, F. J., Daniels, M. J., Hatch, G. E., Graham, J. A., Gardner, D. E., and Selgrade, M. K. 1982. Protein accumulation in lung lavage fluid following ozone exposure. *Environ. Res.* 29:377–388.

Hurst, D. J., and Coffin, D. L. 1971. Ozone effect on lysosomal hydrolases of alveolar macrophages *in vitro*. *Arch. Intern. Med.* 127:1050–1063.

Hurst, D. J., Gardner, D. E., and Coffin, D. L. 1970. Effect of ozone on acid hydrolases of the pulmonary alveolar macrophage. *Res. J. Reticuloendothel. Soc.* 8:288–300.

Ibrahim, A. L., Zee, Y. C., and Osebold, J. W. 1980. The effects of ozone on the respiratory epithelium of mice. II. Ultrastructural alterations. *J. Environ. Pathol. Toxicol.* 3:251–258.

Inoue, H., Sato, S., Hirose, T., Kikuchi, Y., Ubukata, T., Nagashima, S., Sasaki, T., and Takishima, T. 1979. A comparative study between functional and pathologic alterations in lungs of rabbits exposed to an ambient level of ozone: Functional aspects. *Nikkyo Shikkai-Shi.* 17:288–296.

Johnson, D. A. 1980. Ozone inactivation of human alpha 1-proteinase inhibitor. *Am. Rev. Respir. Dis.* 121:1031–1038.

Kavlock, R. J., Daston, G., and Grabowski, C. T. 1979. Studies on the developmental toxicity of ozone. I. Prenatal effects. *Toxicol. Appl. Pharmacol.* 48:19–28.

Kavlock, R. J., Meyer, E., and Grabowski, C. T. 1980. Studies on the developmental toxicity of ozone: Postnatal effects. *Toxicol. Lett.* 5:3–9.

Kenoyer, J. L., Phalen, R. F., and Davis, J. R. 1981. Particle clearance from the respiratory tract as a test of toxicity: Effect of ozone on short and long term clearance. *Exp. Lung Res.* 2:111–120.

Kerr, H. D., Kulle, T. J., McIlhany, M. L., and Swidersky, P. 1975. Effects of ozone on pulmonary function in normal subjects. *Am. Rev. Respir. Dis.* 111:763-773.

Konig, G., Rommelt, H., Kienele, H., Dirnagl, K., Polke, H., and Fruhann, G. 1980. Changes in the bronchial reactivity of humans caused by the influence of ozone. *Arbeitsmed. Sozialmed. Preventivmed.* 151:261-263.

Kyei-Aboagye, K., Hazuka, M., Wyszogrodski, I., Rubinstein, D., and Avery, M. E. 1973. The effect of ozone exposure in vivo on the appearance of lung tissue lipids in the endobronchial lavage of rabbits. *Biochem. Biophys. Res. Commun.* 54:907-913.

Last, J. A., and Cross, C. E. 1978. A new model for health effects of air pollutants: Evidence for synergistic effects of mixtures of ozone and sulfuric acid aerosols on rat lungs. *J. Lab. Clin. Med.* 91:328-339.

Last, J. A., and Kaizu, T. 1980. Mucus glycoprotein secretion by tracheal explants: effects of pollutants. *Environ. Health Perspect.* 35:131-138.

Last, J. A., Jennings, M. D., Schwartz, L. W., and Cross, C. E. 1977. Glycoprotein secretion by tracheal explants cultured from rats exposed to ozone. *Am. Rev. Respir. Dis.* 116:695-703.

Lee, L.-Y., Bleecker, E. R., and Nadel, J. A. 1977. Effect of ozone on bronchomotor response to inhaled histamine aerosol in dogs. *J. Appl. Physiol.: Respirat. Environ. Exercise Physiol.* 43:626-631.

Linn, W. S., Buckley, R. D., Spier, C. E., Blessey, R. L., Jones, M. P., Fischer, D. A., and Hackney, J. D. 1978. Health effects of ozone exposure in asthmatics. *Amer. Rev. Respir. Dis.* 117:835-843.

McKenzie, W. H. 1982. Controlled human exposure studies: cytogenetic effects of ozone inhalation. In *Banbury Report 13: Indicators of Genotoxic Exposure,* eds. B. A. Bridges, B. E. Butterworth, and I. B. Weinstein. Cold-Spring Harbor: Cold-Spring Harbor Laboratory.

McKenzie, W. H., Knelson, J. H., Rummo, N. J., and House, D. E. 1977. Cytogenetic effects of inhaled ozone in man. *Mutat. Res.* 48:95-102.

Mellick, P. W., Dungworth, D. L., Schwartz, L. W., and Tyler, W. S. 1977. Short term morphologic effects of high ambient levels of ozone on lungs of rhesus monkeys. *Lab. Invest.* 36:82-90.

Menzel, D. B. 1970. Toxicity of ozone, oxygen, and radiation. *Annu. Rev. Pharacol.* 10:379-394.

Menzel, D. B. 1976. The role of free radicals in the toxicity of air pollutants (nitrogen dioxide and ozone). In *Free radicals in biology,* Vol. II, ed. W. A. Pryor, pp. 181-202. New York: Academic.

Menzel, D. B. 1980. Environmental toxicants and vitamin E. In *Vitamin E,* ed. E. Machlin, pp. 474-494. New York: Marcel Dekker.

Menzel, D. B., Slaughter, R. J., Bryant, A. M., and Jauregui, H. O. 1975. Prevention of ozonide-induced bodies in human erythroyctes in vitamin E. *Arch. Environ. Health* 30:234-236.

Menzel, D. B., Anderson, W. G., and Abou-Donia, M. B. 1976. Ozone exposure modifies prostaglandin biosynthesis in perfused rat lungs. *Res. Commun. Chem. Pathol. Pharmacol.* 15:135-147.

Miller, F. J., Menzel, D. B., and Coffin, D. L. 1978. Similarity between man and laboratory animals in regional pulmonary disposition of ozone. *Environ. Res.* 17:84-101.

Moore, P. F., and Schwartz, L. W. 1981. Morphological effects of prolonged exposure to ozone and sulfuric acid aerosol in the rat lung. *Exp. Mol. Pathol.* 35:108-123.

Mountain, J. T. 1963. Detecting hypersusceptibility of toxic substances, an appraisal of simple blood tests. *Arch. Environ. Health* 6:357-365.

Mudd, J. B., Leavitt, R., Ongun, A., and McManus, T. T. 1969. Reaction of ozone with amino acids and proteins. *Atmos. Environ.* 3:669-681.

Mudd, J. B., and Freeman, B. A. 1977. Reaction of ozone with biological membranes. In *Biochemical Effects of Environmental Pollutants,* ed. S. D. Lee, p. 97:133. Ann Arbor, Mich.: Ann Arbor Science.

Mustafa, M. G., and Lee, S. D. 1976. Pulmonary biochemical alterations resulting from ozone exposure. *Ann. Occup. Hyg.* 19:17-26.

Mustata, M. G., DeLucia, A. J., York, G. K., Arth, C., and Cross, C. E. 1973. Ozone interaction with rodent lung. II. Effects of oxygen consumption of mitochondria. *J. Lab. Clin. Med.* 82:357–365.

Mustafa, M. G., Elysayed, N. M., Graham, J. A., and Gardner, D. E. 1983. Effects of ozone exposure on lung metabolism: Influence of animal age, species, and exposure conditions. In *Biomedical Effects of O_3 and Oxidants,* eds. S. D. Lee, M. G. Mustafa, and M. A. Mehlman, pp. 57–73. Princeton, N.J.: Princeton Scientific.

Osebold, J. W., Gershwin, L. J., and Zee, Y. C. 1980. Studies on the enhancement of allergic lung sensitization of inhalation of ozone and sulfuric acid aerosol. *J. Environ. Pathol. Toxicol.* 3:221–234.

Peterson, M. L., Rummo, N., House, D., and Harder, S. 1978. The effect of ozone on human immunity: *In vitro* responsiveness of lymphocytes to phytohemmagglutinin. *Arch. Environ. Health* 36:59–63.

Phalen, R. F., Kenoyer, J. L., Crocker, T. T., and McClure, T. R. 1980. Effects of sulfate aerosols in combination with ozone in elimination of tracer particles inhaled by rats. *J. Toxicol. Environ. Health* 6:797–810.

Plopper, C. G., Dungworth, D. L., Tyler, W. S., and Chow, C. K. 1979. Pulmonary alterations in rats exposed to 0.2 and 0.1 ppm ozone: A correlated morphological and biochemical study. *Arch. Environ. Health* 34:390–395.

Posin, C. I., Clark, K. W., Jones, M. P., Buckley, R. D., and Hackney, J. D. 1979. Human biochemical response to ozone and vitamin E. *J. Toxicol. Environ. Health* 5:1049–1058.

Prat, R., Nofre, C. L., and Cier, A. 1968. Effects de l'hypochlorite de sodium, de l'ozone et des radiations ionisantes sur les constituants pyrimidiques d'Escherichia coli. *Annals de Institut Pasteur* 114:595–607.

Previero, A., Signor, A., and Bezzi, S. 1964. Tryptophan modification in poly-peptide chains. *Nature* 204:687–688.

Pryor, W. A., Dooley, M. M., and Church, D. F. 1983. Mechanisms for the reaction of ozone with biological molecules, the source of the toxic effects of ozone. In *Biomedical Effects of O_3 and Photochemical Oxidants,* eds. S. D. Lee, M. G. Mustafa, and M. A. Mehlaman, pp. 7–19. Princeton, N.J.: Princeton Scientific.

Raub, J. A., Miller, F. J., and Graham, J. A. 1983. Effects of low-level ozone exposure on pulmonary function in adult and neonatal rats. In *Advances in Modern Environmental Toxicology,* Vol. 5, eds. S. D. Lee, M. G. Mustafa, and M. A. Mehlman, pp. 363–367. Princeton, N.J.: Princeton Scientific.

Reasor, M. J., Adams III, G. K., Brooks, J. K., and Rubin, R. J. 1979. Enrichment of albumin and IgG in the airway secretions of dogs breathing ozone. *J. Environ. Sci. Health* 13:335–346.

Revis, N. W., Major, T., and Dalbey, W. E. 1981. Cardiovascular effects of ozone and cadmium inhalation in the rat, pp. 171–179. (USEPA 600/9-81-001, PB81 178832).

Roehm, J., Hadley, J. G., and Menzel, D. B. 1971a. Oxidation of unsaturated fatty acids by ozone and nitrogen dioxide: A common mechanism of action. *Arch. Environ. Health* 23:142–148.

Roehm, J., Hadley, J. G., and Menzel, D. B. 1971b. Antioxidants vs. lung disease. *Arch. Intern. Med.* 129:88–93.

Roeham, J. N., Hadley, J. G., and Menzel, D. B. 1972. The influence of vitamin E on the lung fatty acids of rats exposed to ozone. *Arch. Environ. Health* 24:237–242.

Sato, S., Kawakami, M., Maeda, S., and Takishima, T. 1976a. Scanning electron microscopy of the lungs of vitamin E-deficient rats exposed to a low concentration of ozone. *Am. Rev. Respir. Dis.* 113:809–821.

Sato, S., Kawakami, M., Maeda, S., and Takishima, T. 1976b. Electron microscopic studies on the effects of ozone on the lungs of vitamin E-deficient rats. *Jpn. J. Thorac. Dis.* 14:355–365.

Sato, S., Shimura, S., Kawakami, M., Hirosa, T., Maeda, S., Takishima, T., Kimura, S., Yashiro, M., Okazaki, S., and Ito, M. 1978. Biochemical and ultrastructural studies on the effects of long-term exposure of ozone on vitamin E-depleted rats (author's translation). *Nippon Kyobu Shikkan Gakkai Zasshi* 16:260–268.

Sato, S., Shimura, S., Hirosa, T., Maeda, S., Kawakami, M., Takishima, T., and Kimura, S. 1980.

Effects of long-term ozone exposure and dietary vitamin E in rats. *Tohoku. J. Exp. Med.* 120:117-128.

Savino, A., Peterson, M. L., House, D., Turner, A. G., Jefries, H. E., and Baker, R. 1978. The effects of ozone on human cellular and humoral immunity: Characterization of T and B lymphocytes by Rosette formation. *Environ. Res.* 15:65-69.

Schwartz, L. W., Dungworth, D. L., Mustafa, M. G., Tarkington, B. K., and Tyler, W. S. 1976. Pulmonary responses of rats to ambient levels of ozone: Effects of 7-day intermittent or continuous exposure. *Lab. Invest.* 34:565-578.

Shinriki, M., Ishizaki, K., Ikehata, A., Yoshizaki, T., Nomura, A., Miura, K., and Mizuno, Y. 1981. Degradation of nucleic acids with ozone, degradation of yeast RNA, yeast phenylalamine TRNA and tobacco mosaic virus RNA. *Biochim. Biophys. Acta* 655:323-328.

Stephens, R. J., Freeman, G., Stara, J. F., and Coffin, D. L. 1973a. Cytologic changes in dog lungs induced by chronic exposure to ozone. *Am. J. Pathol.* 73:711-726.

Stephens, R. J., Sloan, M. F., Evans, M. J., and Freeman, G. 1973b. Early response of lung to low levels of ozone. *Am. J. Pathol.* 74:31-58.

Stephens, R. J., Sloan, M. F., Evans, M. J., and Freeman, G. 1974. Alveolar type 1 cell response to exposure to 0.5 ppm O_3 for short periods. *Exp. Mol. Pathol.* 20:11-23.

Thomas, G. B., Fenters, J. D., Ehrlich, R., and Gardner, D. E. 1981. Effects of exposure to ozone on susceptibility to experimental tuberculosis. *Toxicol. Lett.* 9:11-17.

Tice, G. B., Bender, M. A., Ivett, J. L., and Drew, R. T. 1978. Cytogenetic effects of inhaled ozone. *Mutat. Res.* 58:293-304.

Watanabe, S., Frank, R., and Yokoyama, E. 1973. Acute effects of ozone on lungs of cats. I. Functional. *Am. Rev. Respir. Dis.* 108:1141-1151.

Williams, S. J., Charles, J. M., and Menzel, D. B. 1980. Ozone induced alterations in phenol red absorption from the rat lung. *Toxicol. Lett.* 6:213-219.

Witz, G., Amoruso, M. A., and Goldstein, B. D. 1983. Effect of ozone on alveolar macrophage function: Membrane dynamic properties. In *Int. Symp. on Biomedical Effects of Ozone and Related Photochemical Oxidants,* eds. S. D. Lee and M. G. Mustafa, pp. 263-272. Princeton, N.J.: Princeton Scientific.

Yokoyama, E. 1969. A comparison of the effects of SO_2, NO_2, and O_3 on the pulmonary ventilation. Guinea pig exposure experiments. *Sangyo Igaku* 11:563-568.

Zelac, R. E., Cromroy, H. L., Bolch, Jr., W. E., Dunavant, B. G., and Bevis, H. A. 1971a. Inhaled ozone as a mutagen. Chromosome aberrations induced in Chinese hamster lymphocytes. *Environ. Res.* 4:262-282.

Zelac, R. E., Cromroy, H. L., Bolch, Jr., W. E., Dunavant, B. G., and Bevis, H. A. 1971b. Inhaled ozone as a mutagen. II. Effect on the frequency of chromosome aberrations observed in irradiated Chinese hamsters. *Environ. Res.* 4:325-342.

TOXICOLOGICAL DATA ON NO_x: AN OVERVIEW

Paul E. Morrow

Department of Radiation Biology and Biophysics, School of Medicine and Dentistry, University of Rochester, Rochester, New York

This overview is based on experimental and epidemiological studies of NO_X toxicity during the past decade. Approximately 130 published studies are cited and about one-fourth of these are discussed briefly under one of the following headings: acute and subacute studies, chronic low-level studies, human studies, and special studies. The latter section examines a selection of comparatively unique investigations, including several devoted to the pulmonary uptake and retention of NO_2, and several examining the potential tumorigenicity of NO_2. For each major section of the overview, a critical evaluation is attempted in terms of the impact of the appropriate studies on the extant NO_X toxicological data base and on the current and planned air quality standards for NO_X.

INTRODUCTION

Enormous natural sources of nitrogen oxides exist, especially for N_2O, and to a lesser extent for other forms, including particulate nitrates. Decaying organic matter, forest fires, volcanism, and atmospheric lightning are perhaps the largest sources of the oxides of nitrogen. In both the urban and rural areas of the world, these natural sources produce an average background level of <5 ppb gaseous NO_x and nitrate aerosols. The most significant species of these nitrogen oxide compounds are listed in Table 1.

Energy utilization in urban areas produces far less total NO_x and nitrates than do natural sources, but this determines average urban air levels to a far greater degree.

The oxides of nitrogen present a complex, not fully understood picture of their chemical origins, interactions, and speciation. A simplified depiction of some of these features is given in Fig. 1 (Guidotti, 1978; NRC, 1977; Pitts et al., 1983; Malatinsky et al., 1973; Lee, 1980).

Partial support from the Department of Energy under contract DE-AC02-76EV03490 with the Department of Radiation Biology and Biophysics, and from the Health Effects Institute are acknowledged.

This paper has been assigned report number UR-3490-2298.

The author greatly appreciates the helpful discussions with Dr. Richard Mavis and Dr. Mark Utell, the illustration done by Ms. Valerie Farnham, the stenographic skills of the department's Word Processing Unit, and the invaluable assistance of Ms. Harriet Beiter.

TABLE 1. Forms of Nitrogen Oxides of Potential Interest to Considerations of Air Pollution Effects

Form	Name	Form	Name
N_2O	Nitrous oxide	HNO_2	Nitrous acid
NO	Nitric oxide	HNO_3	Nitric acid
NO_2	Nitrogen dioxide	MNO_2	Inorganic nitrites
NO_3	Nitrogen trioxide	MNO_3	Inorganic nitrates
N_2O_2	Nitrogen peroxide		
N_2O_3	Dinitrogen trioxide (sesquioxide)	R—NO	Nitroso compounds
			Nitroxy compounds
N_2O_4	Dinitrogen trioxide (dimeric NO_2)	$R—NO_2$	Organic nitrates
N_2O_5	Dinitrogen pentoxide	$R—C(=O)—OONO_2$	Peroxyacyl nitrates (PAN)

The role of the land mass in decaying vegetation and the decomposition of diverse organic matter is implicit, but the breakdown of mineral nitrates and nitrites to NO, N_2O, and NO_2 is also a part of that scheme (Fishbein, 1976; NRC, 1978).

It is a generally accepted practice to designate the many atmospheric forms of the nitrogen oxides as NO_x. It is also generally believed that from a toxicological viewpoint, NO_2 is the dominant urban atmospheric form. Hence, the tendency to use NO_2 and NO_x interchangeably has developed when biomedical considerations of air pollution are discussed. The nongaseous states of the nitrogen oxides are ambiguously excluded (e.g., inorganic nitrate salts) and included (e.g., nitric acid) in the NO_x accounting; presumably, this is more in line with what is analyzed by accredited methods than what is present to be breathed. Analytically this appears to be a small "error," but it may prove biologically to be more important than we recognize presently.

In the past decade, greater attention has been given to indoor pollution levels, and to indoor NO_2 in particular (Melia et al., 1978;

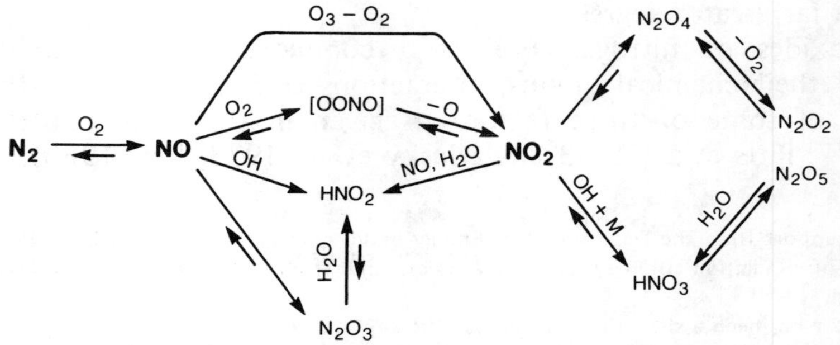

FIGURE 1. Interconversions of NO_X. This depiction, adapted from Guidotti (1978), shows the major interconversions of the atmospheric oxides of nitrogen. Where equilibria are depicted, the length of the arrow represents the favored reaction.

Goldstein et al., 1979; Speizer et al., 1980; Spengler, 1983). Families living in rural or low-pollution areas are exposed to 10–15 μg NO_2/m^3 typically, whereas in homes using gas for cooking the average levels of NO_2 are 3–5 times greater. About 10% of the rural homes assessed have had measurements that exceed the annual NAAQS of 100 μg NO_2/m^3. Moreover, brief peak levels exceeding 1000 μg NO_2/m^3 (0.5 ppm) have also been recorded. This circumstance has presented epidemiologists with an unusual opportunity to evaluate NO_2-related health effects, and they have responded (*op. cit.*).

Another major concern of air-pollution investigators and regulators is the interplay and interdependence of NO_x, ozone, carbonaceous material (e.g., hydrocarbons, carbon monoxide, formaldehyde), and other air pollutants. This concern stems from our limited understanding of the atmospheric chemistry of these pollutants, the belief that some secondary pollutants, e.g., nitrous acid and peroxyacyl nitrate (PAN), are toxicologically important but usually excluded or inadequately considered, and that ozone is probably 10 times more toxic to humans than NO_x, *inter alia*. These background concerns and considerations impinge on NO_2 toxicity evaluations and affect the development of NO_x standards. However, an assessment of progress and needs in many of these areas is outside the purview of this report.

Several review articles on nitrogen oxides' toxicity and health-related effects have been published since the Air Quality Criteria for Nitrogen Oxides was promulgated (U.S. EPA, 1971)—e.g., Morrow (1975), Menzel (1976), Mustafa and Tierney (1978), and Dawson and Schenker (1979). Three major publications (NRC, 1977; Lee, 1980; Schneider and Grant, 1982) have summarized the NO_x data base from both environmental and biomedical perspectives. The World Health Organization's Criteria 4 Document (1977) provided a review of NO_x toxicity, while proposing a new, short-term ambient air standard. A related major assessment on nitrates directed primarily at their environmental impact (NRC, 1978) was also published. Collectively these publications reveal a continued, even intensified, interest in NO_x toxicity; a greater emphasis on NO_x toxicity mechanisms, especially at the biochemical level; indications that intermittent or short-term exposures are of greater interest in standard setting; and indications that important new studies at near ambient levels have been accomplished. In the description that follows, a critical evaluation will be made of this more current information and a synthesis will be attempted that will indicate what we know and need to know about NO_2-induced health effects.

ACUTE AND SUBACUTE TOXICITY STUDIES WITH NO_2 (NO_x)

Although acute intoxication by NO_2 is usually regarded as a condition of limited direct relevance to ambient air quality, there has been some important information developed using this approach. The NRC report (NRC, 1977), for example, reveals a wide variability in acute mortality among different species. Book (1982), elaborating on the work of Gardner et al. (1977), has shown that these acute mortality data can be scaled (minute volume and body size) and correlated as a power function of exposure time. This has led to a 1-h exposure LD50 estimate for humans of 174 ppm NO_2; this estimate agrees well with extant information (NRC, 1977) that acute exposures to >150 ppm NO_2 are generally fatal.

Gardner et al. (1977) utilized subacute NO_2 dose-response information to indicate the plausibility of predicting chronic long-term effects. Coffin et al. (1977) presented evidence that intermittent NO_2 exposures with short-term peak levels may be more toxic than continuous NO_2 exposures. For example, they found that the mortality from a 79-h continuous exposure of mice to 3.5 ppm NO_2 was similar to that obtained from mice given 4 7-h exposures to the same concentration over a 4-d period. Greater differences in responses were ascribed to 1 ppm NO_2 exposures. Differences in the dose-response data related to exposure pattern were also seen to change with postexposure time and to vary depending upon response criteria—e.g., spontaneous mortality vs. mortality—following bacterial challenge.

Azoulay-Dupuis and co-workers (1983), using both changes in lung superoxide dismutase and histology, reported on the greater resistance of newborn guinea pigs and rats to NO_2 compared to their adult counterparts. They also showed that rats were far more tolerant of 10 ppm NO_2 than guinea pigs, suggesting a factor of ~7 on an acute basis. Stephens and co-workers (1982) have again demonstrated the remarkable natural resistance to NO_2 manifested by neonatal rats. They have shown, for example, that continuous exposure to 50 ppm NO_2 is tolerated, whereas 60-d-old rats cannot survive 25 ppm NO_2 for more than 8 h. Cabral-Anderson et al. (1977) found that older rats experienced greater mortality than younger adult rats when exposed to 12–14 ppm NO_2 for up to 15 d. Young, mature, and old adult rats showed similar injury to bronchial ciliated cells and alveolar type I cells, but the oldest rats were slower to repair. The time between injury and repair was effectively the acute edematous phase; the longer it persisted, the greater the mortality. It is well documented that the acute lethality of NO_2 is due almost entirely to pulmonary edema. Bronchiolitis and pneumonia contribute to the subacute mortality. Many subacute effects tend to be persistent and some are progressive, e.g., fibroblastic proliferation to fibrosing bronchiolitis to bronchiolitis obliterans. Other effects tend to be reversible and self-limiting, e.g., bronchial and type II cell hyperplasia. There also appear to be important differences in delayed effects: these may run from 24 h to 1 wk following NO_2 exposure.

Pickrell et al. (1982) pointed out that the subacute toxicity of NO_2 presents qualitative as well as quantitative distinctions among species;

consequently, they question the validity of the rat model and indicate that the dog pattern of injury and repair may be much closer to that of humans.

Although many relatively high-level NO_2 exposures were performed to induce major pulmonary dysfunctions and pathology (e.g., Guidotti, 1978, 1980; Stephens et al., 1972, 1978; Ranga and Kleinerman, 1977; Kleinerman, 1977; Rynbrandt and Kleinerman, 1977; Johnson et al., 1982), most were performed to affect biochemical phenomena. DeNicola et al. (1981), Crapo et al. (1978a,b), Sagai et al. (1982), Williams et al. (1971), Sevanian et al. (1979), Wright and Mavis (1981), Mustafa and Tierney (1978), and Menzel (1976) conducted studies on lung lipids and enzymes. Evans et al. (1973, 1975, 1976, 1978a,b), Freeman et al. (1969), and Hackett (1979) are typical of the studies directed at type II and Clara cell progenitor roles in NO_2-induced epithelial cell injury. The effects of antioxidants on NO_2 toxicity were investigated by several groups, including Elsayed and Mustafa (1982), Csallany and Ayaz (1978a), and Kim (1978).

Collectively, these studies attempted to clarify such basic issues as the importance of lipid peroxidation; the role(s) of antioxidant molecules in NO_2 exposure; the bases of tolerance and cross-tolerance to oxidants; and the nature of the proliferative and repair processes induced by NO_2 injury.

CHRONIC, LOW-LEVEL NO_2 STUDIES

The appropriateness of chronic, low-level NO_2 studies, particularly those using a NO_2 level within an order of magnitude of urban levels, would seem to be self-evident, although constrained by the problem of interspecies extrapolations. For example, Evans and Freeman (1980), Fenters et al. (1973), Lewis et al. (1974), Richters et al. (1978, 1980), Erlich et al. (1977), Csallany and Ayaz (1978b), Gardner et al. (1977), Sherwin et al. (1973), Sherwin and Layfield (1976), Dróždž et al. (1977), Kuraitis et al. (1881), Sherwin and Richters (1982), Ichinose and Sagai (1982), Fujimaki et al. (1982), Pickrell et al. (1982), Elsayed and Mustafa (1982), Roy-Burman et al. (1982), and Sherwin et al. (1983a) have conducted studies at 2 ppm or lower NO_2 levels in either mice, rats, guinea pigs or dogs. In fact, most of these studies involved exposure levels below 1 ppm NO_2. Except for very brief exposure studies using a follow-up bacterial challenge (e.g., Erlich et al., 1977), these studies ranged from 1 wk to several years.

Many of the foregoing studies were designed to examine basic mechanisms of NO_2 toxicity and emphasized pulmonary cytological changes. Collectively, they indicate that biological investigation at low, near-ambient exposure levels is entering a new era. Greater sensitivity of measurements and selection of the more susceptible targets of NO_2 action have both reduced the need for work at higher levels. This is not to imply that studies at higher NO_2 levels have become unreasonable.

The low-level studies cited show evidence of type II cell proliferation; increased alveolar membrane permeability, especially to macromolecules and protein; increased "antioxidant" enzyme production; and lipid peroxidation. These are qualitatively the same kinds of changes reported at higher subacute exposure levels: they represent both injury and compensatory changes. If the widely held view that type II cell proliferation in response to NO_2 exposure is due to type I cell injury, we must conclude not only that the type I pneumocyte is highly susceptible to NO_2 action (Menzel, 1976; Evans et al., 1978a,b) but that its injury is difficult to assess directly.

One study (Fujimaki et al., 1982) showed a suppression of the primary antibody response in BALB/c mice after 4 wk exposure to 0.4 ppm NO_2; Csallany and Ayaz (1978a) reported reduced growth and survival in mice at 0.5–1.0 ppm NO_2 after 18 mo, and Roy-Burman and co-workers (1982), using high and low expressor mice (endogenous retrovirus genes), found elevated but fluctuating levels of viral RNA in spleens of 0.3 and 0.8 ppm NO_2-exposed animals. This last observation is characterized as supporting the "emerging notion" that the spleen is a target of chronic NO_2 exposure even at ambient levels. Their comment is based, in part, on earlier work in the same laboratory that reported splenic changes in newborn and adult mice exposed to 0.35 ppm NO_2 for 6 wk (Kuraitis et al., 1981).

The studies by Richters and co-workers (1978, 1980) showed that explants from the lungs of NO_2-exposed mice (0.4 and 0.5 ppm) manifest an extent of macrophage congregation on epithelial cells and spindle cells not seen with control explants. In a recent mouse study conducted at 0.5–0.8 ppm NO_2, with continuous or intermittent exposures for either 3 or 6 wk (Sherwin et al., 1983b), horseradish peroxidase (HRP) was used as a molecular probe for protein leakage. In most of the NO_2 exposure protocols, a greater lung HRP content resulted. This was interpreted by these investigators as indicative of an increase in plasma protein content of the lungs, probably derived from NO_2-induced overload of bidirectional protein transport in the lungs. Electron microscopic localization of the HRP showed its association with the alveolar membrane.

Utilizing simulated patterns of urban NO_2 levels, Graham et al. (1983) reported that the NO_2 dose rate is a critical factor in determining the degree of susceptibility to streptococcal challenge. Using a continuous (2 wk) exposure to 1.5 ppm NO_2 with 2 1-h peaks daily of 4.2 ppm NO_2, mice showed greater mortality if given the bacterial challenge immediately after the morning NO_2 peak level. The bacterial challenge was relatively ineffectual given before the peak level, however.

The study of Fenters and co-workers (1973) on squirrel monkeys exposed continuously to 1 ppm NO_2 for 16 mo failed to show an increase in viral influenza susceptibility over air-exposed controls, but it did reveal slight to moderate emphysematous changes with thickened bronchial and bronchiolar epithelia, in NO_2-exposed, infected monkeys, that were not present in infected monkeys exposed to air. This interesting finding relates to a more recent guinea pig study (Drόżdż et al., 1977), in which a 1.1 ppm NO_2 exposure was conducted 8 h/d for 180 d. In this investigation, total lung collagen decreased along with the development of emphysematous changes in the lungs. Drόżdż and co-workers believe NO_2 inhibits collagen maturation and/or activates its catabolism.

Except for the spleen and explant studies, whose implications are not immediately evident, the other studies clearly demonstrate that in several normal animal species potentially adverse reactions are produced at NO_2 levels that are approximately a factor of 10 greater than the annual NO_2 standard for ambient air quality.

SPECIAL STUDIES OF NO_x

The deposition ("absorption") of NO_2 in the lungs of rhesus monkeys and its subsequent retention were measured by Goldstein et al. (1977). Using 0.3–0.9 ppm NO_2 levels, 50–60% deposition occurred during normal respiration. After cessation of the ^{13}N-labeled NO_2 exposure, 90% of the terminal exposure value was still present in the lungs 20 min later. In contrast to the washout (elimination) characteristics of radioactive xenon gas, the ^{13}N activity of lungs and blood was persistent and indicative of chemical conversion(s).

Postlethwait and Mustafa (1981) examined NO_2 uptake by the isolated perfused rat lung. They concluded that as NO_2 absorption occurs in the blood perfused lung, it is oxidized to nitrate, principally by the RBC.

Saul and Archer (1983) pursued this matter further with their rat studies of low NO_2 exposures (1.2–8.8 ppm) for 24 h. They found urinary nitrate levels could be correlated with exposure, viz., 8.4 ± 1.1 μmol nitrate/ppm NO_2/24 h exposure. Their absorption kinetics for NO_2 were interpreted to mean that water solubility was not a major determinant; conversion to nitrite probably was. Saul and Archer estimate 3.6 mg nitrite would be formed in the human lung each day from breathing air containing 0.1 ppm NO_2.

These three NO_2 studies are nicely complemented by the study of Valentini and co-workers (1983), who showed that the retention half-times of nitrate from the lungs and blood of rats were ~4.5 min and <9.8 min, respectively. This indicates the relatively prolonged ^{13}N retention measured by Goldstein et al. is based not on nitrate, but on other unknown NO_2 products.

Using a general model for gas transport and mixing in the lungs, the regional deposition of NO_2 was examined by Miller et al. (1982; Miller and Graham, 1983) in several species, including humans. The estimates for humans utilized the dichotomous Weibel lung model and indicated that the respiratory tissue of the transitional airways was the main "sink" for

NO_2 uptake. The model of Miller et al. also incorporated Henry's law, which is reported to have limited validity with NO_2 and to vary by more than an order of magnitude in its estimations. Hence, the most troublesome feature of this otherwise useful approach is the inadequate information on NO_2 reactions *in vivo*.

A novel experiment reported by Sone and co-workers (1983) pertains to the effect of NO_2 on alveolar macrophages (AM). Using both *in vivo* and *in vitro* NO_2 exposures, they showed host AM-mediated immune defense is modulated by NO_2. The *in vitro* exposure of AM lavaged from the lungs of control rats resulted in a dose-related, time-dependent enhancement of AM cytotoxicity toward an abnormal adenocarcinoma cell line.

Holt et al. (1979) reported an immunologic study in BALB/c mice exposed to 10 ppm NO_2 or 10 ppm NO for up to 30 wk. These high short-duration exposures to either gas resulted in similar effects on leucocytosis, increased humoral immune response to sheep red blood cells, decreased phytohemagglutinin response with spleen cells, and decreased tumor rejection. Moreover, NO produced more advanced emphysematous changes in the lungs than did NO_2. The apparent assumption of the relative toxicities of NO_2 and NO being of the order of 5 to 1 (implicit in the occupational threshold levels) was certainly not borne out by these mice studies.

At the ultrastructural level, Case et al. (1982) reported that changes in epithelial tight junctions induced in the lungs of Syrian hamsters by a specific NO_2 level (28 ppm) was not proportional to exposure time.

The potential mutagenic and carcinogenic properties of air pollutants have been discussed, e.g., von Nieding (1978). Recently, Iqbal et al. (1981) reported the biosynthesis of nitrosamine in NO_2-exposed mice. This indication that these carcinogenic substances could be induced by NO_2 exposure was not confirmed by the study of Chaudhari and Dutta (1981) in guinea pigs. However, Richters and Kuraitis (1981) presented evidence that a relatively continuous exposure to 0.4 ppm NO_2 affects the dissemination of existing cancer cells without necessarily being carcinogenic; in other words, NO_2 caused, experimentally, the facilitation of metastases. This is being described as a new type of relationship between an air pollutant and cancer (Richters and Richters, 1983).

Graham et al. (1982) reported that a 3-h exposure to 0.25 ppm NO_2 led to a prolongation of pentobarbital-induced sleeping time in female mice. This unique effect in females, attributed to extrapulmonary action of NO_2 on hepatic xenobiotic metabolism, was not found relatable to mixed-function oxidases or cytochrome P-450 levels.

HUMAN STUDIES OF NO_2 EFFECTS

The report of the NRC–NAS Committee on Medical and Biological Effects of Environmental Pollutants, entitled *Nitrogen Oxides*, devotes one chapter to the human health effects reported up to 1977. In the summary provided, the report cites two epidemiological studies where average NO_2 levels were 0.08–0.16 ppm (with other pollutants present), which indicated adverse effects on respiratory functions. Two other studies with lower NO_2 levels did not show this adversity. Even so, the report questioned any epidemiologic correlation between NO_x and acute respiratory disease, inasmuch as other pollutants were present. In addition, the report characterized the evidence for NO_2-associated chronic respiratory disease as "not convincing."

Newer human studies, both experimental and epidemiologic, have been reviewed in the book on nitrogen oxides edited by Lee (1980). Specifically, Chapter 18 by Shy and Love (1880) summarizes the findings of the experimental studies of Horvath and Folinsbee (1975), Beil and Ulmer (1976), Hackney and co-workers (1975a,b; Hackney, 1976), Kerr et al. (1978), and Orehek et al. (1976) and epidemiologic data from Melia et al. (1977), Love et al. (1974), Finklea et al. (1974), and Mitchell et al. (1974).

Shy and Love (1980) conclude that the experimental studies showed no significant alteration of respiratory ventilatory parameters in either normal or bronchitic subjects at NO_2 exposures below 1.5 ppm and of 2 h or less duration. The Orehek study with asthmatic subjects at 0.11 ppm NO_2 reported an enhanced responsiveness to bronchial challenge in 13 of 20 cases. Shy and Love stated that this study would have been more acceptable if the responders and nonresponders had been identified initially from baseline data. In any case, the appropriateness of this study has been challenged insofar as enhanced susceptibility to NO_2 is concerned.

Shy and Love also presented some new data as a follow-up of the earlier Chattanooga study (Love et al., 1974, 1982), and the results of this new assessment were included in their epidemiologic data review. Summarizing the NO_2 epidemiology, Shy and Love indicated a positive association of increased risk of acute respiratory disease with moderately elevated ambient levels of NO_2. This association was more consistently seen in children than in adults. They postulated that repeated, short-term peak NO_2 concentrations may have an important role in the enhanced risk of respiratory disorders.

Smoking in the household has a demonstrable effect on acute respiratory illness in children, but it appears to be independent of the NO_2 effect. This last statement is germane to the study of Melia et al., since their 4-yr longitudinal study of respiratory dysfunction in more than 5,000 school children was subsequently criticized as not having analyzed the impact of parental smoking. Melia et al. had concluded that the greater incidence of respiratory problems with children, particularly in young girls, was based on homes with gas cooking and the associated higher NO_2 levels, compared to homes with non-gas cooking (Melia et al., 1977, 1978; see also Goldstein et al., 1979; Florey et al., 1979).

More recent human experimental studies have been reported: Hackney et al. (1978), Kerr et al. (1979), von Nieding and Wagner (1979), Sackner et al. (1980, 1981), Ahmed and co-workers (1982, 1983), and Hazucha et al. (1983).

Hackney and co-workers (1978) studied normal male subjects during light exercise (200 kg m/min) at 1 ppm NO_2 and found only marginal reduction in forced vital capacity immediately after NO_2 exposure and on 2 consecutive days ($p < 0.05$). Five of the 20 subjects developed respiratory sumptoms, but no other changes in respiratory mechanics could be detected. Gas exchange measurements were not performed.

Kerr et al. (1979) examined 20 subjects with asthma and chronic obstructive pulmonary disease (COPD) exposed to 0.5 ppm NO_2 for 2 h. This study was evaluated by the investigators as not showing a significant decrement in pulmonary function. Kerr et al. did note some significant changes in certain parameters; for example, normal and diseased subjects both showed significant reduction in pulmonary compliance after NO_2 exposure when compared to control values. Seven of the 13 asthmatic subjects experienced symptoms during NO_2 exposure. This study has been viewed as refuting the Orehek study in asthmatics, which was conducted at a lower NO_2 level, but the assessment methodologies were different.

The effects of nitrogen dioxide on 116 chronic bronchitis were investigated by von Nieding and Wagner (1979). Concentrations of NO_2 below 1.5 ppm had no effect on these subjects with chronic nonspecific lung disease. Disturbance of gas exchange was evidenced by brief (15 min) exposures to 4–5 ppm NO_2. The authors cite unpublished data in humans and earlier findings in rats (Thomas et al., 1967) to suggest that NO_2 causes mast-cell degranulation and histamine release. These NO_2-induced actions may contribute to smooth-muscle constriction and edema formation in the lungs.

Sackner and co-workers conducted 4-h, face-mask exposures of normal (1980) and asthmatic (1981) subjects at 4 NO_2 levels between 0.1 and 1 ppm. The results of these two studies were virtually identical. These investigators found no consistent changes in pulmonary resistance or compliance, arterial oxygen saturation, lung volumes, diffusing capacity, or breathing mechanics immediately following or up to 1–2 wk after NO_2 exposure.

Two of the most recent experimental studies reported are by Ahmed et al., who measured bronchial reactivity in normals and in subjects with intrinsic asthma (1982) and allergic asthma (1983) after a 1-h exposure to 0.1 ppm NO_2. For the respective asthmatic subjects, either carbachol or ragweed antigen was used for bronchial challenges. Ahmed and co-workers reported variable effects of NO_2 on bronchial reactivity (1982) and no significant effect on immediate or latent reactivity to ragweed antigen (1983). In their 1982 study, 10 of 20 normal subjects were found to be "responders" by a preliminary assessment of individual dose-responses to carbachol aerosols. In these "responders," 0.1 ppm NO_2 caused an enhancement of the carbachol-induced fall in specific airway conductance (SGaw). Similarly, in 13 of 20 asthmatic "responders," NO_2 significantly enhanced the carbachol effect on SGaw. In the "nonresponder" groups studied, normals and asthmatics, NO_2 exposure induced no such actions in this single-blind experiment with a cross-over design of 1-h exposures to clean air and 0.1 ppm NO_2.

A study showing no effect of 0.1 ppm NO_2 on airway responsiveness of normal and asthmatic subjects after a 1-h exposure, and when followed by a methacholine challenge, was reported by Hazucha and co-workers (1983). In this double-blind, crossover study, following exposure to NO_2 there was a slight increase in mean airway resistance in both normals and asthmatics, but the changes were not significant. All three of these investigations at 0.1 ppm NO_2 were directed at confirming or refuting the Orehek study. Only the study of Ahmed et al. (1982) appears to provide support to the findings of Orehek et al. (1976).

In 1979, Keller et al., Zagraniski et al., and Florey et al. reported epidemiologic studies. Keller et al. examined respiratory illness in homes with gas and with electric cooking, as did Melia et al., but concluded there was no significant difference in the incidence of acute respiratory illness in the two situations. Zagraniski et al. reported a positive correlation between photochemical oxidant levels and eye irritation and cough, especially among smokers and those with predisposing illnesses, but NO_2, per se, could not be deemed causal.

In a study of 808 school children, Florey and co-workers found a higher prevalence of respiratory illness in homes with gas compared to electric cooking. However, they were unable to correlate prevalence of illness with kitchen levels of NO_2 as well as they could correlate bedroom levels. Both levels were highly variable, and neither correlated with lung function tests, leading the authors to speculate that other factors may be involved than NO_2. However, the later studies of Mostardi et al. (1981) and Speizer et al. (1980) both appeared to incriminate NO_2 with an increased incidence of acute respiratory illness among children. Speizer et al. reexamined homes with gas and electric stoves, since the Melia study was possibly flawed and the smaller Keller study was contradictory. They found that in homes with gas cooking there was increased incidence of respiratory illness in preschoolers and a small but significant loss in pulmonary function among school children. The average NO_2 levels in homes with gas cooking were generally below the 24-h alert level of 0.5 ppm, but brief peak levels of 0.5 ppm or higher were noted.

Mostardi et al. investigated acute respiratory illness in school children next to an industrial site having mean ambient levels of 77.5 $\mu g\ SO_2/m^3$ and 54.5 $\mu g\ NO_2/m^3$ (~0.03 ppm), and in school children at a less polluted size, viz., 21.4 and 36.9 $\mu g\ NO_2/m^3$. Their study concluded that the children from the polluted area had more daily symptoms, more pulmonary functional deficits (e.g., $FEV_{1.0}$ and MMF), more severe respiratory illnesses, and longer recovery times. Like the Zagraniski study, the effects of NO_2 could not be isolated and evaluated.

Chestnut Ridge, a rural area of Pennsylvania with coal-fired power plants, has been the site of an epidemiologic study of 4071 school children. Preliminary information (Schenker et al., 1981, 1982) from multiple logistic analyses indicate positive correlations of air pollutant and parental smoking with respiratory illness, but the presence of gas cooking was only weakly associated with respiratory problems in preschool children. Atmospheric levels of NO_2 correlated strongly with those of SO_2. Consequently, there is a limited expectation that this study can single out the effects of any specific pollutant, e.g., NO_2.

The Six-Cities study (Dockery et al., 1982) is a larger, long-range, prospective epidemiologic effort to assess air pollution and pulmonary function in both adults and children. While specific relationships to NO_2 levels are not reported and unlikely, analyses of pulmonary functions in 4500 children indicate that indoor air pollution (gas stoves and maternal smoking) caused small but significant reductions in pulmonary performance: these are described as "of little medical significance."

CRITIQUE

Nonhuman studies have been the rule, not the exception, in biomedical research of NO_2 toxicology. To better utilize this vast informational base, extrapolation to humans must be more confidently and competently undertaken. Extrapolation of one type of information to another, such as going from the acute to the chronic condition, is another example of this general pursuit where factors of control, capability, and cost dictate the feasibility of undertaking investigations especially protracted ones. And, of course, there is the recognized difficulty of extrapolating from the simplified model, physical or biological, to the "real world." These and other vexing issues will be attended by other papers in this journal volume. For this overview of nitrogen oxide, let us return to the NO_2 data and reflect upon its general suitability and limitations in relation to the current NAAQS for NO_x.

Acute and subacute NO_x studies vary immensely in their apparent relevance to ambient air standards, but relevance, like beauty, is in the eye of the beholder. Nevertheless, it is difficult to evaluate some NO_2 studies because we are still unable to delineate systematically the spectrum of NO_2 toxicity from acute to chronic. There are impressive distinctions between the fulminating pulmonary edema of the acute phase to the more insidious obstructive disease of the chronic. But more and more we see evidence that edema is produced by still lower NO_2 concentrations and the chronicity of NO_2 effects remains uncertain. Is the injury induced in animals surviving acute NO_2 exposures entirely reversible? Does it eventually produce fibrotic or emphysematous changes similar to those occurring with chronic exposures?

The hyperplasia resulting from NO_2 exposure is similarly confusing. Imaging analyses have shown that progenitor cell proliferation is induced by comparatively low NO_2 levels. Are we certain this proliferative response is based exclusively on type I injury? Witschi (1976) has discussed this general problem and raised the same issue in relation to other substances.

Tissue repair, like tissue injury, usually results in changes in cell populations. This can pose two problems: during injury, the responses measured, particularly with subacute and chronic exposures to NO_2, are often explicable as induced effects—e.g., an increase in a specific enzyme—when cell proliferation or hyperplasia can also account for the same change; second, epithelial cell repair may result in altered cell populations, and the original state of function will not be restored exactly. This may be a basis for oxidant tolerance and may also explain the unusual resistance of the neonate to NO_2. Of course, in the latter case, the phenomenon is attributable to cell population changes due to development, not repair.

The mechanistic bases for NO_2 injury remain unclear. Intuitively, the effects of oxidant on membrane lipids are suspect as an initiating step and there are many biochemical data that seem to support this. But the overall scenario of lipid peroxidation remains controversial, as does antioxidant compensation or inhibition of oxidative phenomena.

It is evident that NO_2 can oxidize lipids under various conditions *in vitro*, but the evidence regarding lipid peroxidation as a primary mechanism of NO_2 damage to lung is inconclusive. The possibility of lipid peroxidation as a natural consequence of dying or degenerating tissue makes it impossible to establish it as the *cause* of toxicity rather than the effect in most studies. The exacerbation of NO_2 toxicity by vitamin E deficiency is often interpreted as supporting lipid peroxidation as a mechanism, while in reality such an observation may only demonstrate the cooperative effect of a second stress, vitamin deficiency.

Little is known regarding the chemical interactions of NO_2 with lung tissue *in vivo*. Polyunsaturated lipids and sulfhydryl groups are two likely groups reactant with NO_2, but systematic, consistent information regarding their fate during *in vivo* NO_2 exposures is unavailable.

Measurement of "antioxidant" enzymes in NO_2-exposed lungs is a popular approach, with increases in glutathione reductase, glutathione peroxidase, catalase, and superoxide dismutase being interpreted as the "intelligent" response of the assaulted tissue to oxidant stress. An equally feasible interpretation of such results, however, is simply an increase in cell content due to proliferative or inflammatory response to injury, and usually insufficient information is available to distinguish between these alternatives. Characterization of the various individual lung cell types after

NO_2 insult might permit a more conclusive interpretation of the antioxidant responses and also shed light on the interesting but poorly understood phenomenon of tolerance.

The effects of NO_2 on pulmonary immune responses appear to be dose-dependent in their outcome. Some studies suggest little effect; others suggest activation or suppression depending not only on dosage, but on species and endpoints. In any case, it is premature to evaluate the potential effects of NO_2 exposure on the immune apparatus. Suffice it to say, these effects in concert with infectivity studies suggest something important can occur in host immune defenses of lungs.

The significance of NO_2 in mutagenesis-carcinogenesis and in the facilitation of cancer metastases are obviously matters of considerable concern. The enhanced cytotoxic response of AM in the study of Sone et al. (1983) appears superficially to be contrary to the findings discussed by Richters and Richters (1983) on metastatic facilitation. But there are so many distinctions in the approaches and in the experimental findings that one must hope many investigators will closely and systematically examine these issues as soon as possible.

Although it is not apparent how much of the overall biochemical, immunologic and cytologic effort is directed at discovering a sensitive index of NO_2 exposure—i.e., NO_2-induced injury—it is evidently an objective of some studies, and certainly a matter of considerable importance. It would be especially significant if such an index of pulmonary injury, primary or secondary, were both quantitative and applicable to intact animals and humans. According to an article by Ember (1979), Watanabe and Fukase have proposed reduced glutathione levels as a quantitative indicator of ambient oxidant gas exposure. If this claim is valid, it is a start; unfortunately, this analysis cannot be applied to the intact animal.

Despite the emergence of more experimental studies in humans, there continues to be a remarkable inconsistency in findings. Moreover, all of the recent experimental studies at 1.0, 0.5, and 0.1 ppm NO_2 were deemed negative by the respective groups of investigators. This development is even more significant when phrased in relation to susceptibility: none of the studies indicate any significantly enhanced susceptibilities toward NO_2 among COPD and mild asthmatic patients. However, it should be further noted that all of these human assessments were based primarily on bronchial responsiveness, a criterion that might be regarded as less than ideal for evaluating NO_2 toxicity.

Inconsistency in results may be partly a problem of variations in techniques particularly in the case of human experimental studies. It may be due in part to the use of inappropriate and insensitive methods of testing, and very probably to the variable presence of subpopulations, i.e., of the so-called responders. For example, in Orehek's study, 13 of 20 subjects were retrospectively classified as responders, whereas in the study of Hazucha et al. 3 of 15 were prospectively identified as such. Conversely, Hazucha et al. applied more demanding criteria to the selection of asthmatics initially. According to these investigators, their selection criteria would have eliminated two-thirds of Orehek's asthmatic subjects from consideration. How the normal and asthmatic "responders" in Ahmed's study (1982) would fit into this controversy is unclear. But their study clearly indicates the value of prospective testing and identification of susceptible subpopulations.

In the realm of NO_2 epidemiology, there has also been a disconcerting inconsistency in findings. Epidemiologic studies seem to be more prone to having a flawed design than do the experimental studies. However, the extensive studies of Melia et al. and of Speizer et al. both appear to provide a strong case for the increased susceptibility of young school and preschool children to low NO_2 levels. Although it was the indoor environmental levels of NO_2 under scrutiny in these studies, not the outdoor levels, it is probably completely appropriate to regard these findings as applicable to environmental air pollution generally. Their uniqueness can be regarded as a matter of special circumstance where NO_2 exposures were convincingly different in the groups of cohorts and where both groups were exposed to air relatively free of other pollutants at concentrations high enough to be suspect.

Neither the weak association of gas cooking with increased respiratory illness among preschool children in the Chestnut Ridge study (Schenker et al., 1981) nor the small but significant decrease in pulmonary performance in school children attributed to indoor pollution in the Six-Cities study (Dockery et al., 1982) can be interpreted simply as NO_2-induced effects. In fact, the newer epidemiologic surveys tend to reduce further the clarity of relationships implied by the initial gas cooking studies. The recent difficulties in reproducing the findings of the earlier epidemiology studies may be due in part to the more complex settings selected for these later studies. In any case, the impact of the newest studies on our appreciation of the health-related effects of NO_2 and/or indoor pollution is far from clear.

The World Health Organization recently proposed a short-term (1-h) ambient air standard for NO_x. Their report describes urban NO and NO_2 levels as having annual means in the range of 49–95 and 20–90 $\mu g/m^3$, respectively. The highest daily means for NO_2 range between 130–400 $\mu g/m^3$, while the highest hourly means range between 240 and 850 μg NO_2/m^3. The WHO Summary (1977) *estimates* that 940 μg NO_2/m^3 (~0.5 ppm) is the lowest level producing adverse health effects on a short-term basis. They also state that a minimum safety factor of 3–5 would be consistent with protection of the public health; consequently, they proposed a 1-h standard of 190–320 μg NO_2/m^3 (0.10–0.17 ppm) with the proviso that this exposure limit is not exceeded more than once a month.

The WHO recommendation has the general character of an occupational standard. The advocacy of a range is tantamount to establishing the higher 0.17 ppm value as the 1-h standard. This level is found to be within the range of highest *daily* means, so in certain communities the expectation would be for occasional, excessive 1-h exposures. The NO_2 level to which these exposures can be exceeded is not delimited. If multiple excessive exposures do occur in any month, and there is such a direct inference to adverse health effects, how will the public, courts, and regulators cope with this standard?

The WHO 1-h standard for NO_x appears to have some merit conceptually, except for the safety factor of 3, which is both unprecedented and indefensible in its meagerness. The WHO 1-h standard is also fraught with ambiguities. Its virtual equivalence to the 24-h NAAQ alert for NO_2 of 0.15 ppm is an unmistakeable relaxation of the NO_2 ambient standards. This relaxation is based on an undocumentable, estimated threshold level for adverse effects in humans and the use of the peculiar safety factor cited. It is indeed remarkable that the WHO, which candidly stated insufficient data were available 10 yr ago to justify an ambient air standard for NO_x, has now proposed an ambient standard using the presumption of an adequate NO_x data base for short-term exposures.

While the WHO action falls outside of a review of biomedical studies of the nitrogen oxides, its possible impact on the assessments of the available NO_x data cannot be ignored.

SUMMARY AND CONCLUSIONS

In an earlier review of NO_x-related health effects information (Morrow, 1975), it was concluded that extant biological data together with conventional toxicological principles indicated that the present annual NAAQS for NO_x was not overly conservative. Thus, the main question to be answered in this review is simply; does the newer NO_x toxicity information refute or support this conclusion? My answer must be that it continues to support that earlier position, and my reasons for this opinion follow.

New experimental data, in general, have been described in terms of lower NO_2 levels. This has had the effect of reducing the difference between levels producing biological injury or change and the ambient air standard.

Attempting to derive or crititicize ambient standards for NO_2 by the use of presently available biochemical information would be, at best, a capricious exercise. Detailed and quantitative data are needed on the chemical interactions of NO_2 and tissue and on these interactions and health-related effects before this mechanistic information can be utilized.

Inconsistencies, particularly in the human experimental data, are very disconcerting and generally attest to the difficulty of that approach, irrespective of its apparent relevance. Were it not for several major epidemiologic studies, which, among other findings, related acute respiratory illness in young children with ambient NO_2 levels, the human experimental data might tend to suggest that present standards are too restrictive. However, most 1- to 2-h short-term experimental studies with exposure at 1.5 ppm NO_2 or higher levels have consistency shown pulmonary effects, albeit effects that may not be completely ideal for evaluating actions on the known NO_2 targets of the deep lungs. This gives only a safety factor of 30 when compared to the annual standard of 0.05 ppm. Of course, this relates a short-term exposure condition to a long-term standard; this impropriety is in the direction of reducing the factor of 30 substantially, although we do not know by how much. So on balance, we have some experimental data that do not convey that there is undue conservatism in the annual standard, and some relatively improved epidemiology that implicates elevated ambient levels of NO_2, indoors and outdoors.

One pervasive weakness in the epidemiology is the poorly documented exposure history to NO_2. Greater attention should have been given, and must be given, to the time-course of intermittent, peak levels. On the basis of the available information, it seems very unlikely that much, if any, of the epidemiology is based on >1.5 ppm NO_2 peak levels, however. So again one can conclude that there are no strong signals to modify the present NAAQS, except that at least one short-term standard is needed. We can all hope that this will be more skillfully achieved than was the proposed WHO short-term standard.

The possibility has been raised that NO_2 levels in houses with gas cooking might be surrogates for other pollutants and factors. Although epidemiologists have shown increased respiratory illness and decreased pulmonary performance in children living in houses with gas cooking, they were only partially successful correlating morbidity with NO_2 levels. This matter should be explored further, with special attention given to the limitation of using mean NO_2 levels whenever there are highly variable atmospheric concentrations (60-fold in one study cited). The use of personal samplers was discussed at a recent meeting on air-pollution measurements (Quackenboss et al., 1983); it is conceivable that these devices may provide a basis for coping with the highly variable NO_2 levels encountered in epidemiologic surveys.

Although the evidence is not compelling, there are reasons to question the viewpoint that among the nitrogen oxides NO_2 is the singular toxin. Also, evidence is accumulating that the oxidant gases ozone, nitrogen oxide, and oxygen need to be regarded as separate toxicologic species having some common traits. An unclear consideration related to the foregoing is the toxicologic status of nongaseous, derivative materials—e.g., nitric acid, nitrites, nitrates, and organic reaction products. This problem is particularly intriguing in view of the known chemical conversions of absorbed NO_2 and the complex chemistry of atmospheric NO_x.

Finally, more attention must be given to the greater utilization of experimental information in the development of a NO_2 toxicity data base. This conference on extrapolation modeling is a testimonial to that need.

REFERENCES

Ahmed, T., Marchette, B., Danta, I., Birch, S., Dougherty, R. L., Schreck, R., and Sackner, M. A. 1982. Effects of 0.1 ppm NO_2 on bronchial reactivity in normals and subjects with bronchial asthma. *Am. Rev. Respir. Dis.* 125:152.

Ahmed, T., Danta, I., Dougherty, R. L., Schreck, R., and Sackner, M. A. 1983. Effect of NO_2 (0.1 ppm) on specific bronchial reactivity to ragweed antigen in subjects with allergic asthma. *Am. Rev. Respir. Dis.* 127:160.

Azoulay-Dupuis, E., Torres, M., Soler, P., and Moreau, J. 1983. Pulmonary NO_2 toxicity in neonate and adult guinea pigs and rats. *Environ. Res.* 30:322–339.

Beil, M., and Ulmer, W. T. 1976. Wirkung von NO_2 in MAK-Bereish auf atemmechanik Acetylcholinempfindlichketi bei Normalpersonen. *Int. Arch. Occup. Environ. Health* 38:31–44.

Book, S. A. 1982. Scaling toxicity from laboratory animals to people: An example with nitrogen dioxide. *J. Toxicol. Environ. Health* 9:719–725.

Cabral-Anderson, L. J., Evans, M. J., and Freeman, G. 1977. Effects of NO_2 on the lungs of aging rats. I. Morphology. *Environ. Mol. Pathol.* 27:353–365.

Case, B. W., Gordon, R. E., and Kleinerman, J. 1982. Acute bronchiolar injury following nitrogen dioxide exposure: A freeze fracture study. *Environ. Res.* 29:399–413.

Chaudhari, S., and Dutta, S. 1981. Possible formation of nitrosamine in guinea pigs following exposure to nitrogen dioxide and dimethylamine. *J. Toxicol. Environ. Health* 7:753–763.

Coffin, D. L., Gardner, D. E., Sidorenko, G. I., and Pinigin, M. A. 1977. Role of time as a factor in the toxicity of chemical compounds in intermittent and continuous exposures. Part III: Effects of intermittent exposure. *J. Toxicol. Environ.* 3:821–828.

Crapo, J. D., Sjostrom, K., and Drew, R. T. 1978a. Tolerance and cross-tolerance using NO_2 and O_2. I. Toxicology and biochemistry. *J. Appl. Physiol.* 44:364–369.

Crapo, J. D., Marsh-Salin, J., Ingram, P., and Pratt, P. C. 1978b. Tolerance and cross-tolerance using NO_2 and O_2. II. Pulmonary morphology and morphometry. *J. Appl. Physiol.* 44:370–379.

Csallany, A. S., and Ayaz, K. L. 1978a. The effects of intermittent nitrogen dioxide exposure on vitamin E-deficient and -sufficient rats. *Toxicol. Lett.* 2:97–107.

Casallany, A. S., and Ayaz, K. L. 1978b. Long-term NO_2 exposure of mice in the presence or absence of vitamin E. I. Effect on body weight and lipofusion in pigments. *Arch. Environ. Health* 33:285–291.

Dawson, S. V., and Schenker, M. B. 1979. Health effects of inhalation of ambient concentrations of nitrogen dioxide. *Am. Rev. Respir. Dis.* 120:281–292.

DeNicola, D. B., Rebar, A. H., and Henderson, R. F. 1981. Early damage indicators in the lung. V. Biochemical and cytological response to NO_2 inhalation. *Toxicol. Appl. Pharmacol.* 60:301–312.

Dockery, D. W., Ware, J. H., Speizer, F. E., and Ferris, B. G. 1982. Preliminary longitudinal analysis of pulmonary function in school children in the Six-Cities study. *Am. Rev. Respir. Dis.* 125:145.

Dróždž, M., Kicharz, E., and Szyja, J. 1977. Effect of chronic exposure to nitrogen dioxide on collagen content in lung and skin of guinea pigs. *Environ. Res.* 13:369–377.

Ehrlich, R., Findlay, J. C., Fenters, J. D., and Gardner, D. E. 1977. Health effects of short-term inhalation of nitrogen dioxide and ozone mixtures. *Environ. Res.* 14:223–231.

Elsayed, N. M., and Mustafa, M. G. 1982. Dietary antioxidants and the biochemical response to oxidant inhalation. I. Influence of dietary vitamin E on the biochemical effects of nitrogen dioxide exposure in rat lung. *Toxicol. Appl. Pharmacol.* 66:319–328.

Ember, L. R. 1979. NO_2: The air standard and health effects. *Environ. Sci. Technol.* 13:642–644.

Evans, M. J., and Freeman, G. 1980. Morphological and pathological effects of NO_2 in rat lung. In *Nitrogen Oxides and Their Effects on Health,* ed. S. D. Lee, pp. 243–266. Ann Arbor, Mich.: Ann Arbor Science.

Evans, M. J., Cabral, L. J., Stephens, R. J., and Freeman, G. 1973. Renewal of alveolar epithelium in the rat following exposure to NO_2. *Am. J. Pathol.* 70:175–190.

Evans, M. J., Cabral, L. J., Stephens, R. J., and Freeman, G. 1975. Transformation of alveolar type 2 cells to type 1 cells following exposure to NO_2. *Exp. Mol. Pathol.* 22:142–150.

Evans, M. J., Johnson, L. V., Stephens, R. J., and Freeman, G. 1976. Renewal of the terminal bronchiolar epithelium in the rat following exposure to NO_2 or O_3. *Lab. Invest.* 35:246–257.

Evans, M. J., Cabral-Anderson, L. J., and Freeman, G. 1978a. Role of the Clara cell in renewal of the bronchiolar epithelium. *Lab. Invest.* 38:648–655.

Evans, M. J., Dekker, N. P., Cabral-Anderson, L. J., and Freeman, G. 1978b. Quantitation of damage to the alveolar epithelium by means of type 2 cell proliferation. *Am. Rev. Respir. Dis.* 118:787–790.

Fenters, J. D., Findlay, J. C., Port, C. D., Ehrlich, R., and Coffin, D. L. 1973. Chronic exposure to nitrogen dioxide. *Arch. Environ. Health* 27:85–89.

Finklea, J. F., French, J. G., Lawrimore, G. R., Goldberg, J., Shy, C. M., and Nelson, W. C. 1974. Prospective surveys of acute respiratory disease in volunteer families: Chicago nursery school study, 1969–1970. In *Health Consequences of Sulfur Oxides: A Report from CHESS, 1970–1971.* Report EPA-650/1-74-004, U.S. Environmental Protection Agency, Research Triangle Park, N.C.

Fishbein, L. 1976. Atmospheric mutagens. I. Sulfur oxides and nitrogen oxides. *Mutat. Res.* 32:309–330.

Florey, C. du V., Melia, R. J. W., Chinn, S., Goldstein, B. D., Brooks, A. G. F., John, H. H., Craighead, I. B., and Webster, X. 1979. The relation between respiratory illness in primary schoolchildren and the use of gas for cooking. III. Nitrogen dioxide, respiratory illness and lung infection. *Int. J. Epidemiol.* 8:347–353.

Freeman, G., Crane, S. C., and Furiosi, N. J. 1969. Healing in rat lung after subacute exposure to nitrogen dioxide. *Am. Rev. Respir. Dis.* 100:662–676.

Fujimaki, H., Shimizu, F., and Kubota, K. 1982. Effect of subacute exposure to NO_2 on lymphocytes required for antibody responses. *Environ. Res.* 29:280–286.

Gardner, D. E., Coffin, D. L., Pinigin, M. A., and Sidorenko, G. I. 1977. Role of time as a factor in the toxicity of chemical compounds in intermittent and continuous exposures. Part I. Effects of continuous exposures. *J. Toxicol. Health* 3:811–820.

Goldstein, E., Peek, N. F., Parks, N. J., Hines, H., Steffey, E. P., and Tarkington, B. 1977. Fate and distribution of inhaled nitrogen dioxide in Rhesus monkeys. *Am. Rev. Respir. Dis.* 115:403–412.

Goldstein, B. D., Melia, R. J. W., Chinn, S., Florey, C. du V., Clark, D., and John, H. H. 1979. The relation between respiratory illness in primary schoolchildren and the use of gas for cooking. II. Factors affecting nitrogen dioxide levels in the home. *Int. J. Epidemiol.* 8:339–345.

Graham, J. A., Illing, J. W., Miller, F. L., and Gardner, D. E. 1982. Influence of nitrogen dioxide on xenobitic metabolism in animals. In *Air Pollution by Nitrogen Oxides,* eds. T. Schneider and L. Grant, pp. 467–475. Amsterdam: Elsevier Scientific.

Graham, J. A., Miller, F. J., Illing, J. W., Davies, D., and Gardner, D. E. 1983. Effect of urban patterns of nitrogen dioxide (NO_2) on susceptibility to pulmonary infection. *Toxicologist* 3:12.

Guidotti, T. L. 1978. The higher oxides of nitrogen: Inhalation toxicology. *Environ. Res.* 15:443–473.

Guidotti, T. L. 1980. Toxic inhalation of nitrogen dioxide: Morphologic and functional changes. *Exp. Mol. Pathol.* 33:90–103.

Hackett, N. A. 1979. Proliferation of lung and airway cells induced by nitrogen dioxide. *J. Toxicol. Environ. Health* 5:917–928.

Hackney, J. D. 1976. Effects of atmospheric pollutants on human physiologic function. Final report, U.S. Environmental Protection Agency, contract R-801396.

Hackney, J. D., Linn, W. S., Mohler, J. G., Pedersen, E. E., Preisacher, P., and Russo, A. 1975a. Experimental studies on human health effects on air pollutants. II. Four-hour exposure to ozone alone and in combination with other pollutant gases. *Arch. Environ. Health* 30:379–384.

Hackney, J. D., Linn, W. S., Law, D. C., Karuza, S. K., Greenberg, H., Buckley, R. D., and Pedersen, E. E. 1975b. Experimental studies on human health effects on air pollutants. III. Two-hour exposure to ozone alone and in combination with other gases. *Arch. Environ. Health* 30:385–390.

Hackney, J. D., Thiede, F. C., Linn, W. S., Pedersen, E. E., Spier, C. E., Law, D. C., and Fischer, D. A. 1978. Experimental studies on human health effects of air pollutants. IV. Short-term physiological and clinical effects of nitrogen dioxide exposure. *Arch. Environ. Health* 33:176–181.

Hazucha, M. J., Ginsberg, J. F., McDonnell, W. F., Haak, E. D., Jr., Pimmel, R. L., Salaam, S. A., House, D. E., and Bromberg, P. A. 1983. Effects of 0.1 ppm nitrogen dioxide on airways of normal and asthmatic subjects. *J. Appl. Physiol.* 54:730–739.

Holt, P. G., Finlay-Jones, L. M., Keast, D., and Papadimitrou, J. M. 1979. Immunological function in mice chronically exposed to nitrogen oxides (NO_x). *Environ. Res.* 19:154–162.

Horvath, S. M., and Folinsbee, L. J. 1975. The effect of NO_2 on lung function in normal subjects. Final contract report, U.S. Environmental Protection Agency contract 68-02-1757.

Ichinose, T., and Sagai, M. 1982. Studies on biochemical effects of nitrogen dioxides. III. Changes of the antioxidative protective systems in rat lungs and of lipid peroxidation by chronic exposure. *Toxicol. Appl. Pharmacol.* 66:1–8.

Iqbal, Z. M., Dahl, K., and Epstein, S. S. 1981. Biosynthesis of dimethylnitrosamine in dimethylamine-treated mice after exposure to nitrogen dioxide. *J. Natl. Cancer Inst.* 67:137–141.

Johnson, W. K., Mauderly, J. L., Hahn, F. F., and Muggenburg, B. A. 1982. Lung function and morphology of dogs after sublethal exposure of nitrogen dioxide. *J. Toxicol. Environ. Health* 10:201–221.

Keller, M. D., Lanese, R. R., Mitchell, R. I., and Cote, R. W. 1979. Respiratory illness in household using gas and electricity for cooking. II. Symptoms and objective findings. *Environ. Res.* 19:504–515.

Kerr, H. D., Kulle, T. J., McIlhany, M. L., and Swedersky, P. 1978. *Effects of Nitrogen Dioxide on Pulmonary Function in Human Subjects. An Environmental Chamber Study.* U.S. EPA Publication EPA-600/1-78-025. U.S. Environmental Protection Agency, Research Triangle Park, N.C.

Kerr, H. D., Kulle, T. J., McIlhany, M. L., and Swedersky, P. 1979. Effects of nitrogen dioxide on pulmonary function in human subjects: An environmental chamber study. *Environ. Res.* 19:392–404.

Kim, J. C. S. 1978. The effect of dietary vitamin A on NO_2 exposure on the hamster lung. *Environ. Res.* 17:116–130.

Kleinerman, J. 1977. Some effects of nitrogen dioxide on the lung. *Fed. Proc.* 36:1714–1718.

Kuraitis, K. V., Richters, A., and Sherwin, R. P. 1981. Spleen changes in animals inhaling ambient levels of nitrogen dioxide. *J. Toxicol. Environ. Health* 7:851–859.

Lee, S. D., ed. 1980. *Nitrogen Oxides and Their Effects on Health*. Ann Arbor, Mich.: Ann Arbor Science.

Lewis, T. R., Moorman, W. J., Young, Y., and Stara, J. F. 1974. Long-term exposure to auto exhaust and other pollutant mixtures. Effects on pulmonary function in the beagle. *Arch. Environ. Health* 29:102–106.

Love, G. J., Cohen, A. A., Finklea, J. F., French, J. G., Lawrimore, G. R., Nelson, W. C., and Ramsey, P. B. 1974. Prospective surveys of acute respiratory disease in volunteer families: 1970–1971 New York studies. In *Health Consequences of Sulfur Oxides: A Report from CHESS, 1970–1971.* EPA-650/1-74-004. U.S. Environmental Protection Agency, Research Triangle Park, N.C.

Love, G. J., Lan, S., Shy, C. M., and Riggan, W. B. 1982. Acute respiratory illness in families exposed to nitrogen dioxide ambient air pollution in Chattanooga, Tennessee. *Arch. Environ. Health* 37:75–80.
Malatinsky, J., Kodlic, T., and Kovacik, V. 1973. Acute poisoning by higher nitrogen oxides. *Anesth. Analg. (Cleve.)* 52:94–99.
Melia, R. J. W., Florey, C. du V., Altman, D. G., and Swan, A. V. 1977. Association between gas cooking and respiratory disease in children. *Br. Med. J.* 2:149–152.
Melia, R. J. W., Darley, S. C., Palmes, E. D., and Goldstein, B. D. 1978. Differences in NO_2 levels in kitchens with gas or electric cookers. *Atmos. Environ.* 12:1379–1381.
Menzel, D. B. 1976. The role of free radicals in the toxicity of air pollutants (nitrogen oxides and ozone). In *Free Radicals in Biology,* vol. III, ed. W. A. Pryor, pp. 181–202. New York: Academic.
Miller, F. J., and Graham, J. A. 1983. Uptake of NO_2 in animals and man. *Toxicologist* 3:120.
Miller, F. J., Overton, J. H., Meyers, E. T., and Graham, J. A. 1982. Pulmonary dosimetry of nitrogen dioxide in animals and man. In *Air Pollution by Nitrogen Oxides,* eds. T. Schneider and L. Grant, pp. 377–386. Amsterdam: Elsevier Scientific.
Mitchell, R. I., Williams, R., Cote, R. W., Lanese, R. R., and Keller, M. D. 1974. Household survey of the incidence of respiratory disease in relation to environmental pollutants. In *WHO International Symposium Proceedings: Recent Advances in the Assessment of the Health Effects of Environmental Pollutants,* Paris, June 24–28.
Morrow, P. E. 1975. An evaluation of recent NO_X toxicity data and an attempt to derive an ambient air standard for NO_X by established toxicological procedures. *Environ. Res.* 10:92–112.
Mostardi, R. A., Woebkenberg, N. R., Ely, D. L., Conlon, M., and Atwood, G. 1981. The University of Akron study on air pollution and human health effects. II. Effects on acute respiratory illness. *Arch. Environ. Health* 36:250–255.
Mustafa, M. G., and Tierney, D. F. 1978. Biochemical and metabolic changes in the lung with oxygen, ozone, and nitrogen dioxide toxicity. *Am. Rev. Respir. Dis.* 118:1061–1090.
National Research Council (NRC). 1977. *Nitrogen Oxides: Medical and Biological Effects of Environmental Pollutants,* pp. 1–333. Washington, D.C.: National Academy of Sciences.
National Research Council (NRC). 1978. *Nitrates. An Environmental Assessment,* pp. 723. Washington, D.C.: National Academy of Sciences.
Orehek, J., Massari, J. P., Gayrard, P., Grimaud, C., and Charpin, J. 1976. Effect of short-term, low-level nitrogen dioxide exposure on bronchial sensitivity of asthmatic patients. *J. Clin. Invest.* 57:301–307.
Pickrell, J. A., Gregory, R. E., Johnson, W. K., Hahn, F. F., Bice, D. E., Henderson, R. F., and Hobbs, C. H. 1982. Respiratory toxicology following inhalation of NO_X. Lovelace Biomedical and Environmental Research Institute Report LMF-101/UC-48, Aug. 1982, pp. 1–29.
Pitts, J. N. Jr., Winer, A. M., Atkinson, R., and Carter, P. L. 1983. Comment on "Effect of nitrogen oxide emissions on ozone levels in metropolitan regions," "Effect of NO_X emission rates on smog formation in the California South Coast air basin," and "Effects of hydrocarbon and NO_X on photochemical smog formation under simulated transport conditions." *Environ. Sci. Technol.* 17:54–57.
Postlethwait, E. M., and Mustafa, M. G. 1981. Fate of inhaled nitrogen dioxide in isolated perfused rat lung. *J. Toxicol. Environ. Health* 7:861–872.
Quackenboss, J. J., Spengler, J. D., Letz, R., Kanarek, M. S., Duffy, C. P., Lindsay, M. D., and Knight, W. C. 1983. Models of human exposure to nitrogen dioxide using personal monitoring data. In *U.S. Environmental Protection Agency Third Annual National Symposium on Recent Advances in Pollutant Monitoring of Ambient Air and Stationary Sources,* pp. 71–72. Raleigh, N.C., May 3–6.
Ranga, V., and Kleinerman, J. 1977. A quantitative study of ciliary injury in the small airways of mice: The effects of nitrogen dioxide. *Environ. Lung Res.* 2:49–55.
Richters, A., and Kuraitis, K. 1981. Inhalation of NO_2 and blood borne cancer cell spread to the lungs. *Arch. Environ. Health* 36:36–39.

Richters, A., and Richters, V. 1983. A new relationship between air pollutant inhalation and cancer. *Arch. Environ. Health* 38:69–75.

Richters, V., Elliott, G., and Sherwin, R. P. 1978. Influence of 0.5 ppm nitrogen dioxide exposure of mice on macrophage congregation in the lungs. *In Vitro* 14:458–464.

Richters, V., Wood, T., and Sherwin, R. P. 1980. The frequency of macrophage congregation in tissue cultures of the mouse lung: Influence of *in vivo* nitrogen dioxide exposure and level of indigenous retrovirus expression. *Fed. Proc.* 39:621.

Roy-Burman, P., Pattengale, P. K., and Sherwin, R. P. 1982. Effect of low levels of nitrogen dioxide inhalation on endogenous retrovirus gene expression. *Exp. Mol. Pathol.* 36:144–155.

Rynbrandt, D., and Kleinerman, J. 1977. Nitrogen dioxide and pulmonary proteolytic enzymes. Effects on lung tissue and macrophages. *Arch. Environ. Health* 32:165–172.

Sackner, M. A., Broudy, M., Friden, A. M., and Cohn, M. A. 1980. Effects of breathing low levels of nitrogen dioxide for four hours on pulmonary function of normal adults. *Am. Rev. Respir. Dis.* 121:254.

Sackner, M. D., Birch, S., Friden, A., and Marchetti, B. 1981. Effects of breathing low levels of nitrogen dioxide for four hours on pulmonary function of asthmatic adults. *Ann. Rev. Respir. Dis.* 123:151.

Sagai, M., Ichinose, T., Oda, H., and Kubata, K. 1982. Studies on biochemical effects on nitrogen dioxide. II. Changes of the protective systems in rat lungs and of lipid peroxidation by acute exposure. *J. Toxicol. Environ. Health* 9:153–164.

Saul, R. L., and Archer, M. C. 1983. Nitrate formation in rats exposed to nitrogen dioxide. *Toxicol. Appl. Pharmacol.* 67:284–291.

Schenker, M. B., Samet, J., and Speizer, F. E. 1981. Biologic and environmental determinants of childhood respiratory disease. *Am. Rev. Respir. Dis.* 123:138.

Schenker, M. B., Samet, J., Batterman, S., and Speizer, F. E. 1982. An epidemiologic study of air pollution effects of childhood respiratory disease. *Am. Rev. Respir. Dis.* 125:145.

Schneider, T., and Grant, L., eds. 1982. *Air Pollution by Nitrogen Oxides.* Amsterdam: Elsevier Scientific.

Sevanian, A., Mead, J. F., and Stein, R. A. 1979. Epoxides as products of lipid autoxidation in rat lungs. *Lipids* 14:634–643.

Sherwin, R. P., and Layfield, L. J. 1976. Protein leakage in the lungs of mice exposed to 0.5 ppm nitrogen dioxide. A fluorescence assay for protein. *Arch. Environ. Health* 31:116–118.

Sherwin, R. P., and Richters, V. 1982. Hyperplasia of type 2 pneumocytes following 0.34 ppm nitrogen dioxide exposure: Quantitation by image analysis. *Arch. Environ. Health* 37:306–315.

Sherwin, R. P., Margolick, J. B., and Azen, S. P. 1973. An automated determination of ratios of type 2 pneumocytes to alveolar wall area using an image analyzer. *Am. Rev. Respir. Dis.* 108:1015–1018.

Sherwin, R. P., Richters, V., and Okimoto, D. 1983a. Type 2 pneumocyte hyperplasia in the lungs of mice exposed to an ambient level of ozone. In *Advances in Modern Environmental Toxicology*, eds. S. D. Lee, M. G. Mustafa, and M. A. Mehlman, vol. 5, Chap. 23, pp. 289–297. International Symposium on Biomedical Effects of Ozone and Related Photochemical Oxidants. Princeton, N.J.: Princeton Scientific.

Sherwin, R. P., Okimoto, D. T., Mundy, D., and Richters, V. 1983b. Serum and lung clearance of exogenous horseradish peroxidase: Influence of low levels of nitrogen dioxide. *J. Am. Coll. Toxicol.* 2:225–236.

Shy, C. M., and Love, G. J. 1980. Recent evidence on the human health effects of nitrogen dioxide oxides. In *Nitrogen Oxides and Their Effects on Health,* ed. S. D. Lee, Chap. 18, pp. 291–305. Ann Arbor, Mich.: Ann Arbor Science.

Sone, S., Brennan, L. M., and Creasia, D. A. 1983. *In vivo* and *in vitro* NO_2 exposures enhance phagocytic and tumoricidal activities of rat alveolar macrophages. *J. Toxicol. Environ. Health* 11:151–163.

Speizer, F. E., Ferris, B., Jr., Bishop, Y. M. M., and Spengler, J. D. 1980. Respiratory disease rates and pulmonary function in children associated with NO_2 exposure. *Am. Rev. Respir. Dis.* 121:3–10.

Spengler, J. D. 1983. Nitrogen dioxide inside and outside 137 homes and implications for ambient air quality standards and health effects research. *Environ. Sci. Technol.* 17:164–168.

Stephens, R. J., Freeman, G., and Evans, M. J. 1972. Early response of lungs to low levels of nitrogen dioxide. Light and electron microscopy. *Arch. Environ. Health* 24:160–172.

Stephens, R. J., Sloan, M. F., Groth, D. G., Negi, D. S., and Lunan, K. D. 1978. Cytologic response of postnatal rat lungs to O_3 or NO_2 exposure. *Am. J. Pathol.* 93:183–200.

Stephens, R. J., Tallent, C., Hart, C., and Negi, D. S. 1982. Postnatal tolerance of NO_2 toxicity. *Environ. Mol. Pathol.* 37:1–14.

Thomas, H. V., Mueller, P. K., and Wright, R. 1967. Response of rat lung mast cells to nitrogen dioxide inhalation. *J. Air Pollut. Control Assoc.* 17:33–38.

U.S. Environmental Protection Agency (EPA), Air Pollution Control Office. 1971. *Air Quality Criteria for Nitrogen Oxides,* p. 188. Air Pollution Control Office Publication No. AP-84. Washington, D.C.: U.S. Government Printing Office.

Valentini, J. E., Deal, D., and Menzel, D. B. 1983. Pulmonary clearance of nitrate aerosols. *Toxicologist* 3:120.

von Nieding, G. 1978. Possible mutagenic properties and carcinogenic action on the irritant gaseous pollutant NO_2, O_3 and SO_2. *Environ. Health Perspect.* 22:91–92.

von Nieding, G., and Wagner, H. M. 1979. Effects of NO_2 on chronic bronchitis. *Environ. Health Perspect.* 29:137–142.

Werthamer, S., Schwarz, L. H., and Soskind, L. 1975. Bronchial epithelial alterations and pulmonary neoplasia induced by ozone. *Pathol. Microbiol.* 35:224–230.

Williams, R. A., Rhoades, R. A., and Adams, W. S. 1971. The response of lung tissue and surfactant to nitrogen dioxide exposure. *Arch. Intern. Med.* 128:101–108.

Witschi, H. 1976. Proliferation of type II alveolar cells: A review of common responses in toxic lung injury. *Toxicology* 5:267–277.

World Health Organization. 1977. *Environmental Health Criteria* 4:*Nitrogen Oxides,* p. 79.

Wright, E. S., and Mavis, R. D. 1981. Changes in pulmonary phospholipid biosynthetic enzymes after nitrogen dioxide exposure. *Toxicol. Appl. Pharmacol.* 58:262–268.

Zagraniski, R. T., Leaderer, B. P., and Stolwijk, J. A. J. 1979. Ambient sulfates, photochemical oxidants, and acute adverse health effects: An epidemiological study. *Environ. Res.* 19:306–320.

Part Two

PULMONARY DOSIMETRY MODELS

INTRODUCTION

Robert F. Phalen

Community and Environmental Medicine
University of California, Irvine, California

Pulmonary dosimetry models provide quantitative information on the locations and amounts of deposition of inhaled toxins, both gaseous and particulate. The specification of location may be quite broad, as in an entire region (head, tracheobronchial airways, or parenchyma), or very specific, for example within each of the many generations of the tracheobronchial tree. The uses for such information include proper selection of laboratory animal species in toxicologic studies, refined interpretation of laboratory animal toxicologic data, and prediction of potential sites of high local-injury from inhaled materials. Further, dosimetry models tend to identify the roles of relevant factors in potential lung injury. Such relevant factors may be physical-chemical (particle size, gas water solubility, etc.), physiologic (air flow patterns, mucus characteristics, etc.), or anatomical (airway dimensions, branch angles, etc.).

Predictive dosimetry models are of two major types, theoretical and empirical. In both types, the ultimate requirements for acceptability are the same; they must be realistic, must be validated, and must be of use in predicting what would occur in unexamined circumstances. Theoretical dosimetry models are constructs which usually begin with simplifications of the real world. Such simplifications occur in the selection of equations describing movement of gases and particles toward airway walls, use of manageable airflow descriptions, and incorporation of anatomical geometries that represent averages of actual lung morphology. Such models are later made more sophisticated by the incorporation of additional complexities.

Empirical modeling usually involves analysis of dosimetric data obtained in constructed hollow models, or less often in actual lungs, which have been exposed to airborne material under conditions simulating breathing. The analysis of such data permits development of mathematical relationships that describe the distribution of material deposited. Such relationships are then used to predict deposition patterns in unstudied circumstances. In addition to their value for predicting dose patterns, the equations that result from theoretical and empirical models can shed light on the mechanisms which control such dose patterns. Knowledge regarding mechanism is a fuel which fires scientific inquiry, stimulating the design of future studies.

Deposition studies using hollow airway models are used to study airflow patterns and deposition sites of airborne particles. Such model studies have led to recognition of secondary airflows within the bronchial tree, the function of the larynx in modifying airflow, and to a better understanding of the detailed deposition patterns for particles. Representative publications in this area include those by Schroter and Sudlow (1969), Olson et al. (1970), Martin and Jacobi (1972), Bell and Friedlander (1973), and Schlesinger et al. (1982).

Validation of any model, theoretical or empirical, is essential to its utility and acceptance by scientists. Validation is a relative term because no model provides an exact description of the great variety of structure and situations that occur in the real world. What is a sufficient validation for one purpose may be insufficient for another. Validation procedures are often more extensive for theoretical models than for empirical ones, and can include testing the predictions in simplified airway models, making airflow measurements in lung or lung-like structures, and performing morphometric measurements on airways. Validation must however eventually include comparison of predicted phenomena with those which occur in living subjects.

The papers presented in this part are examples of current research related to pulmonary dosimetry. The papers by Gehr (1984) and Leith (1984) do not present dosimetric models per se, but rather describe quantitative approaches for generating useful basic anatomical information and air flow information, respectively. In both papers, comparative mammalian biology is stressed. Such a comparative approach is an essential ingredient for the work of extrapolating data acquired in one species to another species.

Because this topic covers a great deal of ground, many similar modeling-oriented anatomical efforts are not included. Among them are the detailed quantitative measurements by Jay Schreider and Otto Raabe on the mammalian nasal cavity and on the respiratory acinus (Schreider and Raabe, 1981a, 1981b), and on mammalian tracheobronchial airways published by a variety of scientists (Horsfield et al., 1971, 1982; Phalen et al., 1978; Weibel, 1963; Yeh and Schum, 1979, 1980; Klimet et al., 1972). Several authors have also published information on comparative mammalian ventilation (Guyton, 1947; Crosfill and Widdicombe, 1961; Boyd and Mangos, 1981; Mauderly, 1974; and Likens and Mauderly, 1979).

The paper on pulmonary dosimetry models by Overton (1984) is representative of recent theoretical dosimetry modeling efforts for inhaled gases. Additional gas (or very fine particle) deposition models have been described by others (Morgan and Frank, 1977; Miller et al., 1978; Hofmann, 1982).

Particle deposition models are not covered, but because of their long history and usefulness they should not be overlooked. Among these models many are quite detailed in their dose pattern predictions (Findeisen, 1935; Landahl, 1950; Morrow et al., 1966; Mercer, 1975; Yeh and Schum, 1980). The foregoing references are merely a small portion of those available.

It is difficult if not impossible to predict the future of scientific research efforts as this future is influenced by the availability of scientific expertise, availability and targeting of research monies, and the ever-changing status of knowledge. Nevertheless, several potentially fruitful research problems are identified. Among them are the following:

1. Additional morphometric information (of the type used as input to theoretical models) is needed on the nasal, tracheobronchial, and parenchymal airways of various mammalian species.
2. The nature of growth of the airways (particularly the human) should be quantitatively defined.
3. Gas dosimetry models require additional validation and sophistication. (For examples, see nos. 4–9 below.)
4. Measurements of gas uptake in whole animals (human and non-human) should be performed and compared to theoretical predictions.
5. Hollow models should be used to study airflow and gas uptake phenomena.
6. Gas uptake models for the nose are in need of further development.
7. Models are needed for the dosimetry of mixtures of gases and particles.
8. The nature of airflow in small airways (terminal bronchioles, respiratory bronchioles, alveolar ducts, and alveolar sacs) must be better defined.
9. The characteristics of normal airway mucus (with respect to dissolution and reaction chemistry) require better definition.
10. Individual variations of dose within a species are important for future research.
11. The persistence of deposited material, gaseous and particulate, in the respiratory tract warrants further study.
12. Alterations in dose and dose patterns during exercise require additional attention.
13. Abnormal situations such as occur in various disease states should be modeled.
14. Doses to specific cell-types within the respiratory tract should be given more attention.
15. Differences between males and females with respect to dosimetry of inhaled materials are potentially important.

REFERENCES

Bell, K. A. and Friedlander, S. K. 1973. Aerosol deposition in models of a human lung bifurcation. *Staub-Reinhalt. Luft.* 33:178–182.

Boyd, R. L. and Mangos, J. A. 1981. Pulmonary mechanics of the normal ferret. *J. Appl. Physiol.: Respir. Environ. Exercise Physiol.* 50:799–804.

Crosfill, M. L. and Widdicombe, J. G. 1961. Physical characteristics of the chest and lungs and the work of breathing in different mammalian species. *J. Physiol. (Lond.)* 158:1–14.

Findeisen, W. 1935. Uber das Absetzen kleiner in der Luft suspendierter Tielchen in der menschlichen Lunge bei der Atmung. *Arch. Gas Physiol.* 26:367–379.

Gehr, P. 1984. Respiratory tract structure and function. *J. Toxicol. Environ. Health* 13:235–249.

Guyton, A. C. 1947. Analysis of respiratory patterns in laboratory animals. *Am. J. Physiol.* 150:78–83.

Hofmann, W. 1982. Dose calculations for the respiratory tract from inhaled natural radioactive nuclides as a function of age. *Health Phys.* 43:31–44.

Horsfield, K., Dart, G., Olson, D. E., Filley, G., and Cumming, G. 1971. Models of the human bronchial tree. *J. Appl. Physiol.* 31:207–217.

Horsfield, K., Kemp, W., and Phillips, S. 1982. An asymmetrical model of the airways of the dog lung. *J. Appl. Physiol.: Respir. Environ. Exercise Physiol.* 52:21–26.

Klimet, V., Libich, J., and Kaudersova, V. 1972. Geometry of guinea pig respiratory tract and application of Landahl's model of deposition of aerosol particles. *J. Hyg. Epidemiol. Microbiol. Immunol.* 16:107–114.

Landahl, H. D. 1950. On the removal of airborne droplets by the human respiratory tract. I. The lung. *Bull. Math. Biophys.* 12:43–56.

Likens, S. A. and Mauderly, J. L. 1979. Respiratory measurements in small laboratory mammals: A literature review, Inhalation Toxicology Research Institute Report LF-68. Available from the National Technical Information Service, Springfield, Va.

Martin, D. and Jacobi, W. 1972. Diffusion deposition of small-sized particles in the bronchial tree. *Health Phys.* 23:23–29.

Mauderly, J. L. 1974. Evaluation of the grade pony as a pulmonary function model. *Am. J. Vet. Res.* 35:1025–1029.

Mercer, T. T. 1975. The deposition model of the task group on lung dynamics: A comparison with recent experimental data. *Health Phys.* 29:673–680.

Miller, F. J., Menzel, D. B., and Coffin, D. L. 1978. Similarity between man and laboratory animals in regional pulmonary deposition of ozone. *Environ. Res.* 17:84–101.

Morgan, M. S. and Frank, R. 1977. Uptake of pollutant gases by the respiratory system. In *Respiratory Defense Mechanisms, Part I*, eds. J. D. Brain, D. E. Proctor, and L. Reid, pp. 157–189. New York: Marcel Dekker.

Leith, D. E. 1984. Mass transportation in mammalian lungs: Comparative physiology. *J. Toxicol. Environ. Health* 13:251–271.

Morrow, P. E., Bates, D. V., Fish, B. R., Hatch, T. E., and Mercer, T. T. 1966. Deposition and retention models for internal dosimetry of the human respiratory tract. *Health Phys.* 12:173–207.

Olson, D. E., Dart, G. A., and Filley, G. F. 1970. Pressure drop and fluid flow regime of air inspired into the human lung. *J. Appl. Physiol.* 28:482–494.

Overton, Jr., J. H. 1984. Physicochemical processes and the formulation of dosimetry models. *J. Toxicol. Environ. Health* 13:273–294.

Phalen, R. F., Yeh, H. C., Schum, G. M., and Raabe, O. G. 1978. Application of an idealized model to morphometry of the mammalian tracheobronchial tree. *Anat. Rec.* 190:167–176.

Schlesinger, R. B., Gurman, J. L., and Lippmann, M. 1982. Particle deposition within bronchial airways: Comparisons using constant and cyclic inspiratory flows. *Ann. Occup. Hyg.* 26:47–64.

Schreider, J. P. and Hutchens, J. O. 1979. Particle deposition in the guinea pig respiartory tract. *J. Aerosol Sci.* 10:599–607.

Schreider, J. P. and Raabe, O. G. 1981a. Anatomy of the nasal-pharyngeal airway of experimental animals. *Anat. Rec.* 200:195–205.

Schreider, J. P. and Raabe, O. G. 1981b. Structure of the human respiratory acinus. *Am. J. Anat.* 162:221–232.

Schroter, R. C. and Sudlow, M. F. 1969. Flow patterns in models of the human bronchial airways. *Respir. Physiol.* 7:341–355.

Weibel, E. R. 1963. *Morphometry of the Human Lung.* New York: Academic Press.

Yeh, H. C., Schum, G. M., and Duggan, M. T. 1979. Anatomic models of the tracheobronchial and pulmonary regions of the rat. *Anat. Rec.* 195:483–492.

Yeh, H. C. and Schum, G. M. 1980. Models of human lung airways and their application to inhaled particle deposition. *Bull. Math. Biol.* 42:461–480.

RESPIRATORY TRACT STRUCTURE AND FUNCTION

Peter Gehr

Department of Anatomy, University of Berne,
3000 Berne 9, Switzerland

The relationship between structure and function of the pulmonary gas-exchange apparatus and its comparative aspects is demonstrated in four sections: (1) the lung as gas exchanger, (2) the gas-exchange structures of the human lung, (3) comparative aspects of structure–function correlation, and (4) analysis of structure–function discordancy. Physical activity and body size are described to influence the structural dimensions of the pulmonary gas exchange apparatus in different ways. As far as the inhalation of toxic substances is concerned, different kinds of effects must therefore be taken into consideration, depending on an animal's size and metabolism; hence, experiments of comparative anatomy and physiology are desirable in order to find an appropriate animal model that permits extrapolation to humans.

INTRODUCTION

Structural alterations of lungs of patients who died from adult respiratory distress syndrome (ARDS) have been described in recent years (Bachofen and Weibel, 1974, 1977, 1982). They are characterized by a dramatic increase in tissue volume, along with a loss of air space and capillary volume by irregular widening of the septal structures due to interstitial edema, fibrosis, and increased cellularity (Fig. 1), and by alveolar fibrosis and septal rearrangements. The diffusing capacity of such lungs is severely impaired, being reduced to only one-tenth of its normal value (M. Bachofen, 1984, personal communication), as estimated by Weibel's model of morphometry (Weibel, 1970/71).

Such structural damage appears to be nonspecific, since similar alterations were also found in experiments where animals inhaled toxicants like ozone (Carrington and Green, 1970; Hesterberg and Last, 1981) and nitrogen dioxide (Evans et al., 1973; Crapo et al., 1978), as well as 100% oxygen (Kapanci et al., 1969; Crapo et al., 1980; Witschi et al., 1981; Steinberg et al., 1983). However, no direct clear relationship between toxic findings from such animal studies and possible extrapolated effects in humans have been established so far, either qualitatively or quantitatively.

I am very grateful to Dr. M. Bachofen and Dr. H. Bachofen for critically reviewing the manuscript. I am also indebted to Mrs. M. Schweizer, Ms. R. M. Frankhauser, Mrs. R. Channi and Mr. K. Babl for excellent technical assistance.

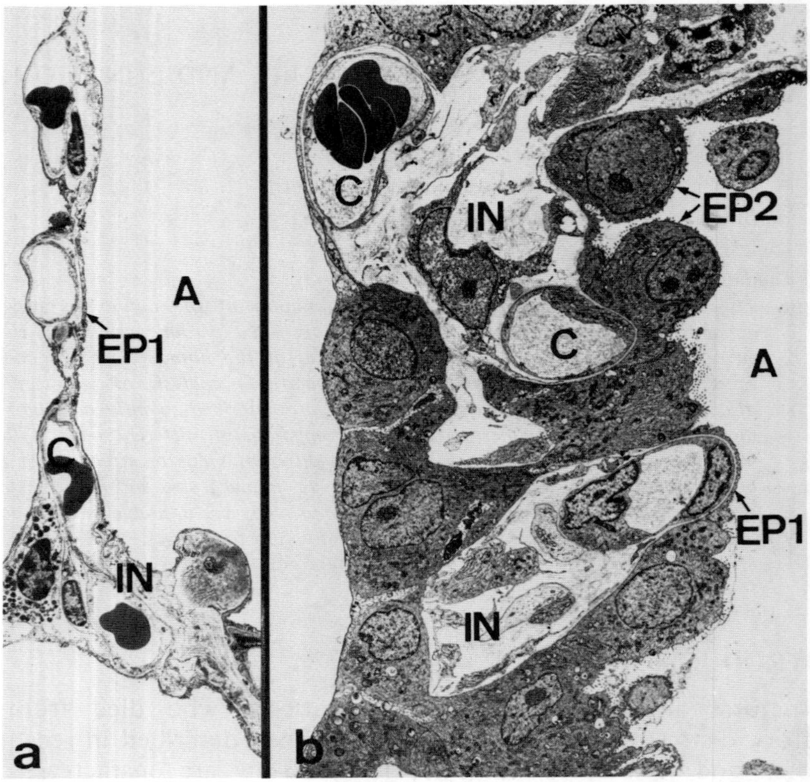

FIGURE 1. Interalveolar septa of human lungs (from Bachofen and Weibel, 1974). (*a*) Slim septum of normal lung. (*b*) Proliferated septum of ARDS patient at same magnification, ×2860. A = alveolar air space, C = capillaries, EP1 = type 1 squamous epithelial cell, EP2 = type 2 granular epithelial cell, IN = interstitium.

The purpose of this presentation is to show how the normal, healthy lung is designed structurally and functionally as a gas exchanger in order to provide the criteria for the structure–function relationship that should eventually be met for an extrapolation from animal to human.

It was found that size and function of the pulmonary gas-exchange apparatus depend on both body size and physical activity (Gehr et al., 1981). It was also found that there exist structural and functional peculiarities in some animals (Gehr et al., 1980).

I propose, therefore, a comparative approach that appears to me to be most easily successful in the search for an appropriate extrapolation model. In the following sections, the gas-exchange structures of the lung, basic aspects of the relationship between gas-exchange structure and function, as well as comparative aspects among various mammalian species are described.

THE LUNG AS A GAS EXCHANGER

The main function of the lung is to supply the organism with oxygen and to remove from it carbon dioxide. For this purpose the mammal lung can be subdivided into functional compartments (Fig. 2). (1) There is a conducting compartment, which comprises the airway tree (bronchi, bronchioli) as well as the pulmonary arteries and veins. They distribute the inspired gas into the terminal airways and the blood into the capillaries or collect expired gas or blood from the capillaries. (2) In the respiratory compartment, air and blood come into close contact to allow the exchange of gases by diffusion. This compartment can be called the actual gas-exchange apparatus. (3) The two compartments are then linked by a transition zone comprising respiratory bronchioli and alveolar ducts. Since gas exchange begins with the respiratory bronchioli, compartments 2 and 3 together are also called gas-exchange or lung parenchyma. The morphologic unit, beginning with the first generation of a respiratory bronchiolus, is the acinus (Weibel, 1973). In Fig. 3, a scanning electron micrograph shows the transition from compartment 1 (nonparenchyma), represented by a terminal bronchiolus as the last generation of the conducting airways (ventilation), to the gas-exchange parenchyma, represented by respiratory bronchioli: 80–90% of the total lung volume consist of gas exchange parenchyma, and only 10–20% is occupied by conducting airways and extraalveolar vessels (nonparenchyma).

The inhalation of a variety of toxic agents predominantly affects the gas-exchange parenchyma. Its fine structure shall therefore be considered in more detail. The structural basis is the interalveolar septum (Fig. 4), where air and blood come into close contact, separated by an air–blood barrier as thin as about 1 μm. This in turn is comprised of three layers: the epithelium, interstitium, and endothelium. Oxygen crosses this thin barrier and reaches the erythrocytes, which are passing by very closely, frequently touching the endothelium. The capillaries are meandering

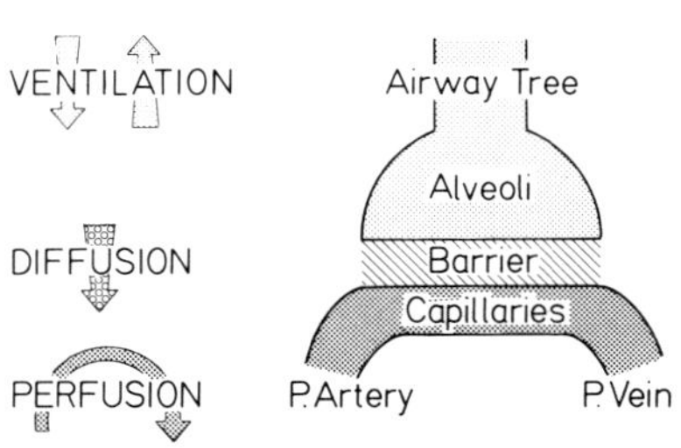

FIGURE 2. Schematic drawing of lung compartments: 90% of the human lung volume is gas-exchange structures, i.e., alveoli, air–blood barrier and capillaries; only 10% is not involved in gas exchange, like the conducting airways (airway tree), larger blood vessels, and the pleura. Compare with Fig. 3. (Courtesy E. R. Weibel.)

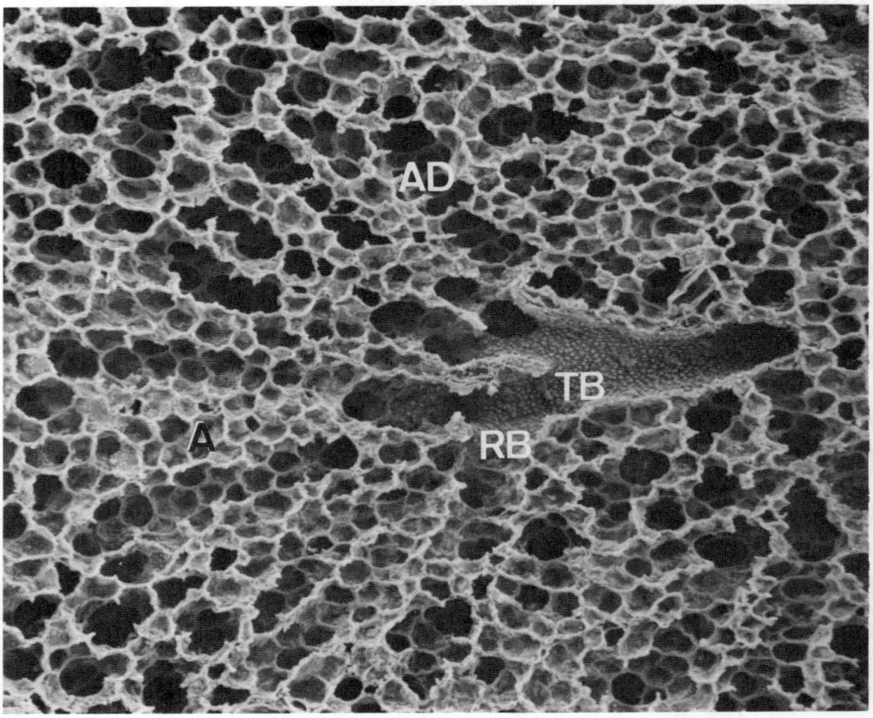

FIGURE 3. Scanning electron micrograph of a bovine lung, showing the terminal piece of the conducting airways, a terminal bronchiole (TB). It ventilates the gas exchange region of the lung, consisting of respiratory bronchioles (RB) and alevolar ducts (AD) with alveoli (A). Compare with Fig. 2. ×100. From Gehr et al. (1981).

around a connective-tissue frame exposing the thin air–blood barrier to alveolar air on either side, i.e., the erythrocytes are loaded with oxygen from two adjacent alveoli (Fig. 4).

According to the model of Weibel (1970/71), the size of the gas-exchange apparatus can be estimated by the morphometric diffusing capacity D_L, which is calculated from the alveolar and capillary surface area (S_a and S_c, respectively), the thickness of the diffusion barrier (τ_h), the capillary volume (V_c), and the hematocrit (Hct). The estimation of these parameters is performed on electron micrographs like the one shown in Fig. 4*b*. This is the basis for structure–function correlations and interspecies comparisons.

The established parameter estimating the capacity of the lung to transfer O_2 from air to blood is the diffusing capacity, D_{L,O_2}, which is defined as the ratio of O_2 flow to driving force (Bohr, 1909):

$$D_{L,O_2} = \dot{V}_{O_2} / \Delta P_{O_2}$$

The lung—i.e., the gas-exchange apparatus (Fig. 4)—can therefore be considered a conductor for O_2. O_2 molecules have to cross three barriers: the tissue (air-blood) barrier t, the plasma layer p (between endothelium and erythrocyte), and part of the erythrocyte (e). The total resistance (reciprocal of conductance) for O_2 hence consists of the three partial resistances in series:

$$\frac{1}{D_L} = \frac{1}{D_t} + \frac{1}{D_p} + \frac{1}{D_e}$$

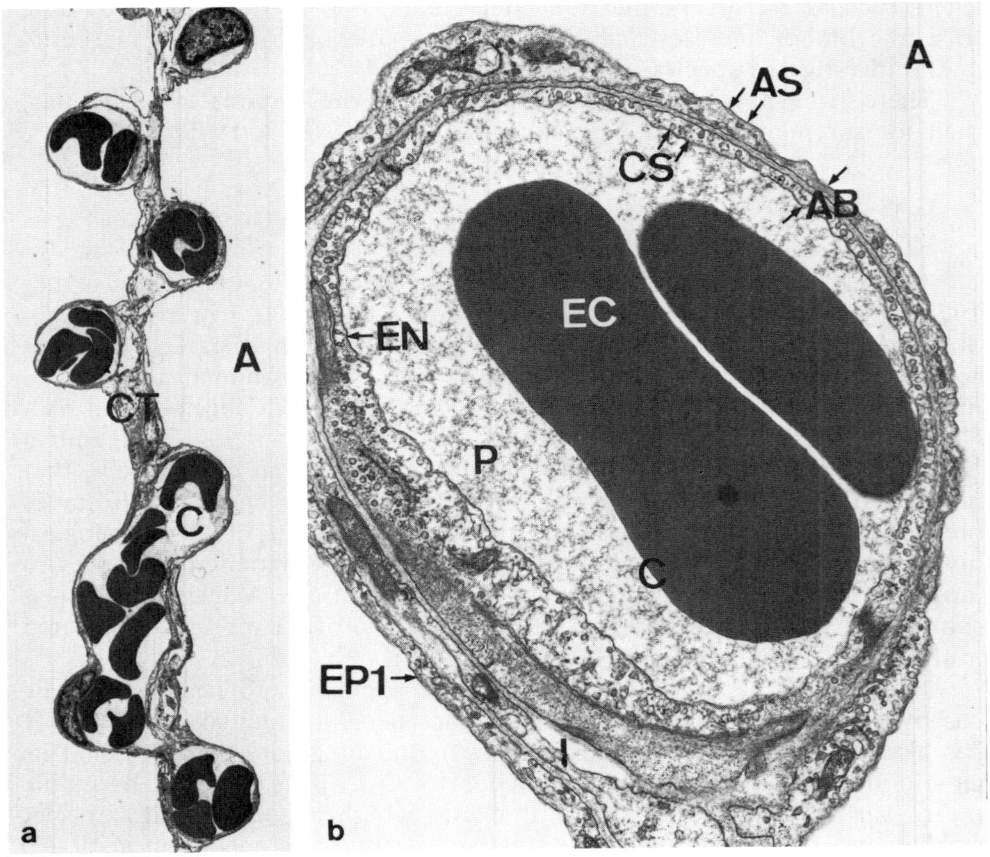

FIGURE 4. (*a*) Interalveolar septum of rabbit lung demonstrating a capillary meandering around a strong connective tissue cord (CT). Courtesy V. Haussener. ×1730. (*b*) Details of interalveolar septum of bovine lung: alveolar air (A), alveolar surface (AS); air-blood barrier (AB), consisting of epithelium (EP1, squamous type 1), interstitium (I), and endothelium (EN); capillary (C), consisting of erythrocytes (EC) and plasma (P); capillary surface (CS). Note thick interstitial space on one side and thin gas-exchange barrier on the other side of septum. ×16,900.

Each conductance is related to some structural parameters mentioned above:

$$D_t = K_t \frac{S_a + S_c}{2\tau_{\mathrm{ht}}}$$

$$D_p = K_p \frac{S_c}{\tau_{\mathrm{hp}}}$$

$$D_e = \theta V_{\mathrm{c}} \frac{\mathrm{Hct}}{\mathrm{Hct}_o}$$

where K_t and K_p are permeation coefficients, θ is the rate of O_2 binding to whole blood, and Hct and Hct_o the estimated and normal hematocrits of the investigated species, respectively.

There have been a number of reviews on the respiratory structures, their measurements, and the gas exchange (Weibel, 1973, 1979, 1983a).

GAS-EXCHANGE STRUCTURE OF THE HUMAN LUNG

The normal human lung has been described in detail (Gehr et al., 1978a). Its gas-exchange apparatus appears to be built like that of any other mammal lung. At least, this is the impression one receives from comparing transmission electron micrographs from animal and human interalveolar septa (Fig. 4, *a* and *b*, as compared with Figs. 1*a* and 5*b*). However, by comparing the scanning electron micrographs from animal (Etruscan shrew) and human gas-exchange parenchyma, respectively, that have been taken at the same magnification (Fig. 8, *a* and *b*), it becomes obvious that the two lungs are not alike. In fact, the stereological investigation of the human lung revealed quite considerable differences to lungs from animals both smaller and larger than man. A selected group is shown in Table 1; a comprehensive consideration of all species investigated in this lab is given elsewhere (Gehr et al., 1981).

Interesting aspects are revealed by such comparisons. The human lung has by far the smallest gas-exchange surface per unit lung volume (S_{Va}); it has also the smallest capillary volume per unit lung volume (V_{Vc}), and its air–blood (tissue) barrier, with respect to O_2 diffusion (τ_{ht}, harmonic mean barrier thickness), is the thickest, whether compared to species smaller or larger than humans or whether compared to species more active or less active than humans. All these parameters are depressing D_{L,O_2}.

Accordingly, the total D_{L,O_2} of the human lung is rather small. Only the cow has a slightly smaller one, relative to body size. This is, however, due to the small lung of the cow, which is only about half the size of a human lung, relative to body size.

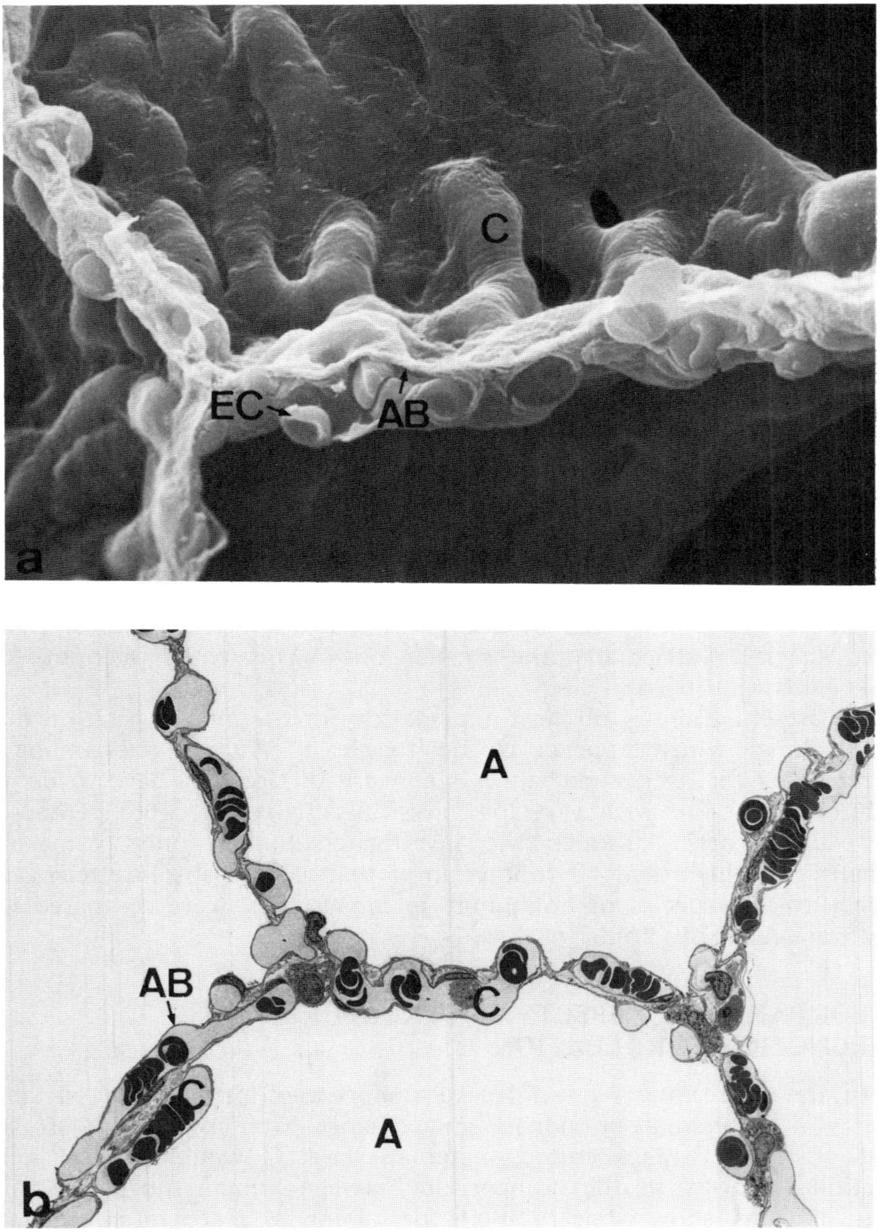

FIGURE 5. Gas exchange structures of human lung (from Gehr et al., 1978). (*a*) Scanning electron micrograph of broken-up interalveolar septa; note the capillaries (C) bulging into the alveolar air space. AB = air–blood barrier, EC = erythrocytes. ×1030. (*b*) Transmission electron micrograph of interalveolar septa: A = alveoli, AB = air–blood barrier, C = capillaries. ×670.

TABLE 1. Steric Parameters Influencing Pulmonary $D_L(O_2)^a$: A Comparison of Lungs of Humans and a Selected Group of Mammals

Mammal	N	W (kg)	τ_{ht} (μm)	V_{Vc} (ml/cm³)	S_{Va} (cm²/cm³)
Etruscan shrew	4	0.0025	0.27	0.123	1831
Laboratory mouse	5	0.023	0.32	0.115	924
Laboratory rat	8	0.140	0.38	0.092	750
Rabbit	6	3.56	0.50	0.107	875
Monkey	6	3.71	0.50	0.093	801
Dog	6	23	0.49	0.079	550
Human	8	74	0.62	0.057	371
Horse	2	510	0.60	0.086	759
Cow	1	700	0.51	0.137	635

[a] Weibel (1970/71).

The human lung deviates a little bit from other, particularly smaller, lungs, in that it has usually much wider interstitial spaces in the interalveolar septa over long distances, which contain cells and many relatively strong collagen and elastic fibers. This fibrous skeleton of the human gas-exchange parenchyma is considerably better developed than in other species, particularly the smaller ones. This could account for a thicker diffusion barrier.

At present there is no clear explanation for the unique position of the human lung, which appears to be the least efficient one among the mammals as far as gas exchange is concerned. Could it have to do with evolution to a life with very low physical activity, or could perhaps the age play a role? Although the investigated human lungs came from patients not older than 40 (not younger than 19), in absolute terms, with respect to the species of nonhuman mammals they were compared with, they were relatively "old."

COMPARATIVE ASPECTS OF STRUCTURE–FUNCTION CORRELATION

In the preceeding section I gave a short description of the structural aspects of the human pulmonary gas-exchange parenchyma only, to show what position it takes among mammals. Here, I would like to include functional aspects in the comparison between human lungs and animal lungs, and also show some problems pertaining to a general consideration of structure–function correlation. Such problems are addressed in particular in several articles on the design of the mammalian respiratory system (Taylor and Weibel, 1981; Gehr et al., 1981; Weibel et al., 1981, 1983b), but they are also mentioned in review articles (Weibel, 1973, 1979, 1983a).

Examining mammals of similar body size but different physical activity, like waltzing mice and laboratory mice, or wild species of bovids that have to move long distances to find food versus domestic bovids that spend most of their life in a stable, a close correlation is found between structure and function of the lung (Geelhaar and Weibel, 1971; Hugonnaud et al., 1977; Gehr et al., 1978b, 1981). The same correlation is revealed by comparing oxygen consumption, $\dot{V}_{O_2}$, and morphometric D_{L,O_2} (representing the size of the pulmonary gas-exchange apparatus) of dog and human, and also of horse and cow, as can be seen in Fig. 6.

Another approach for the study of how structure and function of the lung are related to each other is to investigate the effect of the body size on these parameters, i.e., by examining the scaling of $\dot{V}_{O_2}$ and morphometric D_{L,O_2}. Rather than finding two parallel allometric lines, the scaling was found to be discordant (Fig. 7), indicating that the body size does not influence these two parameters in the same way. Whereas $\dot{V}_{O_2}$ scales with the $\frac{3}{4}$ power of the body weight (Kleiber, 1932, 1961), morphometric D_{L,O_2} scales directly proportional with body weight (Gehr et al., 1981). $\dot{V}_{O_2}$ and morphometric D_{L,O_2}, i.e., function and gas-exchange structure, are hence not directly correlated with each other.

However, there is some structural adaptation, as can be seen in Fig. 8. It shows scanning electron micrographs of the Etruscan shrew and human gas-exchange parenchyma at the same magnification. The alveoli of the Etruscan shrew are extremely small; hence many more can be packed into the same lung volume as compared to man. One human alveolus, in fact, would preserve space for roughly 500–1000 shrew alveoli. This, as a consequence, is providing this tiny animal with much larger gas-exchange surface (Table 1). The Etruscan shrew has a roughly 20-fold higher weight-specific O_2 consumption than man. The structural adaptations

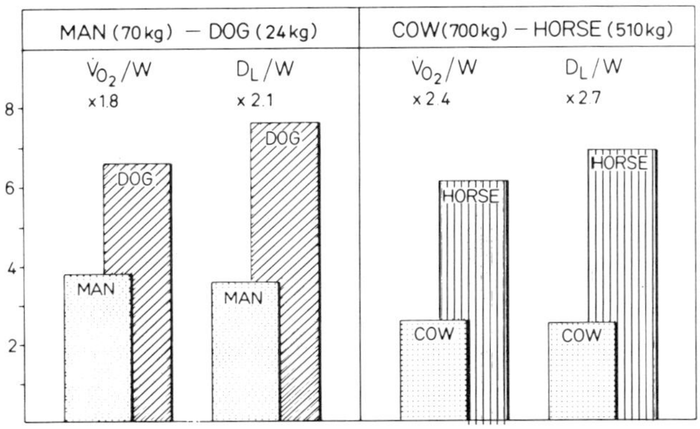

FIGURE 6. Comparison of oxygen consumption and morphometric diffusing capacity (size of gas-exchange apparatus) in species of similar size but different natural physical activity.

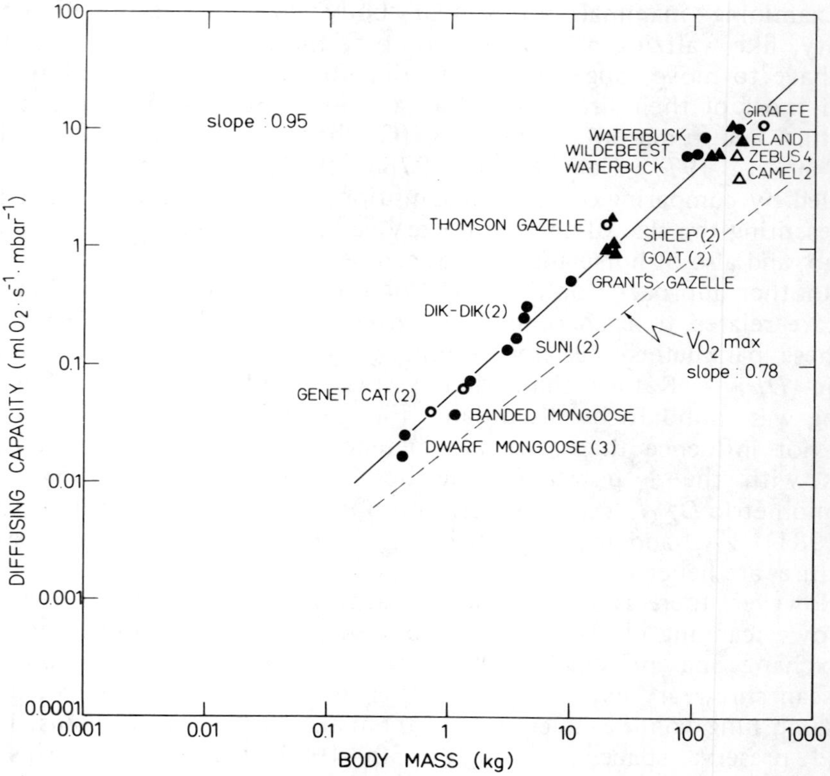

FIGURE 7. Comparison of oxygen consumption and morphometric diffusing capacity (size of gas-exchange apparatus) in species of different body size (scaling). From Gehr et al. (1981).

reflected in the morphometric D_{L,O_2}, however, cannot account for this huge difference in the specific $\dot{V}_{O_2}$. Are there other structural adaptations or additional functional adaptations?

ANALYSIS OF DISCORDANCE OF STRUCTURE–FUNCTION CORRELATION

The problem of discordant scaling has been discussed in detail (Weibel et al., 1981; Gehr et al., 1981). I do not intend to repeat or continue these discussions here, but I would like to give a short summary of the state of the research, since the hypotheses to the solution of this problem could influence the approaches to the extrapolation modeling of inhaled toxicants.

We first ask how $\dot{V}_{O_2}$ and D_{L,O_2} are physiologically related to each other. This relationship is given by the Bohr equation (1909):

$$\dot{V}_{O_2} = D_{L,O_2} \cdot \Delta P_{O_2}$$

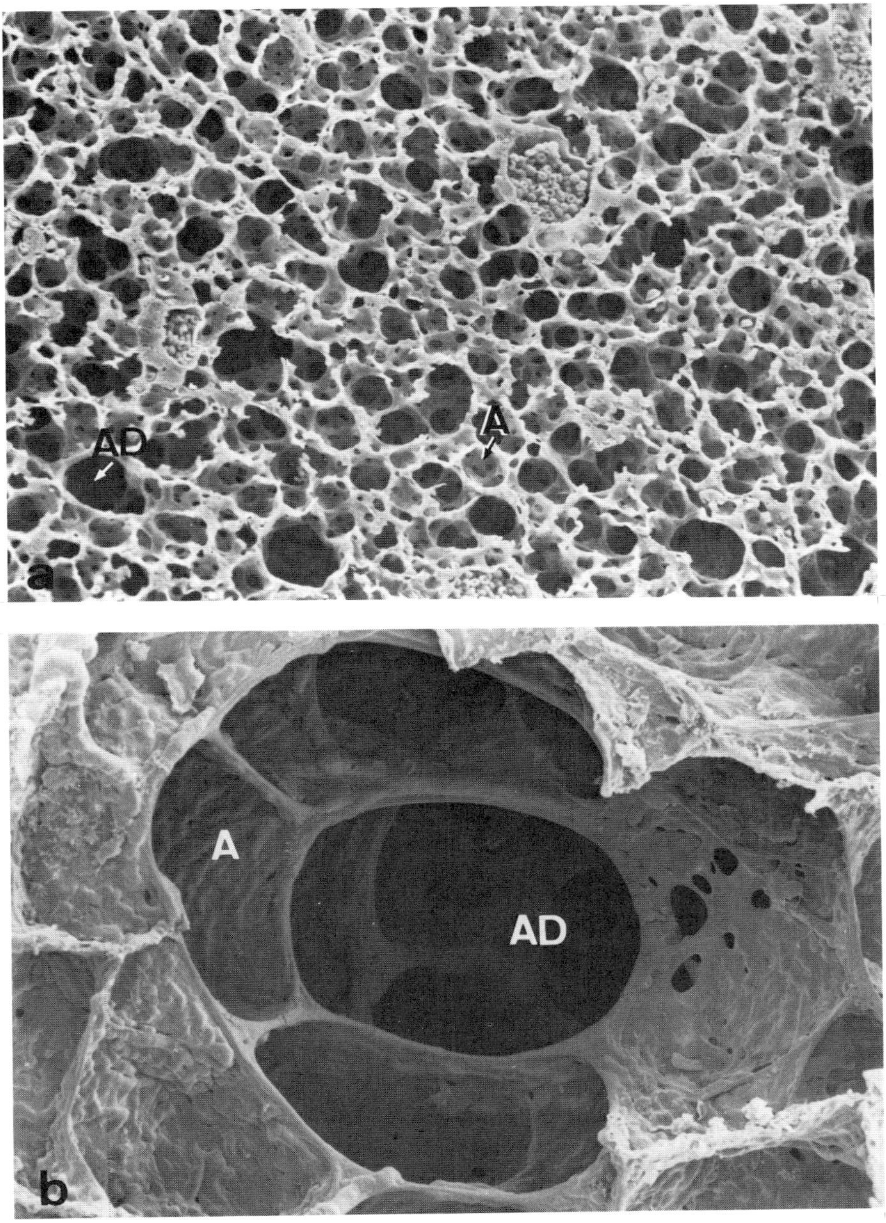

FIGURE 8. Comparison of gas exchange parenchyma of Etruscan shrew (*a*) and human (*b*) at same magnification, X230). A = alveoli, AD = alveolar duct. Pictures by P. Gehr, in Weibel (1979).

From dividing the allometric equations for $\dot{V}_{O_2}$ and morphometric D_{L,O_2} we conclude that ΔP_{O_2} is size-dependent (Weibel et al., 1981):

$$P_{O_2} = \frac{\dot{V}_{O_2}}{D_{L,O_2}} = \frac{1.84\ (W^{0.77})}{0.055\ (W^{0.95})} = 33.5\ (W^{-0.22})$$

This means that the larger the animal, the smaller the partial pressure gradient for O_2 between arterial and mean capillar P_{O_2} —i.e., the smaller is the driving force for oxygen to cross the diffusion barrier. The relationship between $\dot{V}_{O_2}$, D_{L,O_2}, and ΔP_{O_2} is shown in Fig. 9. Any change in ΔP_{O_2} can be due either to a change in the mean capillary P_{O_2} or in the alveolar P_{O_2} or both.

At the moment, there is evidence for a change in the alveolar P_{O_2} only. Alveolar P_{O_2} is defined as the P_{O_2} at the gas-exchange surface. It is maintained by replenishing O_2 there in a two-step process: by ventilation into the entrance part of the acinus, and, from there, by diffusion through the air phase. This diffusion, however, requires a P_{O_2} gradient throughout the acinus, which is called stratification of O_2 by physiologists (Fahri, 1969; Piiper, 1979). Such a stratificational gradient could occur along the axis of an acinus (Fig. 10), and would mean that the longer the distance from the center of the acinus to its periphery, the lower the P_{O_2} at the gas-exchange surface. A size-dependent difference of alveolar P_{O_2} might therefore exist if the size of the acinus would alter with body size. This possibility is discussed in more detail elsewhere (Weibel et al., 1981; Weibel, 1983b), but in fact, from a study currently undertaken by Weibel, the duct length was found to increase with body size, namely from less than 1 mm in the mouse to about 4 mm in man (Weibel, 1984, personal communication). The structural design of the acinus, therefore, appears to be critical for gas exchange. Isn't it also the acinus,

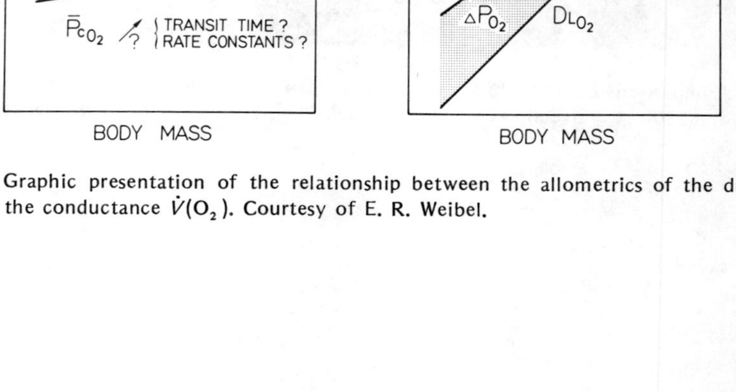

FIGURE 9. Graphic presentation of the relationship between the allometrics of the driving force $\Delta P(O_2)$ and the conductance $\dot{V}(O_2)$. Courtesy of E. R. Weibel.

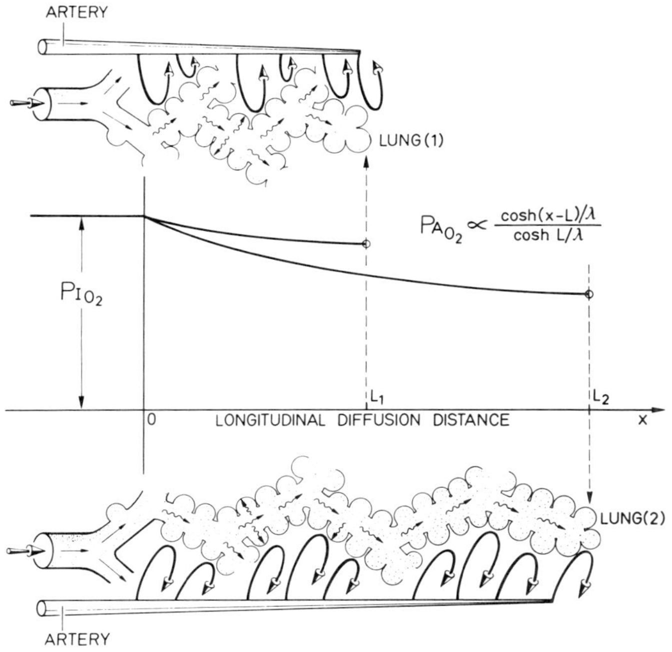

FIGURE 10. Relationship between duct length and ΔP_{O_2}. From Weibel (1983b).

particularly its entrance part, that is the critical structure for inhalation toxicologic effects?

CONCLUSIONS

There is evidence that the structural design of an acinus is a rate-limiting factor for O_2 uptake. Furthermore we have seen that the structural dimensions of the pulmonary gas-exchange apparatus are influenced by the size and the physical activity of an animal. Considering that there is a relationship between design and dimensions of respiratory structures in the lung, and the effects of inhaled toxicants, it can be assumed that, depending on size and activity of an animal, different effects of inhaled toxicants are to be expected. A comparative approach, therefore, is desirable for the evaluation of an appropriate animal model allowing extrapolation to humans.

REFERENCES

Bachofen, M., and Weibel, E. R. 1974. Basic pattern of tissue repair in human lungs following unspecific injury. *Chest* 65:14S–19S.

Bachofen, M., and Weibel, E. R. 1977. Alterations of the gas exchange apparatus in adult respiratory insufficiency associated with septicemia. *Am. Rev. Respir. Dis.* 116:589–615.

Bachofen, M., and Weibel, E. R. 1982. Structural alterations of lung parenchyma in the adult respiratory distress syndrome. *Clin. Chest Med.* 3:35-56.

Bohr, C. 1909. Ueber die spezifische Tätigkeit der Lungen bei der respiratorischen Gasaufnahme. *Scand. Arch. Physiol.* 22:221-280.

Carrington, C. B., and Green, T. J. 1970. Granular pneumocytes in early repair of diffuse alveolar injury. *Arch Intern. Med.* 126:464-465.

Crapo, J. D., Marsh-Salin, J., Ingram, P., and Pratt, P. C. 1978. Tolerance and cross-tolerance using NO_2 and O_2. II. Pulmonary morphology and morphometry. *J. Appl. Physiol.* 44:370-379.

Crapo, J. D., Barry, B. G., Foscue, H. A., and Shelburne, J. 1980. Structural and biochemical changes in rat lungs occurring during exposures to lethal and adaptive doses of oxygen. *Am. Rev. Respir. Dis.* 122:123-143.

Evans, M. J., Cabral, L. J., Stephens, R. J., et al. 1973. Renewal of alveolar epithelium in the rat following exposure to NO_2. *Am. J. Pathol.* 70:175-198.

Fahri, L. E. 1969. Diffusive and convective movement of gas in the lung. In *Circulatory and Respiratory Mass Transport,* ed. by G. E. W. Wolstenholme and J. Knight, pp. 277-293. London: Churchill.

Geelhaar, A., and Weibel, E. R. 1971. Morphometric estimation of pulmonary diffusion capacity. III. The effect of increased oxygen consumption in Japanese waltzing mice. *Respir. Physiol.* 11:354-366.

Gehr, P., Bachofen, M., and Weibel, E. R. 1978a. The normal human lung: Ultrastructure and morphometric estimation of diffusion capacity. *Respir. Physiol.* 32:121-140.

Gehr, P., Bachofen, H., Burri, P. H., Hugonnaud, C., and Weibel, E. R. 1978b. Adaptation of the growing lung to increased $\dot{V}_{O_2}$. III. The effect of exposure to cold environment in rats. *Respir. Physiol.* 32:345-353.

Gehr, P., Sehovic, S., Burri, P. H., Claassen, H., and Weibel, E. R. 1980. The lung of shrews: Morphometric estimation of diffusion capacity. *Respir. Physiol.* 40:33-47.

Gehr, P., Mwangi, D. K., Ammann, A., Sehovic, S., Maloiy, G. M. O., Taylor, C. R., and Weibel, E. R. 1981. Design of the mammalian respiratory system. V. Scaling morphometric pulmonary diffusing capacity to body mass: Wild and domestic mammals. *Respir. Physiol.* 44:113-128.

Hesterberg, T. W., and Last, J. A. 1981. Ozone-induced acute pulmonary fibrosis in rats. *Am. Rev. Respir. Dis.* 123:47-52.

Hugonnaud, C., Gehr, P., Burri, P. H., and Weibel, E. R. 1977. Adaptation of the lung to increased $\dot{V}_{O_2}$. II. Morphometric analysis. *Respir. Physiol.* 29:1-10.

Kapanci, Y., Weibel, E. R., Kaplan, H. P., et al. 1969. Pathogenesis and reversibility of the pulmonary lesions of oxygen toxicity in monkeys. Ultrastructural and morphometric studies. *Lab. Invest.* 20:101-118.

Kleiber, M. 1932. Body size and metabolism. *Hilgardia* 6:315-353.

Kleiber, M. 1961. *The Fire of Life,* p. 454. New York: Wiley.

Piiper, J. 1979. Series ventilation, diffusion in airways, and stratified inhomogeneity. *Fed. Proc.* 38:17-21.

Steinberg, H., Greenwald, R. A., Moak, S. A., and Das, D. K. 1983. The effect of oxygen adaptation on oxyradical injury to pulmonary endothelium. *Am. Rev. Respir. Dis.* 128:94-97.

Taylor, C. R., and Weibel, E. R. 1981. Design of the mammalian respiratory system. I. Problem and strategy. *Respir. Physiol.* 44:1-10.

Weibel, E. R. 1970/71. Morphometric estimation of pulmonary diffusion capacity. I. Model and Method. *Respir. Physiol.* 11:54-75.

Weibel, E. R. 1973. Morphological basis of alveolar-capillary gas exchange. *Physiol. Rev.* 53:419-495.

Weibel, E. R. 1979. Oxygen demand and the size of respiratory structures in mammals. In *Evolution of Respiratory Processes,* eds. S. C. Wood and C. Lenfant, vol. 13, pp. 289-345. New York: Marcel Dekker.

Weibel, E. R. 1983a. How does lung structure affect gas exchange? *Chest* 83:657-665.

Weibel, E. R. 1983b. *The Pathway for Oxygen.* Cambridge, Mass.: Harvard University Press, in press.

Weibel, E. R., Taylor, C. R., Gehr, P., Hoppeler, H., Mathieu, O., and Maloiy, G. M. O. 1981. Design of the mammalian respiratory system. IX. Functional and structural limits for oxygen flow. *Respir. Physiol.* 44:151-164.

Weibel, E. R., Taylor, C. R., O'Neil, J. J., Leith, D. E., Gehr, P., Hoppeler, H., Langman, V., and Baudinette, R. V. 1982. Maximal oxygen consumption and pulmonary diffusing capacity: A direct comparison of physiologic and morphometric measurements in canids. *Respir. Physiol.* 54:173-188.

Witschi, H. R., Haschek, W. M., Klein-Szanto, A. J. P., and Hakkinen, P. J. 1981. Potentiation of diffuse lung damage by oxygen: determining variables. *Am. Rev. Respir. Dis.* 123:98-103.

MASS TRANSPORT IN MAMMALIAN LUNGS: COMPARATIVE PHYSIOLOGY

David E. Leith

Departments of Anesthesia of Brigham and Women's Hospital and Harvard Medical School, Boston, Massachusetts

Comparative physiology of mass transport of gases in mammalian lungs is surveyed in terms of the use of experimental mammals in inhalation toxicology. Principles of similarity, scaling, and the relationship of metabolism to body size are touched on, with reference to the wide variability among mammals of similar size. Mechanisms that influence the magnitude and distribution of pulmonary ventilation are reviewed, including mechanical differences associated with variation in body size. More systematic and complete descriptions and understanding are needed. Recent advances in the understanding of gas mixing and transport in airways and in the pulmonary acinus have applications in comparative physiology and inhalation toxicology that are worth exploring.

INTRODUCTION

When mammals inhale noxious agents, how much is retained, and where?

To help answer those questions, we want to know about ventilation and collection efficiency, including not only their determinants and their regional distributions, but also their variabilities among mammals of interest; and we want to know those things for inhaled materials with a wide range of properties.

That is a tall order, outlined in part in Table 1. Of the items listed there, I think that gas mixing, in airways and at the level of the pulmonary acinus, offers the most interesting opportunities now for joint study by comparative physiologists, physical scientists, and inhalation toxicologists.

In this paper I will assume considerable familiarity with common terms and background physiology, will omit many cautions and qualifications, and will not explain or substantiate all that I say. Speculations are identified as such.

COMPARATIVE PHYSIOLOGY

First I will consider how mammalian size, shape, and metabolic rates are compared, and outline some general relationships among them.

TABLE 1. Some Inhalation Issues

I.	Comparative physiology
	A. Similarity and scaling
	B. Metabolism and body size
	C. Elastic similarity
II.	Gas transport in the lung
	A. $\dot{V}_E$, V_T, and f
	B. Lung volumes
	C. Airways
	D. Distribution of ventilation
	E. $\dot{V}_A/\dot{Q}$
	F. Gas mixing in airways; in acinus

Similarity and Scaling

Some characteristics of mammalian structure and function are invariant with size. For example, horse lungs are bigger than mouse lungs, but the ratio of lung weight to body weight is about the same in these and in other mammals generally. This dimensionless ratio is invariant with size and provides a quantitative criterion of similarity among mammals; that is, if we find a species with lungs substantially more or less than 1% of body weight, knowledge of this rule allows us to recognize that species as unusual. It may represent either a pitfall or an opportunity for experimentalists.

Similarity criteria may also reflect, not invariance, but systematic variation of some characteristics with body size. For example, tracheal radius is, of course, bigger in big animals. One might expect such linear dimensions to vary with the cube root of body mass ($m^{1/3}$, isometric scaling); instead, it appears to be proportional to $m^{3/8}$, an allometric scaling relationship like that expected under the rules for elastic similarity (to be discussed further). We may wonder why tracheas are made in this way, and what implication there may be to respiratory functions of interest to us. Again, departures from this generalized rule may call our attention to specialized structure and function.

These simple examples are merely an invitation to the field of scaling in biology, an old, extensive, and interesting endeavor that is well reviewed elsewhere (McMahon, 1973, 1977, 1983; Stahl, 1962, 1963). Applications of these ideas and techniques in mammalian respiratory biology have been reviewed as well (Heusner, 1983; Leith, 1983a; Stahl, 1967; Spells, 1969–70). An interesting series of papers on scaling and the developmental aspects of mammalian respiratory mechanics is reviewed by Mortola (1983).

Several authors have used dimensional analysis in the study and prediction of inhaled particle deposition in mammals of different size; some of these are briefly reviewed by McMahon and co-workers (McMahon et al., 1977). Book (1982) presents an application of scaling to the extrapolation of toxicological findings from animals to humans. Other aspects of scaling and extrapolation are addressed in reviews of aerosol deposition and toxicology by Brain and co-workers (Brain and Mensah, 1983; Brain and Valberg, 1979).

Heusner (1982) recently challenged some accepted approaches to scaling of metabolism with body weight. Feldman and McMahon responded (1983), clarifying and securing the statistical basis of existing views.

Metabolism and Body Size

Adult mammals range in size from the 2-g Etruscan shrew to great whales weighing around 100 metric tons—over 7 orders of magnitude. Their aerobic metabolism ($\dot{V}O_2$) varies with body mass to the $\frac{3}{4}$ power ($m^{3/4}$) both at rest (Fig. 1) (Benedict, 1938; Kleiber, 1932) and during maximal aerobic exercise (Taylor et al., 1981). Thus aerobic metabolic scope (ratio of maximal to resting $\dot{V}O_2$) is about the same for all

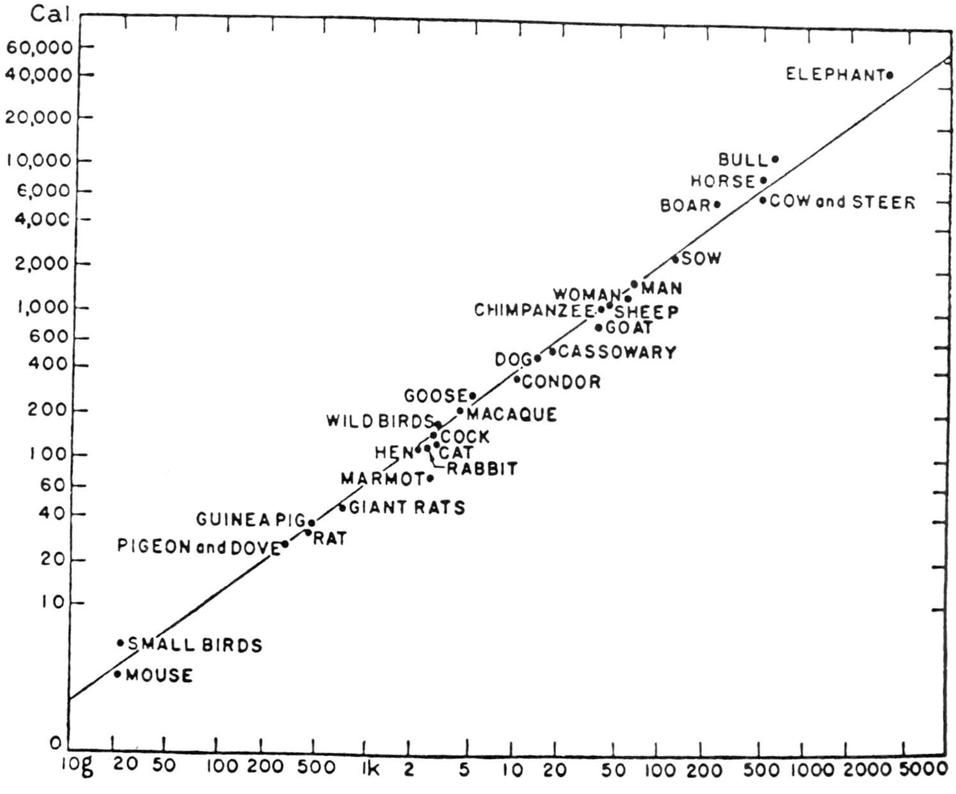

FIGURE 1. "Mouse-to-elephant" curve relating metabolism to body weight (Benedict, 1938).

mammals within the size range observed; values generally range from 10 to 20. These two variables—size and activity—account for much of the variation of $\dot{V}O_2$ among mammals, and to the extent that the species listed in Table 2 follow this generalized scaling rule, we would expect their weight-specific metabolic rates to vary as shown in comparison with that of humans.

But a surprising amount of variation remains. Species of the same size may differ 10-fold in their resting $\dot{V}O_2$, as emphasized in Fig. 2. This idealized representation also shows another interesting scaling rule: within similar groups of mammals ("homomorphic," in Heusner's usage), resting $\dot{V}O_2$ varies with $m^{2/3}$. There is no general agreement about the bases for these relationships, but some implications are clear. One is important for present purposes and has to be emphasized: individual animals and species may depart very widely from "average" mammalian characteristics expressed in scaling rules derived from observations among species over a wide range of body size. Ideally these departures can be used as experimental variables, but if unrecognized they can be a source of apparently conflicting, or at least noisy, experimental results.

Elastic Similarity

McMahon (1975) argues that mechanical considerations in design of mammals (and other biological systems) require scaling according to rules of elastic similarity, rather than geometric similarity. This means that with changes in size, structural elements change their lengths in proportion to $m^{1/4}$, while their diameters vary as $m^{3/8}$. Thus for example among species, cross section of homologous muscles varies with $m^{3/4}$, and with it, for reasons given by McMahon, muscle power and aerobic metabolism.

The requirements of elastic similarity might also be expected to govern trunk dimensions, and indeed there is evidence that thorax circumference (therefore diameter) scales with $m^{3/8}$ in some groups of mammals (Fig. 3).

TABLE 2. Body Mass (m) and Weight-Specific Metabolic Rate ($\dot{V}O_2/m$) Compared with That of Humans

Species	(kg)	Ratio[a]
Mouse	0.025	7.4
Rat	0.25	4.2
Rabbit	2.5	2.3
Dog	15	1.5
Human	75	1.0
Horse	500	0.6

[a] Assume $\dot{V}O_2 \propto m^{3/4}$. Then $(\dot{V}O_2/m) \propto m^{-1/4}$, and for any species x, the ratio $(\dot{V}O_2/m)_x/(\dot{V}O_2/m)_{\text{human}} = (m_x/m_{\text{human}})^{-1/4}$.

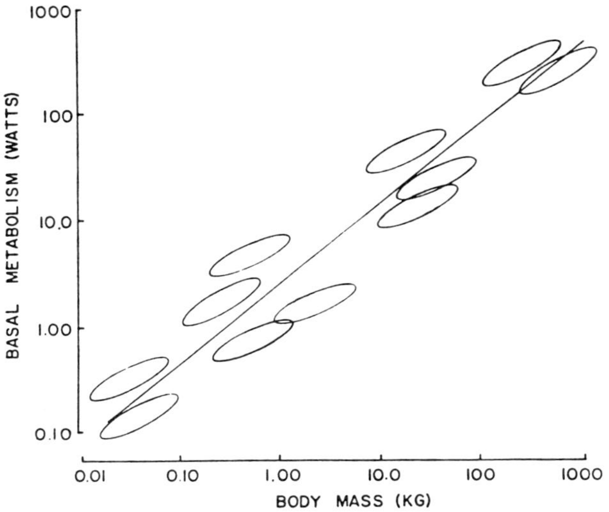

FIGURE 2. Individual species or groups of species may depart widely from the overall regression line for metabolism against body mass. Thus mammals of the same size may vary 10-fold in resting $\dot{V}O_2$ (Feldman and McMahon, 1983).

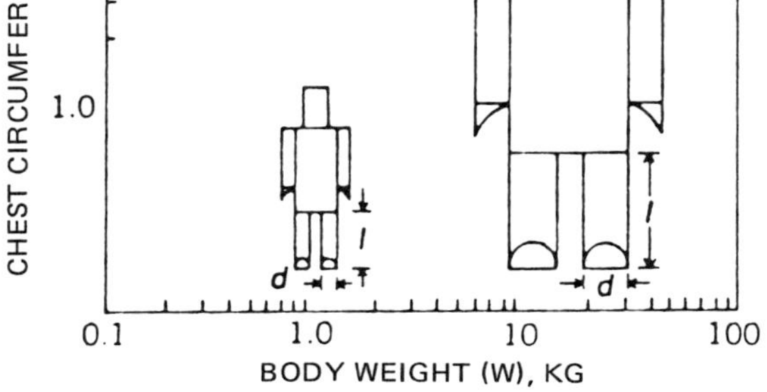

FIGURE 3. Chest circumference in several nonhuman primates varies as $M^{0.37}$ (Stahl and Gummerson, 1967).

RESPIRATORY PHYSIOLOGY

Now I will consider how much mammals breathe and some of the relationships that influence the distribution of that ventilation among regions of the lung.

$\dot{V}_E$, $\dot{V}_T$, AND f

Expired minute ventilation ($\dot{V}_E$) in the steady state can be treated as a function of three variables: (1) metabolism ($\dot{V}O_2$ and respiratory quotient, RQ); (2) control of breathing (P_aCO_2); and (3) gas exchange efficiency (V_D/V_T).

In considering a wide range of body weights, we can regard mammalian RQ, P_aCO_2 and V_D/V_T as invariant with size. Therefore among mammals in comparable physiological states, $\dot{V}_E$ is proportional to $\dot{V}O_2$, which is to say,

$$\dot{V}_E \propto m^{3/4} \tag{1}$$

This is the cornerstone for our thinking about scaling of mass transport of gases in the mammalian respiratory system. These considerations also imply invariance with size of the ventilatory equivalent for oxygen,

$$\dot{V}_E/\dot{V}O_2 = 33 \tag{2}$$

(Stahl, 1967), so that weight-specific $\dot{V}_E$ varies among species in the same ratio (with that of humans) as $\dot{V}O_2$ (Table 2).

Not only lung weight, but also some lung volumes, are directly proportional to body mass. One example is tidal volume:

$$V_T \propto m^{1.0} \tag{3}$$

This implies that the high weight-specific $\dot{V}_E$ of small mammals has to be achieved by high respiratory frequencies:

$$\begin{gathered} f = \dot{V}_E/V_T \\ f \propto m^{3/4}/m^{1.0} \\ f \propto m^{-1/4} \end{gathered} \tag{4}$$

We see here that allometric equations and proportionalities can be multiplied and divided. Note also that I have used approximate exponents for convenience, and that there is variability and uncertainty about reported values. Remember that a particular species may depart from generalized characteristics, and its inclusion or exclusion from the studies that establish the norm will influence our view of what is normal. Finally, individual animals' condition during experimental procedures can influence their behavior and thus the experimental results.

Sweeney and co-workers (1983) explored some of the latter influences in studies of retention and distribution of inhaled 0.45-μm aerosols in hamsters, either awake or deeply anesthetized with pentobarbital. Body temperature in the anesthetized animals fell slowly by 6°C, and there was a much quicker fall of $\dot{V}O_2$ to 36% of awake values, $\dot{V}_E$ to 24%, V_T to 41%, and f to 61%. Associated with the marked reduction in $\dot{V}_E$ there was, of course, a decrease in the inhaled burden to particles and in the total retained.

One might hope to avoid these effects in studies of awake animals, but it is not that simple. Exposure of awake animals to inhaled irritants can lead to similarly profound decreases in $\dot{V}_E$.

Bruce and her colleagues (1979) exposed awake mice to a high concentration of ozone (10 ppm) for 90 min. $\dot{V}_E$ decreased promptly to as little as 20% of what were apparently valid resting control levels, while body temperature fell more slowly to as low as 26°C (Fig. 4). Again, the inhaled ozone dose must have been very much conditioned by the low $\dot{V}_E$.

In somewhat similar studies of awake mice during exposure to inhaled acrylate esters, Silver et al. (1981) observed comparable responses. $\dot{V}O_2$ and $\dot{V}_E$ fell promptly to one-third of resting values or less, in some cases lower

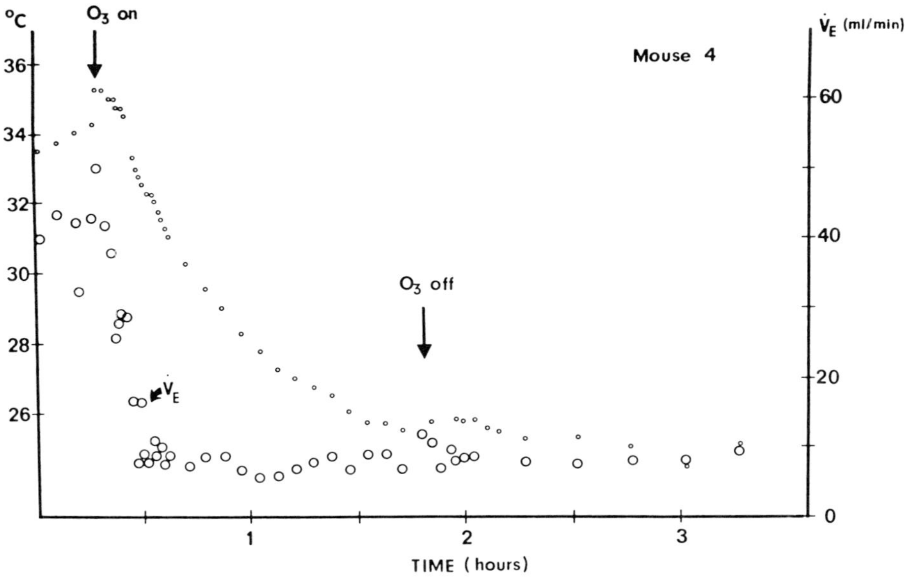

FIGURE 4. Ventilation and rectal temperature in an awake 24-g mouse exposed to ozone at 10 ppm. Ambient temperature 23°C. Note the different time courses. (By permission of M. C. Bruce, unpublished work.)

than anesthetized and sleeping values measured in control animals. Body temperature control was changed, temperature falling slowly by 6–8°C.

These changes are not only attendant on, but also reciprocally influence, the animals' exposure to inhaled oxidants and irritants. If such an interaction is not accounted for (which may require measurement, e.g., of $\dot{V}_E$ and temperature), then the responses cited become confounding variables. More interesting perhaps are the mechanisms by which these profound changes in metabolism are brought about in awake animals, and their species distributions and intensities. They remind one of Scholander's term "the master switch of life," and they raise questions for toxicologists that go beyond inhaled dose and retention. What are the effects of such changes on the response(s) to toxic agents, and on rates of metabolism of anesthetics and other noxious agents?

Lung Volumes (V)

Total lung capacity (TLC) is poorly defined in mammals other than cooperating humans. Reported values consistently scale close to $m^{1.0}$. Respiratory system compliance (C) also scales with $m^{1.0}$. That is to say, elastic pressures P_{el} of the same magnitude are used to operate the respiratory system in mammals from shrew to whale!

$$P_{el} = (1/C)\ V$$
$$P_{el} \propto (1/m^{1.0})\ (m^{1.0}) \qquad (5)$$
$$P_{el} \propto m^{\circ} = k$$

Large mammals have relatively stiff chest walls and small mammals have compliant ones (Gillespie, 1983), but the data are incomplete and our understanding of chest wall mechanics is changing fast right now. There are two consequences of considerable importance to inhalation toxicologists.

First, the relaxation volume of the respiratory system, V_{rel}, is relatively large in large mammals including humans (say, 40–50% of TLC), and the functional residual capacity (FRC) at rest is, in the orthodox view, mostly determined by relaxation mechanics.

Second, residual volume (RV, which like TLC, is poorly defined in large experimental mammals) is also relatively large, say 20–25% of TLC, and is determined (again in the orthodox view) by a static balance between expiratory pressure and the elastic forces of the respiratory system, dominated in this case by the chest wall.

None of these things is true in small (and newborn, and diving) mammals with very compliant chest walls. V_{rel} in such terrestrial mammals is close to RV, which is set (in most experiments) by total airway closure at about 10% of TLC, and resting FRC is maintained at a volume greater than V_{rel} by interrelated mechanisms including passive expiratory mechanics and reflex (active) control of respiratory cycle timing and expiratory flow rate.

We saw in experiments already outlined that anesthesia and irritant exposures decrease $\dot{V}_E$. For that reason, and because of associated reflex changes, FRC diminishes and may approach RV in recumbent small mammals during anestheisa. Figure 5 shows the equal decrements in FRC seen in mice at the instant of rapid intravenous injection of pentobarbital or of pentobarbital and a muscle relaxant. Here is direct evidence that the awake FRC exceeds, and the anesthetized FRC approaches, V_{rel}. Figure 6 shows respiratory flow and volume events, and shows their timing, in awake and anesthetized mice, measured plethysmographically (Vinegar et al., 1979). Anesthesia is attended by loss of expiratory braking, so that expiration becomes an exponential decay; there is also a profound change in frequency and in relative durations of inspiration and expiration.

What does all this mean to distribution of inhaled agents? At low lung volume, airways tend to close. Lung regions served by them are then unventilated—that is, they are no longer exposed to inhaled agents. Nonuniform distribution of inhaled aerosols—that is, the observation of regions with little or no deposition—was accordingly interpreted by Sweeney et al. (1983) as evidence of widespread airway closure in anesthetized hamsters.

I think it likely that exposure to inhaled irritants in awake animals similarly decreases not only $\dot{V}_E$ but also FRC, and therefore influences not only the magnitude of the inhaled dose but also its distribution, by mechanisms including airway closure.

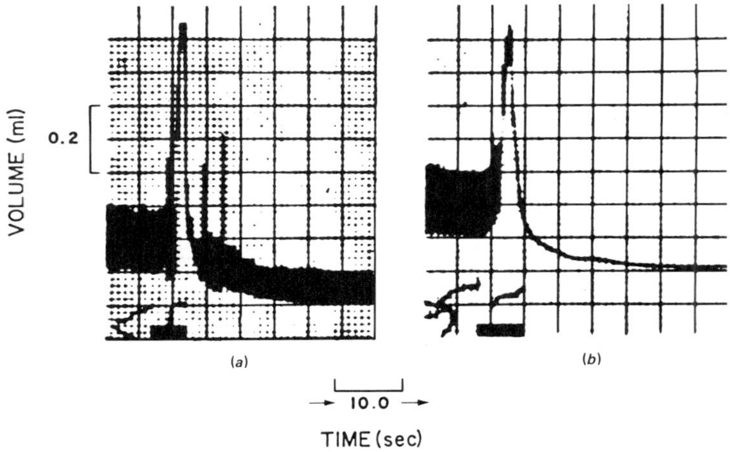

FIGURE 5. Plethysmographic record of volume changes with (*a*) abrupt onset of anesthesia (i.v. pentobarbital) or (*b*) anesthesia plus paralysis in a mouse. Anesthesia alone decreased ventilation by 60%, and functional residual capacity decreased by the same amount as when the animal was paralyzed (Vinegar et al., 1979).

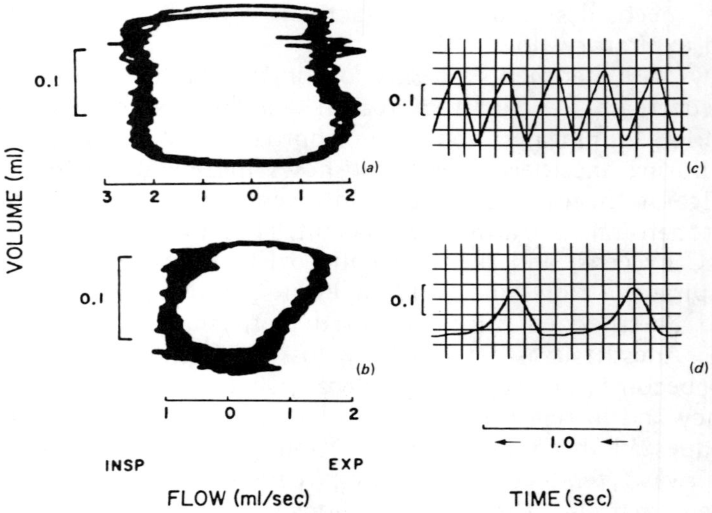

FIGURE 6. Ventilatory patterns in (*a, c*) awake and (*b, d*) anesthetized mouse. In the latter, expiration is unbraked and expiration is prolonged (Vinegar et al., 1979).

Airways

An important limitation to the study of comparative features of gas transport in mammalian lungs has been the lack of quantitative descriptions of airways. A recent National Heart, Lung, and Blood Institute (NHLBI) workshop on Comparative Biology of the Lung suggests that some of the necessary measurements will soon be available (Phalen and Oldham, 1983; Plopper et al., 1983; Tyler, 1983).

Airway size and flow velocity. The relatively high pulmonary ventilation required by small mammals in relation to the lung size is associated with relatively large airways. That is to say, airway size and lung size are dissociated in some sense ("dysanapsis" in humans; Mead, 1980) and presumably respond to different genetic (evolutionary) and developmental constraints.

Tracheal radius (diameter) scales with $m^{3/8}$ (Tenney and Bartlett, 1967), so tracheal cross-section varies with $m^{3/4}$. So does $\dot{V}_E$; therefore in adult mammals, superficial velocities of gas flow in the trachea are presumably invariant with body size. In humans, these are on the order of 1 m/s at rest, 20 m/s in maximal aerobic exercise, and 200 m/s or more (Mach 0.6!) during cough when the trachea is compressed to a fraction of its resting cross-section.

Flow velocities are, however, very much lower in the lung periphery, where the summed cross-section of airways is large. Not enough is known (for example, about airway area-distance functions) to say much about scaling of fluid flow velocities and their distributions in peripheral airways of mammals over a wide range of size.

Our interest in gas flow velocities arises in part because they influence the delivery of inhaled materials to airway walls. Reynolds numbers and boundary layers are touched on further in this paper; related issues are mentioned in the section on $\dot{V}_A/\dot{Q}$, including asymmetry and irreversibility of airway flow profiles.

Here I note that the mechanisms by which intermediate-sized aerosols (0.1–10 μm) are deposited on airway walls are uncertain, specifically the way(s) in which such particles traverse the essentially laminar layer near the wall in regions where flow at the core is turbulent. Davies (1983) suggests that the "stopping distance" of such particles is less than the thickness of the viscous sublayer they must penetrate. He presents calculations and observations to support a new mechanism involving, first, transport toward the wall by eddy diffusivity, and then a series of inertial jumps (down toward the wall) across streamlines each time the flow encounters a wall roughness element; particles finally get close enough to the wall to deposit on roughnesses while moving axially. Like other mechanisms for wall deposition, this one is dependent on flow velocity, Reynolds number (Re), and boundary layer conditions; however, not enough is known about their scaling and distributions in small airways (Schroter, 1978).

Resistance. Resistance of the respiratory system is dominated by that of airways. Reported measurements are quite variable, but it appears that resistance scales approximately as $m^{-3/4}$, that is to say, inversely with $\dot{V}_E$ (flow rate). This implies that the flow-resistive component of pressure (P_{res}) used for breathing is invariant with body size:

$$\begin{aligned} P_{\text{res}} &\propto (R)(\dot{V}_E) \\ P_{\text{res}} &\propto (m^{-3/4})(m^{3/4}) \\ P_{\text{res}} &\propto m^0 = k \end{aligned} \tag{6}$$

Pressure-flow relationships in mammalian lungs are complex, and the factors that govern them are incompletely understood (Pedley et al., 1977; Dorkin et al., 1982; Isabey and Chang, 1981); here again, not enough is known about airway structure to say much about their details in mammals of different size (Schroter, 1978).

Reynolds number and boundary layers. Since air flow velocity, density, and viscosity are invariant with body size, Re in the trachea varies with tracheal radius and therefore

$$\text{Re} \propto m^{3/8} \tag{7}$$

Its value is around 2000 in resting humans and 100 in mice. Thus laminar flow is likely to prevail in the trachea of small mammals, with the usual cautions about the effects of the larynx during inspiration. Velocity and radius of course diminish in peripheral airways, so Re falls to very low values, but its variation with distance, body size, oscillatory flow, and branching patterns are not well worked out.

Similar uncertainty applies to the scaling of boundary layer behavior in the airways. While Womersley's parameter, alpha, scales with $m^{1/4}$, Schroter (1978) concludes that its value does not exceed 1, and that therefore flow can be treated as quasisteady. The relationships among air velocity, tracheal length, and respiratory frequency are such that boundary layer thickness may be a smaller fraction of tracheal diameter in small mammals (Leith, 1983b).

Inertance. Inertance of the respiratory system is dominated by inertance of gas in central airways. Past estimates based on assumptions of airway isometry suggested that inertance scaled with $m^{-1/3}$, but since tracheal area scales with $m^{-3/8}$, I think inertance will vary more like $m^{-1/2}$ (Leith, 1983b). If so, then inertial pressures (P_{in}), like elastic and resistive pressures, would be invariant with body size, since temporal acceleration ($\ddot{V}$) varies with $m^{1/2}$:

$$\begin{aligned} P_{\text{in}} &= I\ddot{V} \\ P_{\text{in}} &\propto (m^{-1/2})(m^{1/2}) \qquad (8) \\ P_{\text{in}} &\propto m^{\circ} = k \end{aligned}$$

Inertial pressures are ordinarily small; only at high frequencies do they become a significant factor in overall respiratory system mechanics, for example, during panting at several cycles per second in large mammals.

The distribution of inertances along and among the branching airways is important in oscillation mechanics at high frequencies. Again, little is known about its scaling and species variation.

Expiratory flow limitation. This mechanism (Hyatt, 1983) is of interest in coughing and forced expirations, and in considering the upper limits to pulmonary ventilation, especially in pathological states. Maximum expiratory flows normally vary among mammals as $m^{3/4}$ (like $\dot{V}O_2$ and $\dot{V}_E$, and with similar wide variability; the evidence is not yet well developed). Expiratory flow limitation involves marked dynamic compression of downstream portions of the airway, with small cross sections and high velocities that markedly influence local clearance and deposition mechanisms.

During severe acute airway obstruction, including the bronchospastic responses to inhaled irritants, and in advanced chronic lung diseases, all or part of the lung can become flow-limited in expiration. This then becomes an important, sometimes dominant, determinant of FRC and its regional distribution, of the distributions of $\dot{V}_A$ and $\dot{V}_A/\dot{Q}$, and of the magnitudes of both V_D/V_T and P_aCO_2. There are good animal models of obstructive lung diseases.

Distribution of Ventilation

Statics. Under ordinary circumstances the time constants for distribution of gas among lung units in parallel are so short in relation to respiratory events that static mechanics dominate (Mead, 1969). Thus regional distributions of FRC and $\dot{V}_E$ ordinarily depend on regional lung elasticity and local differences between alveolar and pleural pressures (unless airways are closed or very nearly so). Assume that lung elastic properties and alveolar pressure are uniformly distributed, and consider the factors influencing regional distribution of pleural pressure, P_{pl}.

Gravity sets up a quasihydrostatic vertical gradient of P_{pl} associated with support of the lung and presumably influenced by overall density of the lung (tissue mass divided by tissue plus gas volume). Because V_{rel} (expressed as a fraction of TLC) increases with body size, influenced by chest wall stiffness, lung density at V_{rel} ranges from around 0.5 in small mammals (tissue and gas volumes about equal, each about 10% of TLC) to about 0.2 in large mammals. If thorax diameter varies with $m^{3/8}$ (elastic similarity), then in horizontal mammals the variation in lung height with body size is much greater than the variation in lung density, and so one would expect a much greater difference in P_{pl} from top to bottom of the lung in large than in small mammals, with associated differences in regional FRC, closure, ventilation, and $\dot{V}_A/\dot{Q}$ values. However, these expectations about distribution of P_{pl}—which I still hold—are not supported by careful work of Agostoni and colleagues (Agostoni, 1972), who found nearly equal top-to-bottom differences in P_{pl}, in animals from rat to ram. The question is of interest to inhalation toxicologists, because if Agostoni is right, one would expect similar top-to-bottom differences in regional FRC (and perhaps ventilation) in large and small mammals; but if my view is correct, those top-to-bottom differences should be negligibly small in small mammals.

Airway closure depends on poorly described distribution functions of small-airway closing mechanics (Macklem, 1971). On a gross regional scale within the lung, these can be considered to be everywhere the same, at least as a first approximation. The relevant factors include airway size, smooth-muscle activity, mucous-membrane and connective-tissue conditions (inflammation, edema, glandularity), and local elastic and surface-tension forces and their hysteresis, among others. Because gravity-dependent regional distribution of lung volume is a major determinant of local elastic recoil forces, the distribution of airway closure responds to gravitational influences. In large mammals, closure is seen first in dependent regions during lung deflation below V_{rel}; this occurs, however, at relatively high overall lung volumes, say 20–25% of TLC, near RV in normal young adult humans. But if gravity's influence is negligible (excised lungs, zero G, small mammals in my opinion) then even at low lung volumes, lung distension is relatively uniform and the distribution of airway closure should reflect only the (regionally uniform?) distribution function of closing mechanics. That is, closure should begin to occur at lower overall lung volumes (near 10% of TLC) and show no gravity-oriented distribution. This view conflicts with what Agostoni would have to predict, I think, but it is compatible with observed patterns of aerosol deposition in anesthetized hamsters (Sweeney et al., 1983).

Changes in P_{pl} during the breathing cycle are sometimes nonuniformly distributed in systematic ways, so that the regional distribution of ventilation becomes nonuniform in other ways than the usual top-to-bottom nonuniformity.

In recumbent mammals, the quasihydrostatic gradient of abdominal pressure influences both the shape of the relaxed diaphragm and its changes of shape upon becoming active during inspiration (Froese and Bryan, 1974). Therefore, the distribution of ventilation, especially near the diaphragm, is different during spontaneous and artificial ventilation. Again, I have to predict that this effect is negligible in small mammals, where the vertical height of the abdomen, and therefore the top-to-bottom difference in abdominal pressure is small. One can appreciate the diaphragm design problem by noting that a mouse's diaphragm is about 1 cm high, while that of a medium-sized whale is about a meter high, exactly according to the scaling rule $d \propto m^{3/8}$. The vertical range of transdiaphragmatic pressure in these two creatures is vastly different, but the implications of this scaling relationship have not been explored much.

A related effect, by contrast, is presumably more influential in small than in large mammals. During anesthesia, the normally coordinated activity of intercostal muscles and diaphragm becomes disorganized; in deep anesthesia, all but the diaphragm may become inactive. The more deformable the rib cage, the greater its paradoxical motion in response to an inspiratory fall in P_{pl}. Again, evidence from the experiments of Sweeney et al. (1983) suggests that ventilation of the lung apex diminishes markedly in hamsters under these conditions.

Dynamics. I said above that the distribution of tidal volume was ordinarily determined by static mechanics. But there are many circumstances of interest here in which that is not true.

Frequency-dependent nonuniformities of the distribution of ventilation are to be expected when regional time constants are both long in relation to respiratory events and nonuniform. These conditions exist in abnormal lungs, during high-frequency events, and when dense or viscous gases or liquids are respired. Frequency-dependent changes in other characteristics are then associated, e.g., the values of lumped mechanical parameters (compliance, resistance, inertance).

Over what ranges of frequencies are such effects to be seen? The frequencies vary systematically in inverse relation to body size. Some of the expected behavior can be predicted (with limited confidence) using standard approaches to alternating current circuit theory and the scaling relationships given above for V and C, $\dot{V}$ and R, $\ddot{V}$ and I. A single example below shows the kinds of things that can be done. Much more grandiose predictions of the scaling of oscillatory mechanics are possible, but not worthwhile here because not enough data now exist over a wide range of body size to confirm or deny them.

Time constants for the respiratory system as a whole can be approximated as the product of resistant and compliance:

$$RC \propto (m^{-3/4})(m^{1.0})$$
$$RC \propto m^{1/4} \tag{9}$$

Such time constants characterize (indeed can be derived from) the quasiexponential time courses of passive expirations, which are systematically faster in small mammals (Bennett and Tenney, 1982). But again, in a useful sense, they are not faster at all, for their time courses (and time constants) scale with body mass in the same way as the respiratory period [inverse of frequency, Eq. (4)] and are, in all mammals, about one-tenth as long as the resting breath duration. Respiratory frequencies in exercise reach about five times the resting value, less than half the frequency at which frequency-dependent behavior might start to become apparent in normal mammals (say 2 cps in humans, 15 cps in mice). If the tachypnea resulting from inhaled irritants approaches that of thermal panting (at the resonant frequency of the respiratory system, about 30 times the resting frequency, where elastic and inertial loads balance one another), then special equipment is needed to follow it: around 50 cps in mice. Finally, respiratory cycling frequency can be used as an experimental variable in studies of deposition of inhaled materials (Valberg et al., 1982), or can introduce interpretive difficulties if such studies are performed at unusually low (or high) frequencies.

$\dot{V}_A/\dot{Q}$ and Gas Exchange with Blood

This section is included only as a note of caution: this review omits considerations of mass transport of gases in the lung having to do with gas exchange and its regional distribution. Nonuniform distributions of $\dot{V}_A/\dot{Q}$ values (and therefore of regional gas exchange, transport, and composition) exist normally in a gravity field and are also characteristic of many diseases. These gravity-dependent nonuniformities, I think, must be negligible in small mammals and marked in large mammals.

These mechanisms are ordinarily relatively unimportant in thinking about distribution and retention of particles and insoluble gases. But soluble and reactive gases are another story entirely; their regional deposition, concentrations, and effects are influenced by regional removal mechanisms, including blood flow. And their presence and exchange can influence the transport and exchange of other gases (and presumably particles) in the respired medium.

Gas Mixing

There are complex geometry-dependent interactions between convection and diffusion in airways and in the pulmonary acinus. Their nature and their importance in gas mixing and gas transport are becoming better understood. They surely influence the transport and deposition of inhaled particles and gases (including soluble and reactive ones). But few studies, either theoretical or experimental, have so far applied these ideas to toxic materials or explored their interesting comparative applications.

I think these topics are worth attention by toxicologists. Levich (1983), summarizing a recent conference on physicochemical hydrodynamics, expressed the views that problems of biological fluid dynamics "are intimately connected with the hydrodynamics of loaded flow and rheology, and that these are among the most attractive and significant problems of modern biophysics."

Airways. As gas flows around curves or corners, more or less complex secondary flow patterns are set up (wall separation, closed eddies, vortices, helices) that may extend for some distance downstream (Pedley et al., 1977; Schroter, 1978). These occur in laminar as well as turbulent flow; for example, vortex shedding occurs at Re values as low as 20. Effects of these kinds are found at changes in cross section and at airway branching points. The flow patterns are quite different during inspiration and expiration. A net effect is enhanced gas mixing, both axial and radial, during oscillatory flow. There may be quantitatively important enhancement of transport of gases (e.g., during high-frequency ventilation) and of airway deposition of aerosols (Scherer and Haselton, 1982). Few comparative data are available, but species- and size-dependent variations in airway geometry, when adequately described, appear to offer experimental opportunities.

Acinus. In the pulmonary acinus, distances and bulk flow velocities are small; convection plays a less important role in gas transport as diffusion becomes more influential. The situation is dramatically shown by Fig. 7. Tracers with very different diffusivities were inhaled simultaneously in a 600-ml tidal volume; their expired concentrations were plotted as a function of volume during subsequent expiration to RV. One tracer was an essentially nondiffusing half-micron aerosol. It (mostly) remained in the 600-ml gas bolus and exited with it; that is, it mixed very little with the resident gas in the lung. The converse was true for the highly diffusive helium, so the mixing "front" for it was more central, and the Fowler deadspace measured by its washout was much smaller.

Note here that we can (ideally) treat helium as completely insoluble. The lung provides minimal sinks for it, so all that goes in can later be accounted for, in either expired or resident gas.

All of the aerosol particles that go in, of course, do not remain in suspension. The respiratory system does provides sinks for these particles,

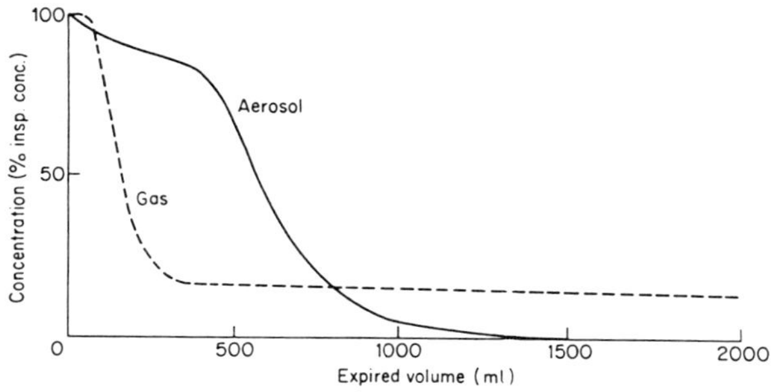

FIGURE 7. Expired concentrations as a function of volume following 600-ml inspirations containing helium or a nondiffusing aerosol (Muir, D.C. E. 1966. *J. Physiol.* 184:73P–74P).

of variable capacities for different particles. The same is true for soluble, reactive, or metabolized materials, including NO_x and O_3, as well as O_2 and anesthetic gases. *Where* these sinks reside and how they behave is, of course, very interesting.

In recent years, new models have provided new, intuitively appealing understanding of the interdependence of convection and diffusion in the acinus, briefly reviewed by Engel (1983). These approaches use symmetrical and asymmetrical two-branched trumpet models like that shown in Fig. 8, and allow calculation of time-dependent distributions of gas composition along airways, as exemplified in Fig. 9. The application of these ideas to soluble gases (Zwart et al., 1983), including those of interest to toxicologists and anesthesiologists, looks worthwhile. Again, size- and species-dependent differences in acinar geometry, once they are adequately

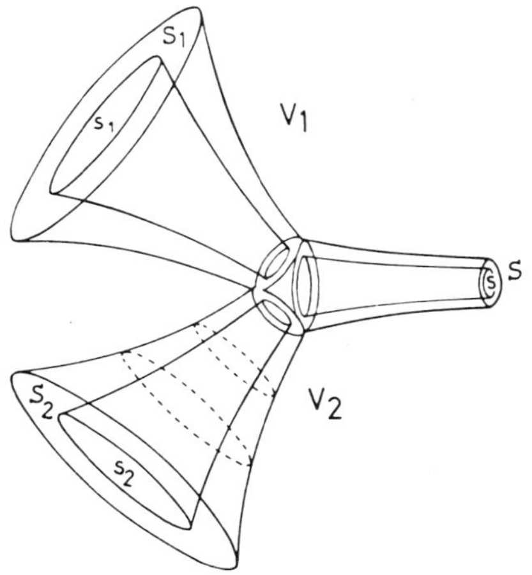

FIGURE 8. Two-branched trumpet model of pulmonary acinus. Asymmetry is introduced by shortening V_2 (dashed lines) (Engel, 1983).

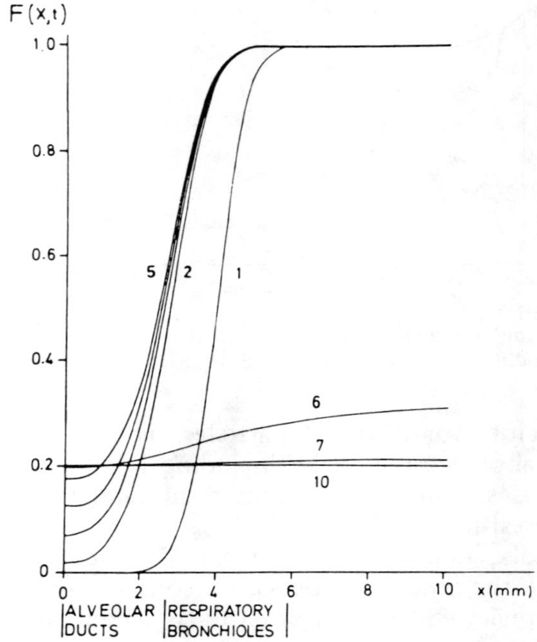

FIGURE 9. Concentration (F) of inspired tracer gas as a function of distance (x) from the "alveolar" end of a symmetrical trumpet model, at 0.2-s intervals (1–10). During inspiration there is a "stationary front" of concentration where bulk flow is balanced by diffusion exchange with resident gas (Engel, 1983).

described, appear to offer theoretical and experimental opportunities. Perhaps there is some insight here into the apparently paradoxical discrepancy between the scaling of maximum $\dot{V}O_2$ (proportional to $m^{3/4}$) and the lung's diffusing capacity for O_2 (proportional to $m^{1.0}$) (Weibel et al., 1981).

There is reason, and some evidence, to think that shape changes in the acinus are not reversible (i.e., they show hysteresis with volume) during the respiratory cycle. There is mechanical coupling between surface-tension forces and the elastic structures that define the boundaries of alveolar ducts (AD) and form the mouths of alveoli (Hoppin and Hildebrandt, 1977). High surface tension tends to retract these partitions, making the AD bigger and the alveoli shallower at the same total volume. Cyclic shape changes of this kind may occur during lung volume cycling, and would be expected to show hysteresis with volume (Wilson, 1982; Wilson and Bachofen, 1982). Preliminary measurements of backscattering of laser light at the pleural surface of excised lungs indeed suggests that mean linear intercepts are greater during stepwise inflation than at the same volumes during stepwise deflation (J. P. Butler et al., personal communication).

If such out-of-phase changes in shape occur during normal cyclic ventilation, they can be expected to contribute to gas mixing, for then even flow of extremely low Re in the acinus need not be reversible.

Even if shape changes in the acinus are reversible, it is not necessarily true that flow is reversible, even though superficial Re value is extremely low. Fluid flowing over a sharp edge has appreciable angular acceleration, which is to say velocity. Inertial forces become significant and local Re values difficult to define.

Laminar vortices can form downstream of abrupt changes in cross section, especially at thin plates, with apparent Re values as low as 5, adding eddy diffusivity to local mixing mechanisms, on both sides of the plate if flow is oscillatory.

In tubes partially obstructed by transverse orifice plates, creeping flow can be irreversible at Re values as low as 0.4 (Jeje, 1983).

If these two mechanisms contribute to mixing in the acinus, then one would expect their effects to be greater in its proximal than in its distal regions, for flow velocities over alveolar mouths are greater proximally.

CONCLUSION

Many features of mammalian respiration influence the distribution and retention of noxious agents in inspired air. Some of these have been reviewed, but systematic description and understanding are not yet available.

Recent advances in knowledge about gas transport and mixing in airways and the pulmonary acinus appear to have applications in comparative physiology and inhalation toxicology that should be explored.

REFERENCES

Agostoni, E. 1972. Mechanics of the pleural space. *Physiol. Rev.* 52:57–128.

Benedict, F. G. 1938. Vital energetics. A study in comparative basal metabolism. Carnegie Inst. Washington Monograph 503, Washington, D.C.

Bennett, F. M., and Tenney, S. M. 1982. Comparative mechanics of mammalian respiratory system. *Resp. Physiol.* 49:131–140.

Book, S. A. 1982. Scaling toxicity from laboratory animals to people: An example with nitrogen dioxide. *J. Toxicol. Environ. Health* 9:719–725.

Brain, J. D., and Mensah, G. A. 1983. Comparative toxicology of the respiratory tract. *Am. Rev. Respir. Dis.* 128:587–590.

Brain, J. D., and Valberg, P. A. 1979. Aerosol deposition in the respiratory tract. *Am. Rev. Respir. Dis.* 120:1325–1373.

Bruce, M. C., Bruce, E. N., Leith, D. E., and Murphy, S. D. 1979. Diethyl maleate and/or ozone (10 ppm) reduce ventilation by 60–80% in awake mice. *Physiologist* 22:16 (Abst.).

Davies, J. T. 1983. A new theory of the deposition of colloidal particles from turbulent fluids. *Ann. N.Y. Acad. Sci.* 404:313–326.

Dorkin, H. L., Jackson, A. C., Strieder, D. J., and Dawson, S. V. 1982. Interaction of oscillatory and unidirectional flows in straight tubes and an airway cast. *J. Appl. Physiol.* 52:1097–1105.

Engel, L. A. 1983. Gas mixing within the acinus of the lung. *J. Appl. Physiol.* 54:609–618.

Feldman, H. A., and McMahon, T. A. 1983. The ¾ mass exponent for energy metabolism is not a statistical artifact. *Respir. Physiol.* 52:1–5.

Froese, A., and Bryan, A. C. 1974. Effects of anesthesia and paralysis on diaphragmatic mechanics in man. *Anesthesiology* 41:242–255.

Gillespie, J. R. 1983. Mechanisms that determine functional residual capacity in different mammalian species. *Am. Rev. Respir. Dis.* 128:574–577.

Heusner, A. A. 1982. Energy metabolism and body size, I. Is the 0.75 mass exponent of Kleiber's equation a statistical artifact? *Respir. Physiol.* 48:1–12.

Heusner, A. A. 1983. Body size, energy metabolism, and the lungs. *J. Appl. Physiol.* 54:867–873.

Hoppin, F. G. Jr., and Hildebrandt, J. 1977. Mechanical properties of the lung. In *Bioengineering Aspects of the Lung,* ed. J. B. West, pp. 83–162. New York: Marcel Dekker.

Hyatt, R. E. 1983. Expiratory flow limitation. *J. Appl. Physiol.* 55:1–8.

Isabey, D., and Chang, H. K. 1981. Steady and unsteady pressure-flow relationships in central airways. *J. Appl. Physiol.* 51:1338–1348.

Jeje, A. A. 1983. On the flow of water in the xylem vessels of living plants. *Ann. N.Y. Acad. Sci.* 404:177–181.

Kleiber, M. 1932. Body size and metabolism. *Hilgardia* 6:315–358.

Leith, D. E. 1983a. Comparative mammalian respiratory mechanics. *Am. Rev. Respir. Dis.* 128:577–582.

Leith, D. E. 1983b. Mammalian tracheal dimensions: Scaling and physiology. *J. Appl. Physiol.* 55:196–200.

Levich, B. G. 1983. Concluding remarks. *Ann. N.Y. Acad. Sci.* 404:527–532.

Macklem, P. T. 1971. Airway obstruction and collateral ventilation. *Physiol. Rev.* 51:386–436.

McMahon, T. A. 1973. Size and shape in biology. *Science* 179:1201–1204.

McMahon, T. A. 1975. Using body size to understand the structural design of animals: Quadrupedal locomotion. *J. Appl. Physiol.* 39:619–627.

McMahon, T. A. 1977. Allometry. In *Yearbook of Science and Technology,* pp. 49–57. New York: McGraw-Hill.

McMahon, T. A. 1984. Effects of scale. In *Muscles, Reflexes, and Locomotion,* pp. 234–296. Princeton: Princeton University Press.

McMahon, T. A., Brain, J. D., and Lemott, S. 1977. Species differences in aerosol deposition. In *Inhaled Particles IV,* ed. W. H. Walton, pp. 23–32. New York: Pergamon.

Mead, J. 1969. The distribution of gas flow in lungs. In *Circulatory and Respiratory Mass Transport,* eds. G. E. W. Wolstenholme and J. Knight, pp. 204–214. London: J. and A. Churchill.

Mead, J. 1980. Dysanapsis in normal lungs assessed by the relationship between maximal flow, static recoil, and vital capacity. *Am. Rev. Respir. Dis.* 121:339–342.

Mortola, J. P. 1983. Some functional mechanical implications of the structural design of the respiratory system in newborn mammals. *Am. Rev. Respir. Dis.* 128:569–572.

Muir, D. C. F. 1966. Bulk flow and diffusion in the airways of the lung. *Br. J. Dis. Chest* 60:169–176.

Pedley, T. J., Schroter, R. C., and Sudlow, M. F. 1977. Gas flow and mixing in the airways. In *Bioengineering Aspects of the Lung,* ed. J. B. West, pp. 163–265. New York: Marcel Dekker.

Phalen, R. F., and Oldham, M. J. 1983. Tracheobronchial airway structure as revealed by casting techniques. *Am. Rev. Respir. Dis.* 128:51–54.

Plopper, C. G., Marriassy, A. T., and Lollini, L. O. 1983. Structure as revealed by airway dissection. A comparison of mammalian lungs. *Am. Rev. Respir. Dis.* 128:54–57.

Scherer, P. W., and Haselton, F. R. 1982. Convective exchange in oscillatory flow through bronchial-tree models. *J. Appl. Physiol.* 53:1023–1033.

Schroter, R. C. 1978. Convective flows in mammalian lungs. In *Comparative Physiology: Water, Ions, and Fluid Mechanics,* eds. K. Schmidt-Nielsen, L. Bolis, and S. H. P. Maddrell, pp. 303–325. New York: Cambridge University Press.

Silver, E. H., Leith, D. E., and Murphy, S. D. 1981. Potentiation by triorthotolyl phosphate of acrylate ester-induced alterations in respiration. *Toxicology* 22:193–204.

Spells, K. E. 1969–70. Comparative studies in lung mechanics based on a survey of literature data. *Respir. Physiol.* 8:37–57.

Stahl, W. R. 1962. Similarity and dimensional methods in biology. *Science* 137:205–212.

Stahl, W. R. 1963. The analysis of biological similarity. *Adv. Biol. Med. Phys.* 9:355–464.

Stahl, W. R. 1967. Scaling of respiratory variables in mammals. *J. Appl. Physiol.* 22:453–460.

Stahl, W. R., and Gummerson, J. Y. 1967. Systematic allometry in five species of adult primates. *Growth* 31:21–34.

Sweeney, T. D., Brain, J. D., and Lemott, S. 1983. Anesthesia alters the pattern of aerosol retention in hamsters. *J. Appl. Physiol.* 54:37–44.

Taylor, C. R., Maloiy, G. M. O., Weibel, E. R., Langman, V. A., Kamau, J. H. Z., Seeherman, M. J., and Heglund, N. E. 1981. Design of the mammalian respiratory system III. Scaling maximum aerobic capacity to body mass; wild and domestic mammals. *Respir. Physiol.* 44:25–37.

Tenney, S. M., and Bartlett, D. Jr. 1967. Comparative quantitative morphology of the mammalian lung: Trachea. *Respir. Physiol.* 3:130–135.

Tyler, W. S. Comparative subgross pulmonary anatomy of lungs. Pleuras, interlobular septa, and distal airways. *Am. Rev. Respir. Dis.* 128:532–536.

Valberg, P. A., Brain, J. D., Sneddon, S. L., and LeMott, S. R. 1982. Breathing patterns influence aerosol deposition sites in excised dog lungs. *J. Appl. Physiol.* 53:824–837.

Vinegar, A., Sinnett, E. E., and Leith, D. E. 1979. Dynamic mechanisms determine functional residual capacity in mice, *Mus musculus*. *J. Appl. Physiol.* 46:867–871.

Weibel, E. R., Taylor, C. R., Gehr, P., Hoppeler, H., Mathieu, O., and Maloiy, G. M. O. 1981. Design of the mammalian respiratory system IX. Functional and structural limits for oxygen flow. *Respir. Physiol.* 44:151–164.

Wilson, T. A. 1982. Surface tension-surface area curves calculated from pressure-volume loops. *J. Appl. Physiol.* 53:1512–1520.

Wilson, T. A., and Bachofen, H. 1982. A model for mechanical structure of the alveolar duct. *J. Appl. Physiol.* 52:1064–1070.

Zwart, A., Luijendijk, S. C. M., and de Vries, W. R. 1983. Expiratory and arterial partial pressure relations under different ventilation perfusion conditions. *J. Appl. Physiol.* 54:1745–1753.

PHYSICOCHEMICAL PROCESSES AND THE FORMULATION OF DOSIMETRY MODELS

John H. Overton, Jr.

Northrop Services, Inc., Research
Triangle Park, North Carolina

The major physical and chemical processes involved in the transport and absorption of O_3 or NO_2 in the lower respiratory tract are discussed. This includes the development of respiratory tract models, flow patterns, and transport in tube networks, the mucous, surfactant, and tissue layers, and chemical reactions and transport of O_3 or NO_2 within these layers. Descriptions of the individual processes are simplified and integrated to illustrate the formulation of dosimetry models. Data from a dosimetry model, formulated from the concepts discussed, are used to illustrate the types of information obtained by modeling.

INTRODUCTION

A lung and the mechanisms for transporting and absorbing gases such as O_3 or NO_2 are very complex. Ideally, we would like to completely define a given lung's dimensions and the geometry of its air spaces and airways and characterize the transport and absorption of gases by applying the equations of mass transport and fluid mechanics. Not only is this impractical, but our present knowledge precludes this approach. However, we can create simple models that effectively simulate the structure and performance of the lungs.

This paper discusses some of the physical and chemical processes that are involved in the transport and absorption of O_3 and NO_2 in the lower respiratory tract (tracheobronchial and pulmonary regions) and the methods by which descriptions of these processes can be simplified to formulate dosimetry models.

Transport and absorption occur in the lower respiratory tract. This paper illustrates how the many branching and different sized airways can be reduced to a simple respiratory tract model that is useful for modeling gaseous transport and absorption in this region. Airflow in the lower respiratory tract is illustrated by the results of investigations of convection in model bifurcating tubes. The work of Taylor (1953) and of Scherer et al. (1975) indicates how transport in the intricate three-dimensional airway network can be simplified to a one-dimensional description of transport in a single tube.

O_3 or NO_2 absorption takes place in the fluid lining the airways and in the underlying epithelial layer. The degree of O_3 or NO_2 absorption and penetration into these layers depends on their physical and chemical properties. Although little is known about these properties, present knowledge allows us to create models of these layers and of the chemical and transport processes that occur within them. A possible approach to modeling this region is introduced, and the results of absorption modeling based on the concepts and processes described in the paper are discussed.

TRACHEOBRONCHIAL AND PULMONARY MODEL

The lungs' airways form a very intricate system. Since these spaces define the region through which gases are transported, their dimensions, geometry, and structure must be quantified before equations describing transport can be applied.

Many schemes of airway quantification that are compatable with mathematical transport equations represent airways as sequential sets of right circular cylinders. Figure 1 illustrates the reduction of several airways to right circular cylinders. A method is chosen to identify airway location in the respiratory tract and to measure airway lengths, diameters, and branching angles (Horsefield and Cumming, 1967; Phalen et al., 1978). Vertical orientation, important in particle deposition, may also be determined. Location identity, diameter, and length are considered the relevant measurements for gaseous transport. The three airways illustrated (Fig. 1) have been reduced to three right circular cylinders, and each will have an identification number that allows for an ordering or sequencing of the cylinders corresponding to the order in the actual airway.

Each airway is associated with a generation number, which is the number of branchings that must be passed on traveling a defined path from an origin to the airway. In Fig. 2, the origin is the "plus" in the left-most airway. The dashed lines define two possible paths. The number associated with each airway is the airway's "generation." Because daughter branches generally do not have the same lengths, the distances from the origin to airways of the same generation will depend on the specific path or airway. Likewise, two paths of the same length do not always end in airways of the same generation.

Traditionally, lung models have been generation based. A characteristic length and diameter is assigned to each generation. For one method, the average terminal bronchiole generation is defined as *the* model terminal bronchiole generation (Yeh et al., 1979; Yeh and Schum, 1980). For each proximal generation, the charcteristic dimensions are the average of the dimensions of all of that generation's airways. These dimensions, along with the number of airways in each generation, define a simple and usable model of the airways in the conductive zone (Fig. 3) that has frequently been used in investigating transport in the lower respiratory tract.

A generation-based model is not unique. A generation other than the

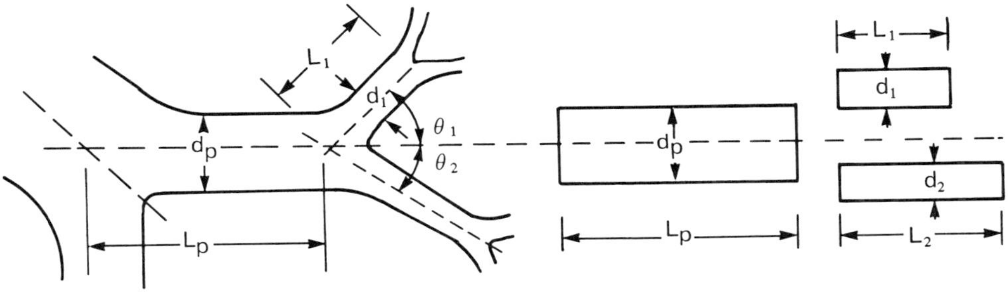

FIGURE 1. Reduction of several airways to right circular cylinders.

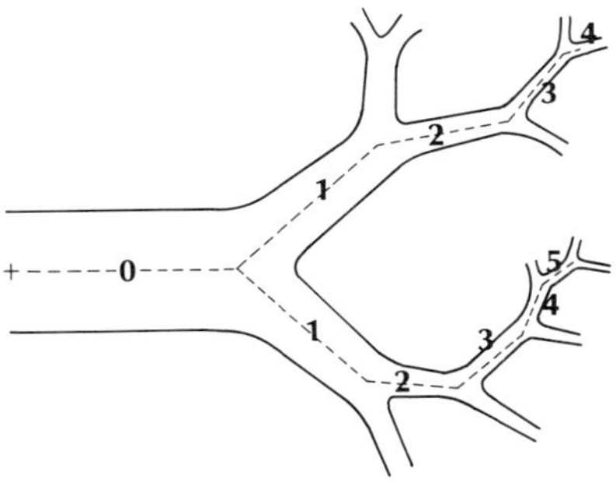

FIGURE 2. The concept of generation and pathways.

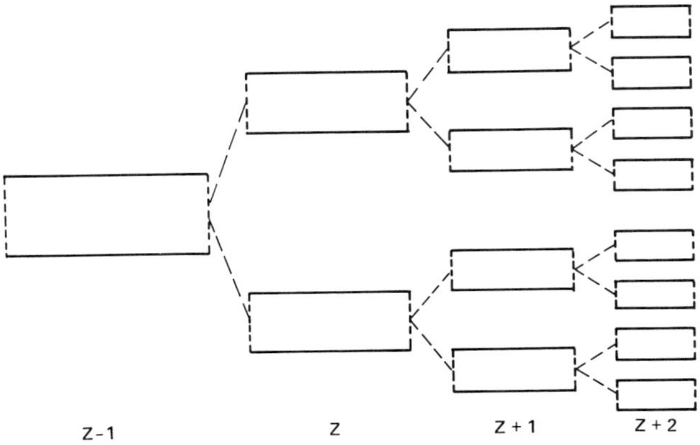

FIGURE 3. Idealized conductive zone airway model based on the concept of sequential sets of right circular cylinders (z = generation index).

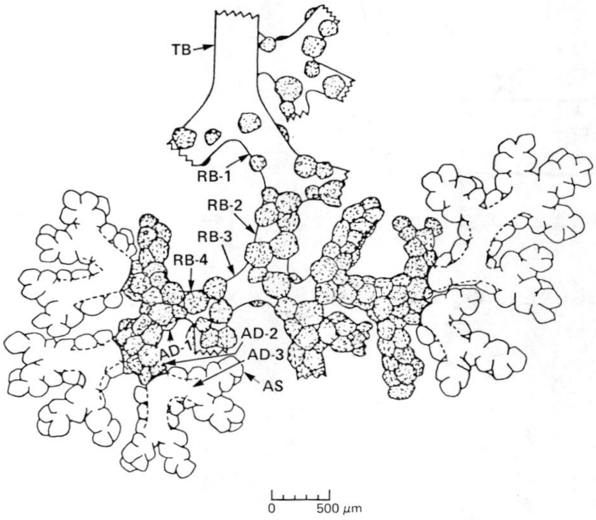

FIGURE 4. Human respiratory acinus (Schreider and Raabe, 1981).

terminal bronchioles could be chosen as the defining generation. For example, Weibel (1963) chose the alveolar sacs. Horsfield and Cumming (1967) reversed the procedure and counted up from bronchi (with diameters greater than 0.7 mm) to the trachea. Although this model is not generation-based, it is like the generation-based models because it is based on branching. However, the mathematical equations generally used to describe gas transport in the respiratory tract are based on distance along a single path. Thus, a distally based airway model may be more appropriate to describe the transport and absorption of gases.

The methods described above can be extended into the pulmonary region, where, although alveoli (Fig. 4) complicate the procedure, generation and dimensions can be assigned to air spaces. The length and diameter of the air spaces or channel about which alveoli cluster can be defined. The number of alveoli, their surface area, air volume, and the fraction of the channel surface area covered by alveoli are also needed to characterize

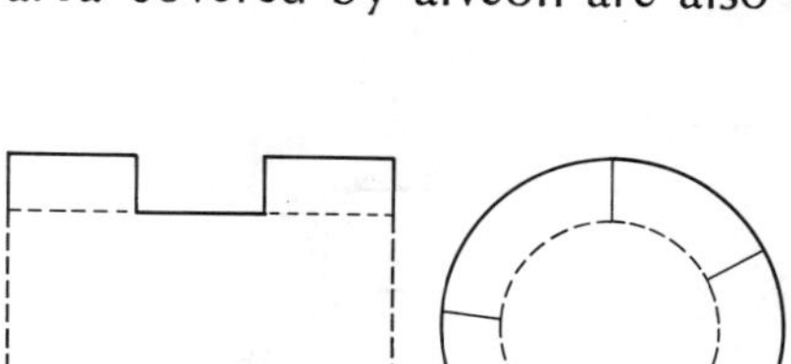

FIGURE 5. Idealized model of an alveolated airway.

this region. With this information, a concentric two-cylinder model of the type illustrated in Fig. 5 can be constructed.

The lower respiratory tract model is completed by augmenting the conductive zone airway model with the model of the pulmonary region (Fig. 6*a*). A very useful feature of this model is that the sequence and dimensions of cylinders describing any path from the trachea to any airway or air space of a given generation are the same. Thus, all these paths can be replaced by a single path model (Fig. 6*b*) that is representative of any path taken from the trachea to the "end" of the respiratory tract.

TRANSPORT OF GASES IN THE AIRWAYS

Convection (airflow), molecular diffusion, and absorption affect the transport of O_3 or NO_2. Convection is often described in terms of the average local velocity and the random variations about the average (the turbulent component). Molecular diffusion due to O_3 or NO_2 concentration gradients creates a diffusion velocity that adds to convection. Absorption removes gases from the lumen and affects concentration gradients, which in turn affect the transport rate.

The intricate structure of the respiratory tract combined with the breathing pattern creates complex gas flows (Schroter and Sudlow, 1969; Schreck and Mockros, 1970; Olson et al., 1973). Figure 7, based on Schroter and Sudlow (1969), illustrates the following observations about the air flow in ideal bifurcating tubes: (1) inhalation and exhalation flow patterns differ; (2) velocity profiles are affected by the orientation of the

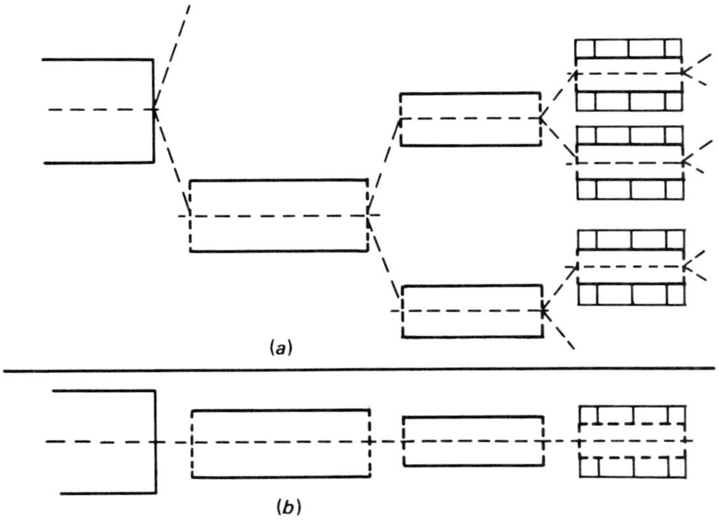

FIGURE 6. (*a*) Idealized tracheobronchial and pulmonary airway model. (*b*) The idealized tracheobronchial and pulmonary airway model reduced to a single-path model.

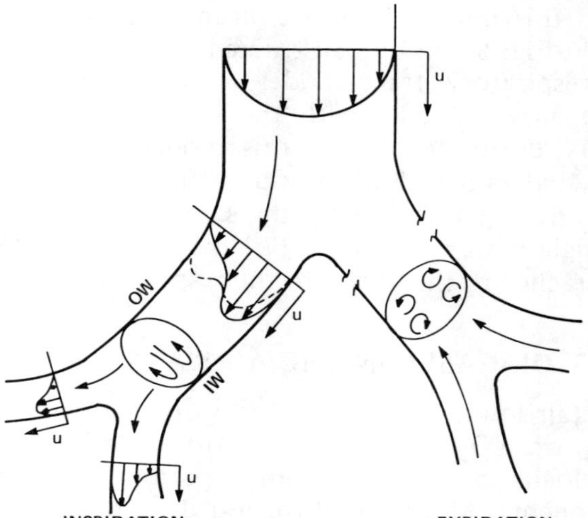

FIGURE 7. Airflow in experimental bifurcating tubes (based on Schroter and Sudlow, 1969).

plane of bifurcation; and (3) the velocity profiles are not axially symmetric.

Nonturbulent flow established in the parent tube (upper tube in Fig. 7) creates an approximately parabolic velocity profile, with zero (or near zero) velocity at the tube walls and maximum velocity at the tube's center. The velocity at the walls of the daughter airways must also be zero or near zero. In the plane of bifurcation and on entrance to a daughter tube, the highest velocities are near the second tube's inner wall (IW, Fig. 7) and rapidly fall to zero at the inner wall. The lower velocities from the larger tube are transferred toward the outer-wall side (OW, Fig. 7). Thus, higher velocities exist at the inner walls than at the outer wall. Just as there is geometric symmetry in the plane perpendicular to the bifurcation plane, the velocity profile in the perpendicular plane (dotted curve, Fig. 7) is also symmetric about this plane.

Proceeding distally to the next bifurcation, we would expect a greater flow in the third generation tube in line with the higher velocities of the proceeding tube if the second plane of bifurcation is the same as the first. If the planes are perpendicular and we invoke physical symmetry, then equal flow can be expected in both branches of the third generation.

Upon simulating inhalation, Schroter and Sudlow (1969) observed two vortices in the branches, while reversed flow produced four vortices in the parent generation. In other words, airflow during inhalation is not the reverse of airflow during exhalation. It will be shown later that this asymmetry may affect gas transport. The respiratory tract, with more than one branching airway, varying orientations of bifurcation planes, differing branching angles, changing dimensions, and alternating flow directions, produces much more complex flows than those illustrated.

Investigators often use the single-path lung model for describing respiratory tract transport. Airflow in a single tube whose diameter decreases distally is not the same as airflow in the respiratory tract. Nevertheless, theoretical and experimental work has demonstrated that the single-path approach is viable (Scherer et al., 1975; Ultman and Blatman, 1977). To understand this, the investigations of Taylor (1953) and Scherer et al. (1975) are discussed.

Taylor (1953) theoretically and experimentally demonstrated that the behavior of the cross-sectional average concentration of a substance in a tube in which parabolic flow was established could be described by one-dimensional convection and dispersion. Specifically, a pulse of substance moves down the tube with an average velocity equal to the medium's average velocity; its spread in the axial direction is governed by a disperson coefficient that is larger than the molecular diffusion coefficient.

Figure 8 illustrates some of the aspects that produce these results. Figure 8*a* is the velocity profile of the tube outlined in Fig. 8*b*. Initially, a cross-sectionally uniform pulse of trace gas (hatched region of Fig. 8*b*) is inserted into the tube's entrance. Some time later, due to convection and molecular diffusion, the pulse has moved down the tube. If molecular motion did not exist, only convection would occur. In this case, the molecules at the tube walls would have no motion because the convective velocity is zero there. Velocity is fastest along the axis, and that portion of the gas moving along the axis penetrates farthest into the tube. The penetration depth of the pulse depends on radial distance from the axis and would appear as region 2 (Fig. 8*b*), with no molecules in regions 1 and 3. Even without molecular diffusion, the average concentration of the pulse spreads as a result of convection. Diffusion moves the molecules of the pulse to regions of lesser concentration (1 and 3 in Fig. 8*b*), as indicated by the arrows in Fig. 8*b*. The overall result is that in region 1,

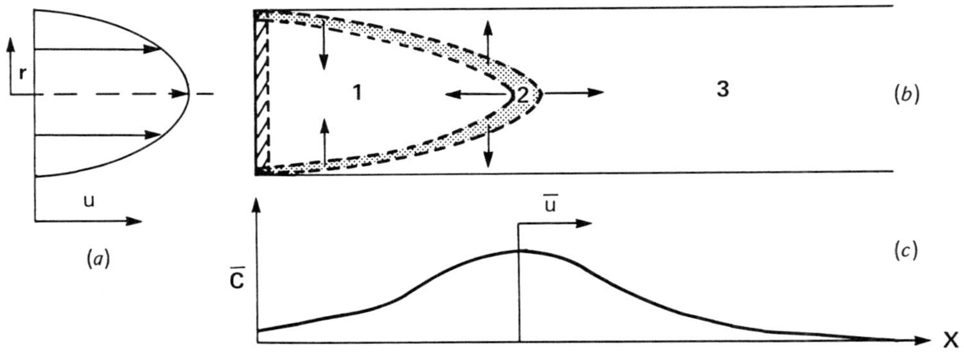

FIGURE 8. Taylor dispersion. (*a*) Velocity profile. (*b*) Distribution of molecules in the tube. (*c*) The cross-sectional average concentration of the molecules in the tube.

the concentration increases as the radius increases; i.e., concentrations are higher along the wall than along the axis of the tube. The reverse is true in region 3, where the concentration decreases radially.

Figure 8*c* shows a plot of the cross-sectional average concentration of the molecules in Fig. 8*b*. Figure 9 shows the behavior of the average concentration as the pulse moves down the tube by illustrating the average concentration profile at various times (indicated by t_1, t_2, t_3, and t_4, where $t_1 < t_2 < t_3 < t_4$) and its positions during movement through the tube. The pulse moves with an average velocity $\bar{u}$, which is the same as the average medium velocity. In addition, it spreads in a manner that can be described by Ficks' law of diffusion. Thus, the average concentration satisfies the simple convective-diffusion equation,

$$\frac{\partial c}{\partial t} + \bar{u}\,\frac{\partial c}{\partial x} = D_e\,\frac{\partial^2 c}{\partial x^2} \tag{1}$$

where D_e is the effective dispersion coefficient (usually called Taylor's dispersion or diffusion coefficient). D_e depends on the average velocity, tube radius, and the molecular diffusion coefficient.

The same principles that apply to an initially narrow pulse travelling down a tube also apply to a pulse with an arbitrary initial shape (as expected in respiration). An initial concentration profile can be thought of as being composed of a sequential series of very narrow pulses along the *x* axis. If the transport of each narrow pulse is described by Eq. (1), the transport of the sum of the concentrations of all the pulses at a given distance *x* and time *t* is also described by Eq. (1). The sum of the concentrations at a given *x* and *t* is the concentration of the arbitrary-shaped pulse at *x* and *t*. That is, the principle of superposition is valid for Eq. (1); if not, the results of Taylor and Sherer et al. (1975; described below) could not be applied describe transport in the respiratory tract.

This result appears to be "Taylor"-made for application to the single path respiratory tract model illustrated in Fig. 6*b*. However, Taylor's formulation does not consider the effects of the turbulence, branchings,

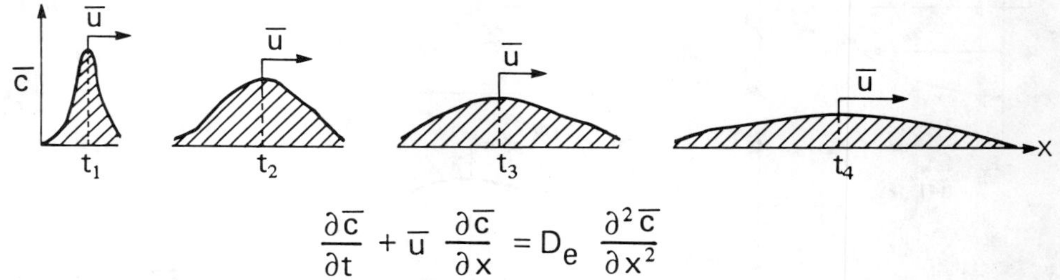

FIGURE 9. The concentration profile of a trace gas pulse as it moves down the tube. This average concentration is a solution to the convective-dispersion equation.

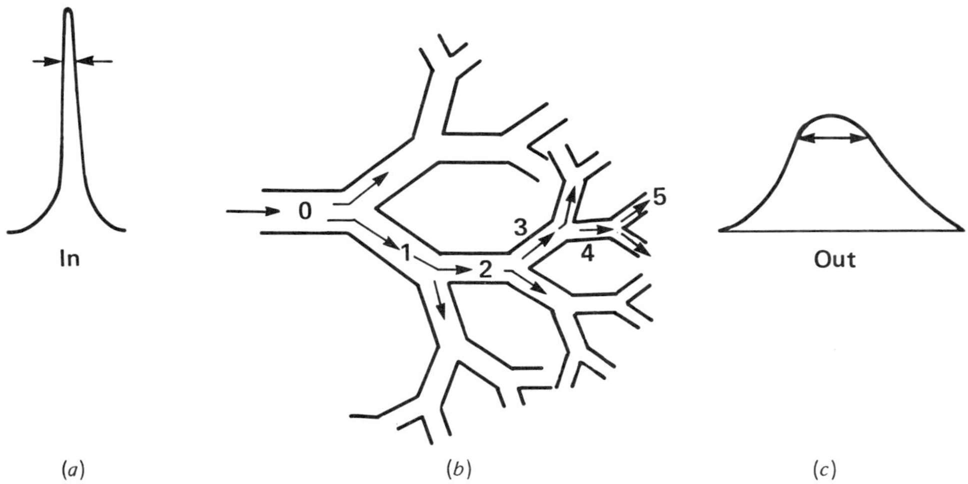

FIGURE 10. The idealized glass tube lung model experiment of Scherer et al. (1975).

secondary flows, inhomogenous ventilation, or gas absorption that occur in the respiratory tract.

Nevertheless, the more recent work of Scherer et al. (1975) and of Ultman and Blatman (1977) indicates that the description of the convection and dispersion of the average concentration in the complex lung system is essentially correct, at least in the absence of absorption.

Scherer et al. (1975) constructed a six-generation glass lung model (Fig. 10*b*) based on the airway dimensions of Wiebel (1963) and the branching angles of Horsefield and Cumming (1967). A trace gas was injected into the model's entrance (generation 0) and its shape was recorded (Fig. 10*a*). The pulse shape was also recorded as it exited the model in generation 5 (Fig. 10*c*). This information, along with the initial pulse width, transition time, flow rates, etc., allowed Scherer et al. (1975) to determine the effective dispersion coefficient of a pulse passing through five bifurcations with its complex flows. Scherer et al. also simulated expiration with airflow in the opposite direction (right to left, Fig. 10*b*) and found the dispersion coefficient to be about one-third of that for inspiration. The difference between inspiration and expiration may be related to the difference in airflow pattern cited previously.

In either case, the functional dependence of the dispersion coefficient on the physical and dynamic parameters was found to be different than that determined by Taylor (1953) for single tubes with simple laminar flow. This result is not surprising given that the single-tube system and the six-generation system differ. Nevertheless, Ultman and Blatman (1977) theoretically demonstrated that the results of Scherer et al. (1975) could be understood in terms of the single-path model.

Taylor (1953) had pointed out that his formulation was not valid until after a pulse had traveled a critical distance. Gill and Sankarasubramanian (1970) later described a pulse's behavior before the critical distance. They showed that by replacing Taylor's effective dispersion coefficient with a theoretically derived, time-dependent coefficient, a pulse could be described by convection and dispersion at all times throughout the tube. Taylor's formulation is a "long" time or asymptotic approximation that is valid after the pulse moves the critical distance. Ultman and Blatman (1977) showed that the results of Scherer et al. (1975) could be explained by pulses that had not reached the critical distance, or in other words, the effective dispersion coefficient was not "fully developed."

A complete description of transport in the air spaces of the pulmonary region must include the effects of alveoli. Fortunately, transport in this region can also be modeled in terms of convection and dispersion in the axial direction (Paiva, 1973; Scherer et al., 1972). In this region, molecular diffusion is the dominant form of transport (Wilson and Lin, 1970). However, the reduction of the three-dimensional molecular process to one dimension creates (as in the Taylor–Scherer case) a dispersion process. The dispersion coefficient is less than the molecular coefficient in the pulmonary region, which is just the opposite of that for the conductive zone.

Figure 11 illustrates an idealized alveolated channel in the pulmonary region. There are two basic transport directions, one along the channel axis and one perpendicular to this axis. Perpendicular transport carries an O_3 or NO_2 molecule into the alveoli, but due to the alveoli walls there is no net axial transport as there is in the central channel. Thus, the average axial transport is slowed because only a fraction of the molecules in the cross-sectional average can move axially, resulting in a dispersion process with a dispersion coefficient less than the molecule coefficient. The coefficient is a function of the molecular diffusion coefficient, the total air volume, and the generation's alveolar airspace volume.

The loss of a gas to the walls of the airway is not considered in the above-mentioned experiments. A molecule of O_3 or NO_2 (or any other

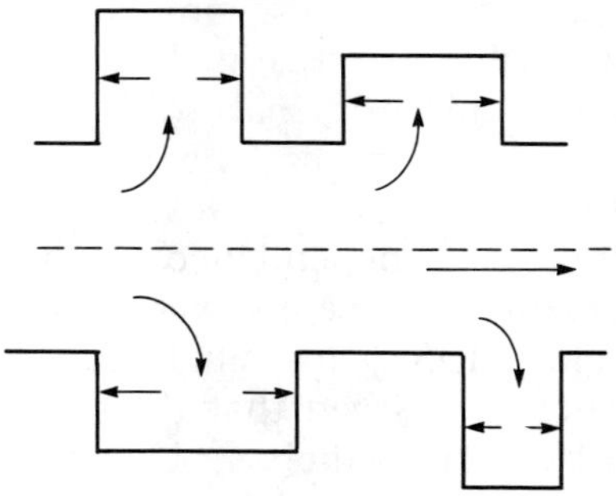

FIGURE 11. Transport in an alveolated airway.

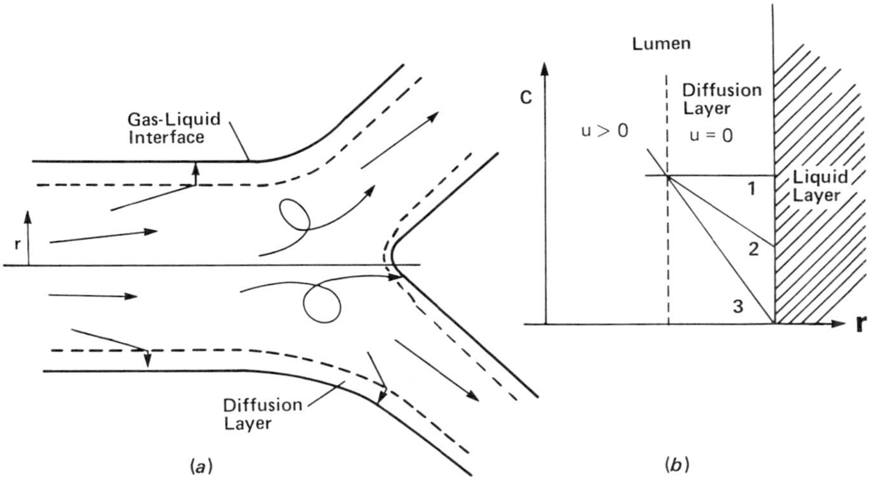

FIGURE 12. (*a*) Possible molecule paths in a bifurcating airway. (*b*) Possible concentration profiles in the diffusion layer.

species) entering the lumen of an airway (see Fig. 12) can pass through the lumen into another airway, strike the surface, and be absorbed and lost to the lumen, or be reemitted to the airstream.

Molecules are transferred from the flowing gas into the liquid layer lining the airway wall via molecular diffusion (Danckwerts, 1970). Several theories describe the process [see discussions by Skelland (1974) and Danckwerts (1970)]. One of the simplest descriptions is the film model, which assumes that the air velocity becomes very small as the air–liquid interface is approached. In a thin layer called the diffusion layer (Fig. 12*a*), convection is insignificant on the gas side next to the interface. A thin stagnant layer or film is postulated, where, by random molecular motion or diffusion, the molecules from the moving air enter the diffusion layer and are transported to the liquid surface by molecular diffusion. Transfer through this film depends on the gas-phase diffusion coefficient, film thickness, and the O_3 or NO_2 concentrations at the film boundaries. Film thickness depends on the system's hydrodynamic properties. The boundary concentrations are determined by properties of the gas phase and the liquid phase's absorption capacity. If moelcules are not absorbed, there is little effect on the concentration in the diffusion layer (curve 1, Fig. 12*b*). Conversely, if they are absorbed, then the concentration of the particular gas in the diffusion layer is reduced at the liquid interface relative to the concentration at the other boundary (curves 2 and 3, Fig. 12*b*). As the ability of the liquid to remove the gas increases, the relative concentration at the gas-liquid interface decreases and the mass transfer from the gas phase to the liquid phase increases. Transfer is also enhanced by a reduction in diffusion layer thickness that is dependent on the nearby rate of airflow; the higher the flow velocity, the thinner the film. Because velocities are relatively higher at the inner wall's branchings (e.g., carina), we can expect relatively higher potential for absorption there.

Figure 13 summarizes the simple descriptions of the physical processes or aspects involved in the transport of O_3 or NO_2 in the airspaces of the respiratory tract. The complex airway system has been reduced to a single-path model in which characteristic airway dimensions (diameter, d_z; length, L_z; alveolar volume, alveolar surface area, etc.) have been assigned to each generation (indicated by z). The transport dynamics of O_3 or NO_2 are expressed in terms of their average concentrations, which is a solution to a one-dimensional dispersion and convection partial differential equation that includes the effects of wall loss (S in Fig. 13). Values for the average convective velocity U_z and the dispersion coefficient D_{ez} depend on characteristic dimensions of each generation and the breathing pattern and can be based on the results of theoretical or experimental investigations.

The wall-loss absorption, or "sink" term, depends not only on the hydrodynamics of the airspaces but also on the transport properties of O_3 or NO_2 in the liquid layer (characterized in the following section) that coat the airway wall.

TRANSPORT AND CHEMISTRY IN THE LIQUID AND TISSUE LAYERS

The respiratory tract is lined with a liquid that is important in protecting the lung. In the conductive zone, this fluid, mucus, has

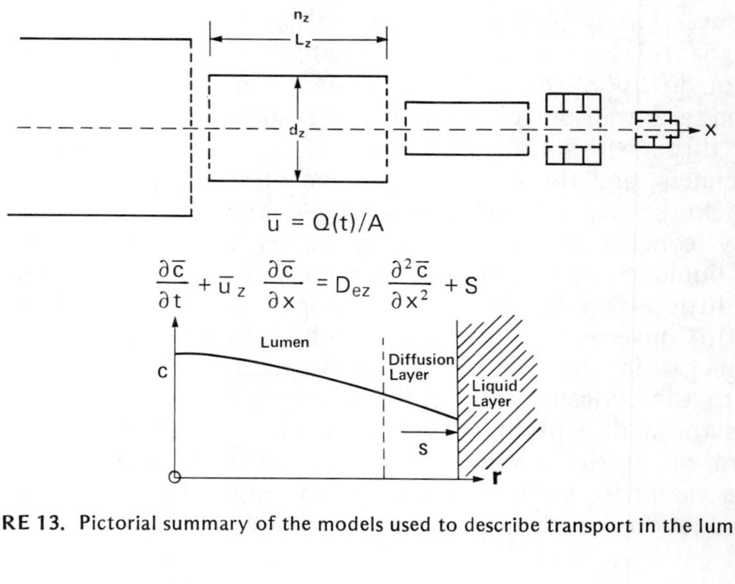

FIGURE 13. Pictorial summary of the models used to describe transport in the lumen.

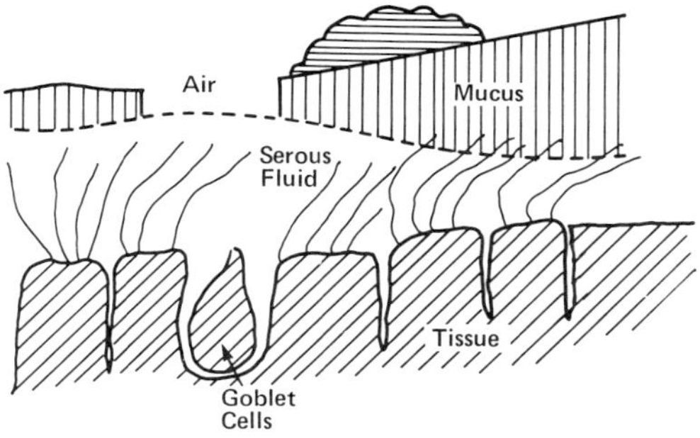

FIGURE 14. The mucous, serous, and tissue layers.

properties different than the "surfactant" that lines the tissue of the pulmonary region. Ozone or NO_2 from the gas phase must pass through the liquid layer before it can reach the tissue layer, where it may cause damage. These layers influence the quantity of O_3 and NO_2 absorbed. Thus, their characterization is very important in any transport and absorption model that predicts doses of O_3 and NO_2 or other toxic molecules delivered to the tissue.

Figure 14 illustrates the physical features of the liquid-tissue complex in the tracheobronchial region that may play a role in absorption. The liquid layer is thought to be composed of a mucous layer that floats on a film of serous fluid (Lucas and Douglas, 1934). The serous layer coats the underlying tissue cells. The mucous layer is nonuniform in thickness (Luchtel, 1978; Yoneda, 1976; Adler et al., 1973) and discontinuous (Van As and Webster, 1974; Iravani and Van As, 1972), with lumps of matter on its surface (Luchtel, 1978). Cilia in the serous layer propel the mucous layer up the respiratory tract. Two important epithelial cells are the goblet cell, which is the source of the mucin that accumulates in the upper layer, and the ciliated cell, which is the origin of the cilia.

For modeling purposes, the liquid-tissue system must be simplified. The simplification depends on the purpose of the modeling, the properties of the system, and the extent of our knowledge about the properties. Figure 15*a* is a model of the liquid-tissue complex (illustrated in Fig. 14), which contains mucus, serous, and tissue layers. Within each generation of the respiratory tract, each layer is uniformly thick, continuous, and horizontally homogeneous. If the transport properties of O_3 or NO_2 are the same in any two adjacent layers, then those two layers can be combined (as in Fig. 15*b*) for modeling transport. This simplifcation may also be performed because of incomplete data, i.e., there is no information concerning any differences between the two layers regarding transport.

In the pulmonary region, the airway walls are covered by a fluid called surfactant. This liquid layer is made up of two layers, a thin ($\sim$0.005 μm) surfactant layer coating a thicker (0.1 to 0.2 μm) serous fluid (Clements

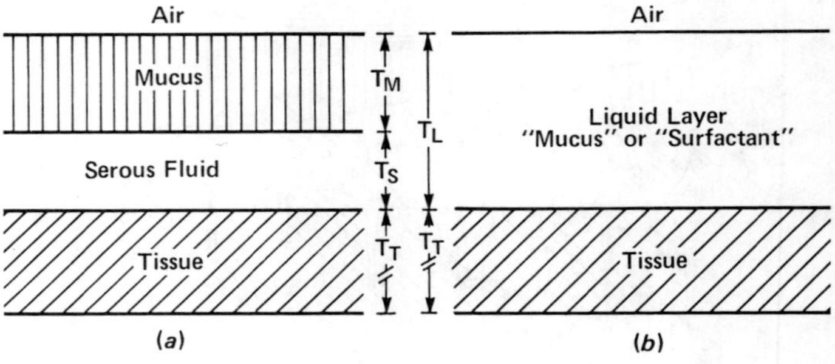

FIGURE 15. (*a*) Idealized model of the mucous, serous, and tissue layers. (*b*) An alternate idealized model of A and of the "surfactant" tissue layers.

and Tierney, 1965). Less is known about the physical properties of this layer than about the mucous layer and, at present, a single-liquid-layer model (Fig. 15*b*) is appropriate. However, surfactant can affect solubility.

In addition to their physical properties, the layers' chemical properties must be considered. Ozone or NO_2 react with many types of the molecules that make up the liquid and tissue layers. As a result, O_3 or NO_2 absorption from the lumen increases, whereas liquid layer reactions can reduce (relative to no reactions) the delivery of O_3 or NO_2 to the tissue. If O_3 or NO_2 are the only toxic molecules (i.e., reaction products are not toxic), then the liquid layer can protect the tissue (Miller et al., 1978). Conversely, if reaction products are toxic, then reactions could increase the delivery of toxic molecules to the tissue layer by increasing the absorption of O_3 or NO_2.

Figure 16 illustrates some of these processes. An O_3 molecule from the gas-phase diffusion layer enters the liquid layer. Here, ozone can pass through the liquid layer to the tissue or it can react with a component of the liquid layer; in this case it does not reach the tissue. Reactants (compounds that react with O_3) are produced by tissue layer cells and are mixed in with the liquid layer by diffusional or ciliar action. Reaction

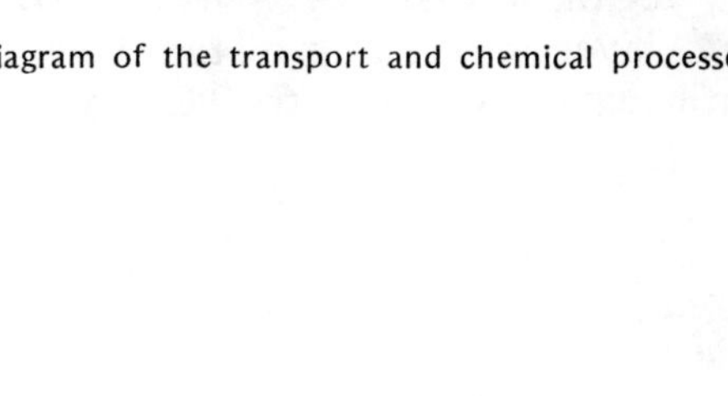

FIGURE 16. Schematic diagram of the transport and chemical processes involved in the mucous–serous–tissue complex.

with an O_3 molecule produces one or more new molecules, which may be transported back to the tissue to be absorbed or removed by mucociliary transport. The effectiveness of the latter process in removing any toxic products before they reach the tissue is unknown.

Strictly speaking, modeling the effects of chemistry would require a knowledge of all the significant reactions occurring in each layer, their rate constants, and the concentrations of the components. Equation (2) illustrate a hypothetical set of reactions in which O_3 reacts with organic molecules (HC) with resulting products (P).

$$\begin{aligned} O_3 + HC_1 &\xrightarrow{k_1} P_{11} + P_{12} + \cdots \\ O_3 + HC_2 &\xrightarrow{k_2} P_{21} + \cdots \\ &\vdots \\ O_3 + HC_n &\xrightarrow{k_n} \cdots \\ O_3 + P &\longrightarrow \cdots \\ &\vdots \end{aligned} \tag{2}$$

The actual reactions or rate constants are not completely known. Ozone or NO_2 reacts (not indicated in these equations or in Fig. 16) with species other than organic compounds. Ozone decomposes in water, and NO_2 reacts with itself (as well as with O_3) and initiates other inorganic nitrogen compounds and further chemical reactions. The importance of these reactions in the liquid layer is unknown. Presently all these uncertainties can be dealt with by simplifying the mechanism. For example, Eq. (3) illustrates a one-step surrogate mechanism in which O_3 reacts with the total reactant concentration with one effective rate constant (k).

$$O_3 + HC \xrightarrow{k} \text{Products} \tag{3}$$

The pseudo-products are assumed not to react with O_3 or HC.

A one-dimensional (perpendicular to tissue layer) diffusion process plus the simple chemical reaction model is at present sufficient for modeling transport and absorption in the liquid–tissue complex:

$$\frac{\partial[O_3]}{\partial t} = D_{O_3} \frac{\partial^2 [O_3]}{\partial z^2} - k\,[HC][O_3] \tag{4a}$$

$$\frac{\partial[HC]}{\partial t} = D_{HC} \frac{\partial^2 [HC]}{\partial z^2} - k\,[HC][O_3] \tag{4b}$$

So far, the descriptions of transport in the several different phases (gas, liquid, and tissue) have been considered independently. For a given compartment, adjacent phases can be a source or sink for O_3 or NO_2 (as well as for the reactants). Figure 17 illustrates hypothetical O_3 or NO_2 concentrations in the compartments or phases that have been discussed. The concentration C of O_3 or NO_2 versus the radial distance r from an airway's center or axis is plotted. Moving radially from the center, a gas phase is encountered that can be divided into a convective region and a diffusion layer, followed by the mucous–serous or surfactant layer and then the tissue compartment. For some regions, such as the pulmonary region where tissue is thin, a compartment after the tissue that represents blood may be needed. For illustration, however, the O_3 or NO_2 concentration is assumed depleted by chemical reactions before it enters the blood.

To solve mathematical equations that describe transport and absorption processes [e.g., see Fig. 13 and Eq. (4)], we must relate or couple the equations that describe the processes in each compartment. This is accomplished at the compartment boundaries, where physical properties dictate the relationships between the variables at the boundary. Two relationships are usually required: Henry's law, which relates concentrations, and the conservation of matter flowing across the interface. The Henry's law constant H is a physically measureable parameter and is the constant of proportionality between two interfacial concentrations (i.e., $C_{gi} = HC_{li}$; C_{gi} and C_{li} = gas- and liquid-phase interfacial concentrations, respectively; a similar relationship is usually assumed at each interface). Conservation of matter requires that the current in each phase at the boundary be equal, a requirement easily represented mathematically.

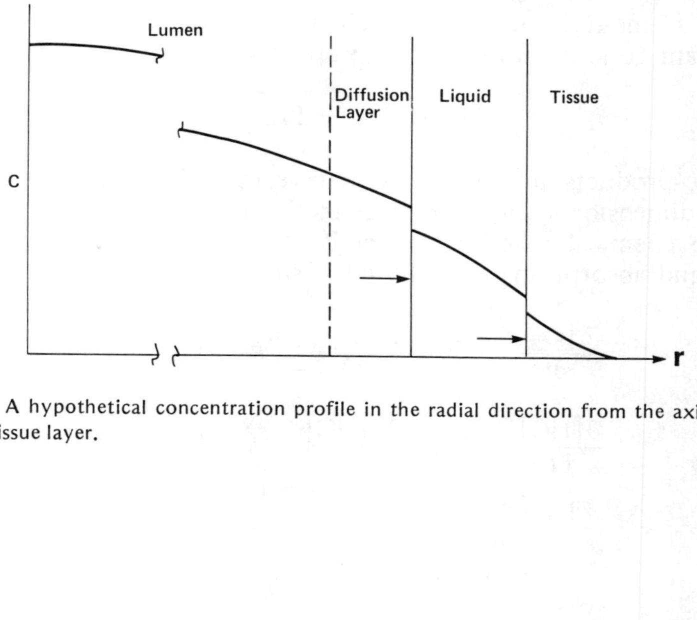

FIGURE 17. A hypothetical concentration profile in the radial direction from the axis of an airway through the tissue layer.

EXAMPLES OF MODELING RESULTS

The simulation model (Miller, 1977; Miller et al., 1978, 1982) used to generate the data for the examples in this section is based on the concepts and the representations of the processes discussed in prior sections. The single-path model is used (Fig. 13), and dimensions are based on the data of Weibel (1963) for humans and on the data of Kliment (1973) for the rat, guinea pig, and rabbit. Transport in the gas phase of the airways is considered by using a one-dimensional effective convection and dispersion equation (Fig. 13). A one-layer film represents the mucous-serous or the surfactant layer (Fig. 15*b*), and film thickness decreases distally from 10 μm in the trachea to 0.125 μm in the alveolar ducts and sacs. A tissue layer is not considered because in this model the O_3 or NO_2 tissue reactivity is assumed to be so high that free O_3 or NO_2 cannot exist inside the tissue compartment. In this case, the O_3 or NO_2 concentrations is set to zero at the tissue–liquid interface. This does not imply zero delivery of O_3 and NO_2 to the tissue. On the contrary, the current is the maximum (relative to a nonzero boundary condition) for the given physical and chemical conditions. For chemical reactions and transport in the liquid layer, the "instantaneous reaction regime" (Astarita, 1967) is used. This is a bimolecular reaction [Eqs. (3) and (4)] in which the rate constant is assumed to be very large. The degree of reactivity decreases distally, which is based on the observation that goblet-cell density decreases distally (Reid, 1970) and on the assumption that reactivity is correlated with goblet-cell density.

A consequence of the instantaneous reaction regime is that for the given physical and chemical parameters, the absorption or loss of O_3 or NO_2 from the lumen is a maximum and the delivery of O_3 or NO_2 to the tissue is a minimum. This model of the O_3 or NO_2 reactions in the liquid layer is too simplistic. Nevertheless, we can use the model to obtain estimates of the upper limits of absorption and the lower limits (relative to the upper limit) of O_3 or NO_2 delivery to the tissue.

The model has been used to investigate the sensitivity of results to a variety of variables and parameters such as liquid-layer thickness, lung solubility, ambient concentrations, and the dispersion coefficient. For the examples, however, only the effect of chemical reactivity, exercise, and the application of the simulation model to the respiratory tract models of several animals will be considered.

The example illustrate the uptake of NO_2; however, the simulation model was developed with O_3 in mind. The physical properties of the two gases are essentially the same, but the chemical properties are different. At this time we consider the results presented to be qualitatively correct for both NO_2 and O_3.

Figure 18 shows patterns for the uptake (or loss from the lumen) and

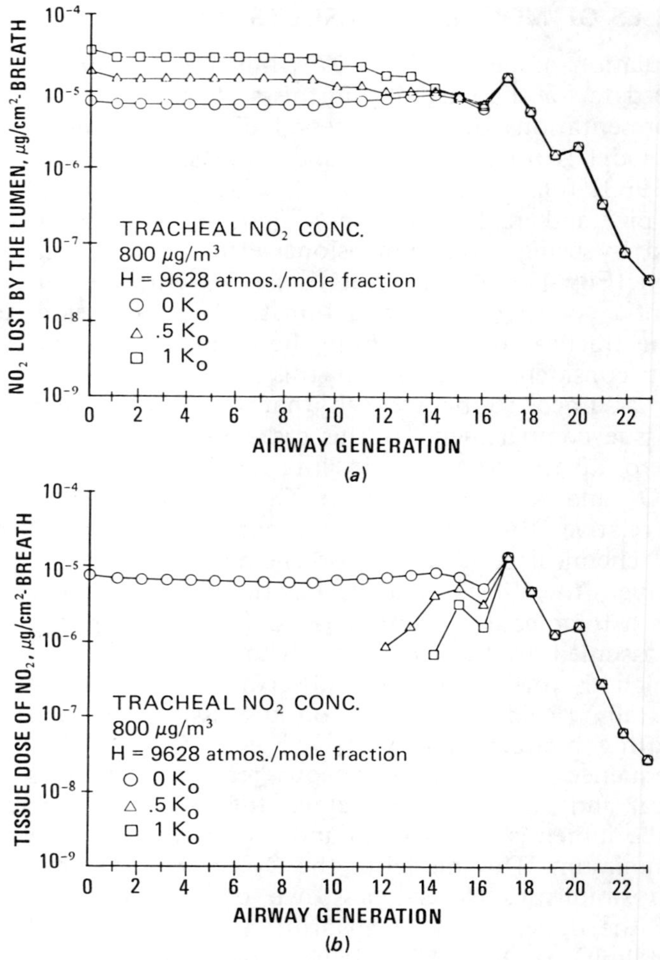

FIGURE 18. (*a*) NO_2 lost by the airway lumen as a function of airway generation in the human lung. (*b*) Predicted NO_2 tissue dose curves in human lungs. Simulations used a tidal volume of 500 ml and a respiratory frequency of 15 breaths/min (Miller et al., 1982).

tissue absorption, respectively, for several levels of chemical reactivity. The airway model is based on the dimensions of Weibel (1963). In each of the two figure parts, three curves are plotted. One is for no reactions and the other two are for reactivity in which the second is half the third ($0K_o$ = no reactivity, K_o symbolizes the maximum reactivity used, and $0.5K_o$ is half the maximum). The vertical axis (Fig. 18*a*) represents the quantity of NO_2 lost/cm^2·breath·generation; the horizontal axis is the generation number. The $0K_o$ curve is for physical absorption only, and the quantity/unit area delivered to the liquid is the same as that delivered to the tissue (Fig. 18, *a* and *b*); in this case, the layer offers the tissue no protection from NO_2. As the reactivity increases (from $0K_o$ to $0.5K_o$ to $1K_o$), the amount absorbed in each generation (Fig. 18*a*) increases in the tracheobronchial region, but does not change in the pulmonary region. Conversely, the dose of NO_2 to the tissue (Fig. 18*b*) decreases in the tracheobronchial region as the reactivity increases, again with no effect in the pulmonary region. Further, the tissue does in the pulmonary region is the same as that in the case of only physical absorption.

With increased reactivity (beyond that illustrated), the uptake and the dose in the pulmonary region will be affected. For sufficiently high reactivity, much of the NO_2 (or O_3) will be absorbed in the tracheobronchial region and only a relatively small quantity will reach the pulmonary region (Miller et al., 1978).

The effect of NO_2 (or O_3) uptake in different regions of the respiratory tract due to exercise is illustrated in Fig. 19. The percent of uptake (based on total inspired NO_2) versus tidal volume is plotted. The shaded areas represent tracheal concentrations from 200 to 800 mg/m^3 (0.11 to 0.43 ppm) of NO_2. The tissue in the pulmonary region (RESP, Fig. 19) receives the greatest fraction of inhaled NO_2 (or O_3), and this increase as tidal volume (or exercise) increases. Exercise results in the delivery of a greater quantity of NO_2 (or O_3) to the tissue of the pulmonary region than that delivered in the resting state or in the tissue or liquid of the tracheobronchial region (COND, Fig. 19). The figure does not indicate that for the same time period, 15 times more NO_2 (or O_3) is

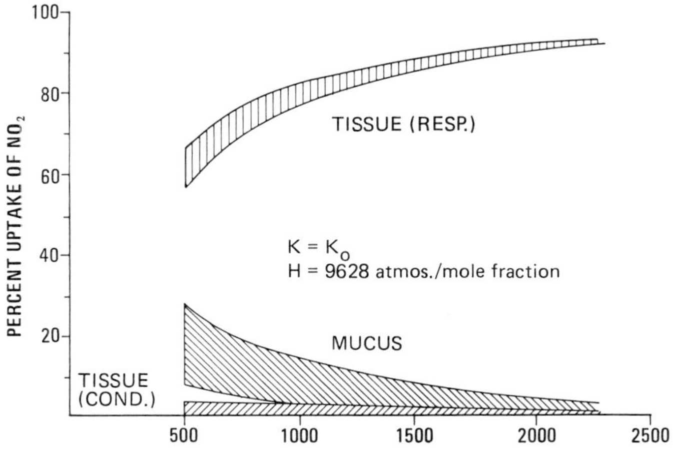

FIGURE 19. Uptake of NO_2 by mucus and airway walls of humans as a function of tidal volume. Results from simulation using tracheal NO_2 concentrations of 200 and 800 μg/m^3 form the lower and upper bounds, respectively, for the uptake by tissue regions, while the reverse is true for the uptake by mucus (Miller et al., 1982).

delivered to pulmonary tissue at the maximum tidal volume than during the resting tidal volume.

Figure 20 illustrates the results of simulating uptake in the rat, guinea pig, and rabbit. The airway model of Kliment (1973), based on "morphometric zones," was used. The zones, which have a length and diameter, were treated like airway generations for obtaining simulations. Figure 20 plots the tissue dose versus zone. The curves can be compared to the human tissue doese curve ($0.5K_0$) in Fig. 18*b*, keeping in mind the following approximate relationship between zone and generation: zones 0 through 4 = generations 0 through 11, zones 5 and 6 = generations 12 through 19, zone 7 = generations 20 through 22, and zone 8 = generation 23. The tissue-dose patterns for the guinea pig and rabbit are relatively similar. Both differ from the rat curve in two ways: the rat curve is "flatter" and peaks in the 7th zone, whereas the dose-curve of the other two animals peaks in the 6th zone. The location of the peak depends on the modeler's choice of where the mucous layer ends and where the surfactant layer begins. For the rat simulation, a mucous layer was assumed for zones 0 to 6, with a surfactant coating in zones 7 and 8. For the other two animals, the surfactant layer began one zone "sooner." In all three animals as well as in the human (Fig. 18*b*), the peak tissue does occurs in the first surfactant-lined zone or generation. Miller et al. (1978) discuss this and cite the results of experimental investigations that show maximum damage in the junction of the conductive and respiratory zones.

SUMMARY

The major factors that influence the transport and absorption of O_3 and NO_2 in the lower respiratory tract and how they can be simplified and integrated to formulate dosimetry models have been discussed. This discussion was developed by (1) considering how the complex branching system of the lower respiratory tract could be reduced to a simple model for modeling O_3 or NO_2 uptake, (2) discussing flow patterns in model bifurcating tubes, (3) describing the works of Taylor (1953) and Scherer et al. (1975), which suggest that transport in the intricate airway network can be simplified to a one-dimensional description of transport in a single tube, (4) discussing the physical, chemical, and biological properties of the liquid layer and tissue layers that make up the walls of airways, and (5) considering transport and chemical reactions in these layers.

By considering all the relevant physical, chemical, and biological factors, a conceptual model of transport and absorption can be developed. Mathematical equations that describe the processes implied by the conceptual model can then be applied to the model. Usually, the resulting mathematical equations are too complex to solve without a computer, and the equations must be suitably reformulated for programming by using numerical techniques.

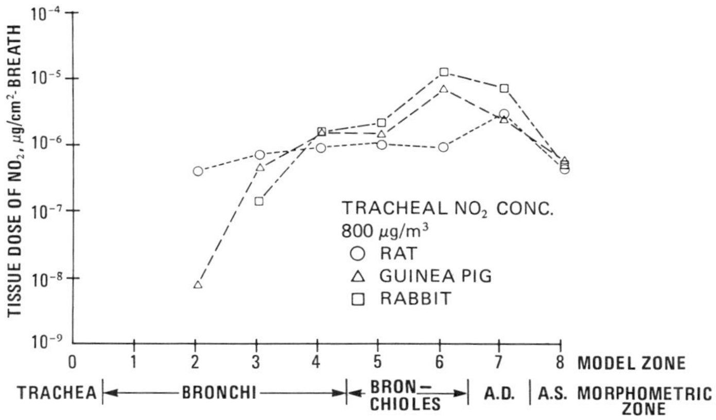

FIGURE 20. Predicted NO_2 tissue curves in the lungs of rats, guinea pigs, and rabbits for various morphometric regions. For rats, guinea pigs, and rabbits, tidal volumes of 0.95, 1.8, and 21 ml and breathing frequencies of 105, 90, and 50 breaths/min, respectively, were used. Alveolar ducts and sacs are represented by A.D. and A.S., respectively (Miller et al., 1982).

Mathematical and numerical aspects also influence the development of dosimetry models. Conceptual model processes may be too complex to describe mathematically; hence, conceptual models need to be simplified as discussed in this paper. The resulting equations can still be complex and impractical to solve with a computer, and further simplifications are required.

Some results of dosimetry modeling were presented. The model that generated the data via computer takes into account many of the factors, processes, and simplifications discussed in this paper. In the development of this model, the major limiting feature has been the lack of physical, chemical, and biological data on the liquid and tissue layers, and not mathematical or programming aspects. Nevertheless, sufficient data exist to allow reasonable estimates of necessary parameters, and the illustrations are considered qualitatively correct as well as in agreement concerning the location of major O_3 or NO_2 damage in the respiratory tract.

REFERENCES

Adler, K. B., Wotten, O., and Dulfano, J. M. 1973. Mammalian respiratory mucociliary clearance. *Arch. Environ. Health* 27:364–369.

Astarita, G. 1967. *Mass Transfer with Chemical Reaction,* p. 188. New York: Elsevier.

Clements, J. A., and Tierney, D. G. 1965. Alveolar instability associated with altered surface tension. In *Handbook of Physiology,* eds. W. D. Fenn and H. Rahn. Section 3, *Respiration,* Vol. II, pp. 15–84. Washington, D.C.: American Physiological Society.

Danckwerts, F. R. S. 1970. *Gas-Liquid Reactions,* p. 276. New York: McGraw-Hill.

Gill, W. N., and Sankarasubramanian, R. 1970. Exact analysis of unsteady convection diffusion. *Proc. R. Soc. Lond. A* 316:341–350.

Horsfield, K., and Cumming, G. 1967. Angles of branching and diameters of branches in the human bronchial tree. *Bull. Math. Biophys.* 29:245.

Iravani, J., and Van As, A. 1972. Mucus transport in the tracheobronchial tree of normal and bronchitic rats. *J. Pathol.* 106:81–93.

Kliment, V. 1973. Similarity and dimensional analysis, evaluation of aerosol deposition in the lungs of laboratory animals and man. *Folia Morphol.* 21:59–64.

Lucas, A. M., and Douglas, L. C. 1934. Principles underlying ciliary activity in the respiratory tract. *Arch. Otolaryngol.* 20:528–541.

Luchtel, D. L. 1978. The mucous layer of the trachea and major bronchi in the rat. *Scanning Electron Microscopy* 11:1089–1098.

Miller, F. J. 1977. A mathematical model of transport and removal of ozone in mammalian lungs. Ph.D. thesis, North Carolina State University, Raleigh, N.C.

Miller, F. J., Menzel, D. B., and Coffin, D. L. 1978. Similarity between man and laboratory animals in regional pulmonary deposition of ozone. *Environ. Res.* 17:84–101.

Miller, F. J., Overton, J. H., Myers, E. T., and Graham, J. A. 1982. Pulmonary dosimetry of nitrogen dioxide in animals and man. In: *Air Pollution by Nitrogen Dioxide*, eds. T. Schneider and L. Grant, Proceedings of U.S.-Dutch International Symposium on NO_x, Maastrict, Netherlands, pp. 377–386. Amsterdam: Elsevier.

Olson, D. E., Sudlow, M. F., Horsfield, K., and Filley, G. F. 1973. Convective patterns of flow during inspiration. *Arch. Intern. Med.* 131:51–57.

Paiva, M. 1973. Gas transport in the human lung. *J. Appl. Physiol.* 35:401–410.

Phalen, R. F., Yeh, H. C., Schum, G. M., and Raabe, O. G. 1978. Application of an idealized model to morphometry of the mammalian tacheobronchial tree. *Anat. Rec.* 190:167–176.

Reid, L. 1970. Evaluation of model systems for study of airway epithelium, cilia, and mucus. *Arch. Intern. Med.* 126:428–434.

Scherer, P. W., Shendalman, L. H., and Greene, N. M. 1972. Simultaneous diffusion and convection of a single breath lung washout. *Bull. Math. Biophys.* 34:393–412.

Scherer, P. W., Shendalman, L. H., Greene, N. M., and Bouhuys, A. 1975. Measurement of axis diffusivities in a model of the bronchial airways. *J. Appl. Physiol.* 38:719–723.

Schreck, R. M., and Mockros, L. F. 1970. Fluid dynamics in the upper pulmonary airways. American Institute of Aeronautics and Astronautics; paper no. 70; p. 788.

Schreider, J. P., and Raabe, O. G. 1981. Structure of the human respiratory acinus. *Am. J. Anat.* 162:221–232.

Schroter, R. C., and Sudlow, M. F. 1969. Flow patterns in models of the human bronchial airways. *Resp. Physiol.* 7:341.

Skelland, A. H. P. 1974. *Diffusional Mass Transfer.* New York: Wiley.

Taylor, G. 1953. Dispersion of soluble matter in solvent flowing slowly through a tube. *Proc. R. Soc.* (Lond.) A219:186–203.

Ultman, J. S., and Blatman, H. S. 1977. Longitudinal mixing in pulmonary airways. Analysis of inert gas dispersion in symmetric tube network models. *Resp. Physiol.* 30:349–367.

Van As, A., and Webster, I. 1974. The morphology of mucus in mammalian pulmonary airways. *Environ. Res.* 7:1–12.

Weibel, E. R. 1963. *Morphometry of the Human Lung.* New York: Academic.

Wilson, T. A., and Lin, K.-H. 1970. Convection and diffusion in the airways and the design of the bronchial tree. In *Airway Dynamics. Physiology and Pharmacology*, ed. A. Bouhuys, pp. 5–19. Springfield, Ill.: Charles C Thomas.

Yeh, H.-C., and Schum, G. M. 1980. Models of human lung airways and their application to inhaled particle deposition. *Bull. Math. Biophys.* 42:461–480.

Yeh, H.-C., Schum, G. M., and Duggan, M. T. 1979. Anatomic models of the tracheobronchial and pulmonary regions of the rat. *Anat. Rec.* 195:483–492.

Yoneda, K. 1976. Mucous blanket of rat bronchus. *Am. Rev. Respir. Dis.* 114:837–842.

Part Three

EXTRAPOLATION OF STRUCTURAL AND FUNCTIONAL CHANGES FROM ANIMALS TO HUMANS

INTRODUCTION

Philip A. Bromberg

Center for Environmental Health and Medical Sciences, University of North Carolina School of Medicine, Chapel Hill, North Carolina

The goals of the process defined as "extrapolation modeling" have been presented previously by Miller et al. (1983). They focussed on the need to establish relative dosimetry for inhaled gases and particles for different portions of the respiratory tract in human vs. experimental animals. They indicated that the factors determining dosimetry are relatively well understood—although there are significant information gaps—and that adequate "realistic" mathematical models relating the composition of inspired air to respiratory system deposition, or uptake, can be developed and tested experimentally.

The paper presented by Leith in this book (Leith, 1984) suggests to me that despite the enormous range of lung size and respiratory system geometry among experimental animals, children, and adults, remarkable similarities in dosimetry may emerge if the proper normalization procedures are developed.

Given a better quantitative approach to comparative dosimetry, how can one deal with the vagaries of interspecies differences in metabolism and responsitivity? This is again a difficult problem in toxicology.

The papers by Autor et al. (1984) and by Rasmussen (1984) support the usefulness of well-controlled studies in isolated systems, including cultures of single cell types from different species. One may also underline the possibility of studying interactions among two or more cell lines under *in vitro* conditions. For example, oxidant-induced alteration in endothelial cell function may result in the release of factors that activate the proliferation and differentiation of adjacent smooth muscle precursor cells. The idea that the respiratory epithelial cell is a primary target of oxidant pollutants is well supported by dosimetry considerations and by morphologic observation. Primary culture of respiratory epithelial cells is now feasible and such cultures could be used for toxicologic and mechanistic studies.

The use of nasal epithelium as an *in vivo* model of airways epithelium in man is worthy of consideration since the nose is most directly exposed to nasally inhaled pollutants, and is also accessible for a variety of relatively non-invasive studies examining functional, morphologic, and inflammatory responses of epithelium.

The review of O_3 toxicology by Menzel (1984) and the paper by Crapo et al. (1984) suggest that similar patterns of epithelial cell injury are observed across various species exposed to a modest range of ambient O_3 concentrations. These morphologic patterns are compatible with the predicted sites of absorption of a highly-reactive trace gas of limited aqueous solubility. Furthermore, the particular susceptibility of ciliated airway cells and Type I acinar alveolar cells to injury is seen in multiple species. The potential for cell-cell interactions following injury to epithelium is brought out by Crapo who shows recruitment of macrophages and an interstitial fibrotic response in the proximal acinar region.

In the review of lung function changes associated with diffuse disease in animals and in humans, Mauderly (1984) suggests that substantial parallelism exist across species lines, despite the variability in size and geometry of the respiratory system. Thus, functional testing is probably useful for extrapolation modeling of pollutant effects. One must, however, be aware that focal changes—e.g., centri-acinar airspace enlargement and de-septation—might be undetectable by such testing in spite of involvement of many lobular units. It is also worth noting that a characteristic physiologic response seen in human subjects exposed to ozone, i.e., the reduction in inspiratory capacity, may be very difficult to test for in an experimental animal.

It is appropriate to recall that animal *disease* models can be useful in studying mechanisms of toxicant-induced injury and also in considering effects of pollutants on potentially sensitive diseased human subjects.

Green's review of host defense mechanisms against pulmonary infections (Green, 1984) outlines the multiple inter-species similarities which support extrapolation modeling efforts to elucidate the suspected interaction between pollutant exposure and enhanced susceptibility to infection. This theme is expanded in the reviews of Goldstein (1984) and Gardner (1984). Certainly, many of the components of this defense system can be studied in humans as well as in animals.

Although substantial difficulties stand in the way of quantitative extrapolation modeling, the papers generally provide reasons for thinking that these difficulties are not insuperable. Furthermore, a number of specific approaches and experiments are suggested to contribute to the achievement of this goal. In my opinion, the extrapolation of effects of chronic experimental exposures from animals to humans will only become convincing when the mechanisms of the observed effects are relatively well understood, and when the experimental work is developed in the context of the broader framework of lung biology.

REFERENCES

Autor, A. P., Bonham, A. C., Thies, R. L. 1984. Toxicity of oxygen radicals in cultured pulmonary endothelial cells. *J. Toxicol. Environ. Health* 13:387–395.

Crapo, J. D., Barry, B. E., Chang, L., Mercer, R. R. 1984. Alterations in lung structure caused by inhalation of oxidants. *J. Toxicol. Environ. Health* 13:301–321.

Gardner, D. E. 1984. Oxidant-induced enhanced sensitivity to infection in animal models and their extrapolations to man. *J. Toxicol. Environ. Health* 13:423–439.

Goldstein, E. 1984. An assessment of animal models of testing the effect of photochemical oxidants on pulmonary susceptibility to bacterial infection. *J. Toxicol. Environ. Health* 13:415–421.

Green, G. M. 1984. Similarities of host defense mechanisms against pulmonary infectious diseases in animals and man. *J. Toxicol. Environ. Health* 13:471–478.

Leith, D. E. 1984. Mass transport in mammalian lungs: Comparitive physiology. *J. Toxicol. Environ. Health* 13:251–271.

Mauderly, J. L. 1984. Respiration function responses of animals and man to oxidant gases and to pulmonary emphysema. *J. Toxicol. Environ. Health* 13:345–361.

Menzel, D. E. 1984. Ozone: An overview of its toxicity in man and animals. *J. Toxicol. Environ. Health* 13:183–204.

Miller, F. J., Graham, J. A., Gardner, D. E. 1983. The changing role of animal toxicology in support of regulatory decisions. *Environ. Health Perspect.* 52:169–176.

Rasmussen, R. E. 1984. *In vitro* systems for exposure of lung cells to NO_2 and O_3. *J. Toxicol. Environ. Health* 13:379–411.

ALTERATIONS IN LUNG STRUCTURE CAUSED BY INHALATION OF OXIDANTS

James D. Crapo, Brenda E. Barry, Ling-Yi Chang, Robert R. Mercer

Departments of Medicine and Pathology, Duke University Medical Center, Durham, North Carolina

Morphometric and morphologic methods have been used to evaluate changes in rat lungs caused by the inhalation of a variety of oxidants. Exposure to 100% oxygen causes diffuse pulmonary injury and leads to death after 66–72 h of exposure. The primary insult leading to death in rats exposed to hyperoxia is injury to pulmonary capillary endothelium. Sublethal exposure to hyperoxia was found to cause diffuse injury to all major components of the alveolar septum and was associated with destruction of approximately 50% of the pulmonary capillary endothelial cells. A corresponding decrease in pulmonary capillary surface area and capillary lumen volume also occurred. Exposure to ozone and to nitrogen dioxide in low concentrations did not cause a diffuse injury throughout the alveolar region of the lung, but rather led predominantly to structural alterations in terminal bronchioles and in their adjacent alveoli. Morphometric evaluation of animals exposed to 0.25 ppm ozone and to 2 ppm NO_2 demonstrated quantitatively and qualitatively similar lesions. These lesions primarily involve injury and remodelling of the alveolar epithelium. These changes in the alveolar epithelium were also associated with the recruitment of increased numbers of alveolar macrophages to the proximal alveolar region. The different types of lung injury caused by various oxidants are most likely to be related to differences in their reactivity with tissue components and to differences in concentration, distribution, and diffusion characteristics of the oxidant gases.

INTRODUCTION

Exposures to hyperoxia at pressures of less than 1 atm can cause extensive pulmonary injury. Most animals die after prolonged exposure to 100% O_2. The primary injury is to the lung, and the animals die with signs of progressive respiratory distress (Clark and Lambertson, 1971). Previous morphologic and morphometric studies have demonstrated that hyperoxic injury to the lung causes a diffuse injury involving all major

This work was supported in part by National Institute of Health grant 5P01 HL07896, Environmental Protection Agency cooperative agreement CR 807255010, and Southern California Edison contract C-1522903.

Although the research described in this article has been funded wholly or in part by the United States Environmental Protection Agency through cooperative agreement CR 807255010, it has not been subjected to the Agency's required peer and policy review and therefore does not necessarily reflect the views of the Agency, and no official endorsement should be inferred.

components of alveolar septum including the epithelium, interstitium, and endothelium (Kistler et al., 1967). Extensive alveolar type II epithelial cell proliferation occurs in some species of animals. This may be preceeded by frank destruction of the alveolar type I epithelium (Kapanci et al., 1969). Extensive destruction of capillary endothelium has been shown to occur. The time-course of this capillary injury relates closely to the progressive hypoxemia and subsequent death of the animal (Crapo et al., 1980). Interstitial changes include interstitial edema, inflammatory cell infiltration, fibroblast proliferation and deposition of increased amounts of connective tissue (Crapo et al., 1978b, 1980).

Oxygen represents a gas in high concentrations whose uptake from the lung, both by the blood stream and by lung tissue metabolism is relatively low compared to gas delivery. Since delivery of oxygen in hyperoxic atmospheres far exceeds its consumption, the tissue and blood concentrations in distal alveoli will equilibrate with the gas in the small airways. No significant stratified inhomogeneity will be expected to exist in the lungs of animals breathing 85 or 100% O_2 (Laforce and Lewis, 1970). Because of the uniform distribution of the gas, the primary site of O_2 toxicity will not depend on distribution but rather on the relative sensitivity of the various cells which are exposed (Fig. 1).

Gases like O_3 and NO_2 that are more highly reactive and occur in relatively low concentrations are known to cause an entirely different pattern of lung injury. The alveolar region adjacent to terminal bronchioles, which we term "proximal alveolar regions," has been reported to be one of the major sites of injury after inhalation of less than 1 ppm ozone (Freeman et al., 1968a,b; Plopper et al., 1979; Crapo et al., 1978a). Alveolar epithelial type I cell injury, including frank destruction of type I epithelium, has been found to occur. Type II cells have been found to proliferate, presumably to replace damaged type I cells (Evans et al.,

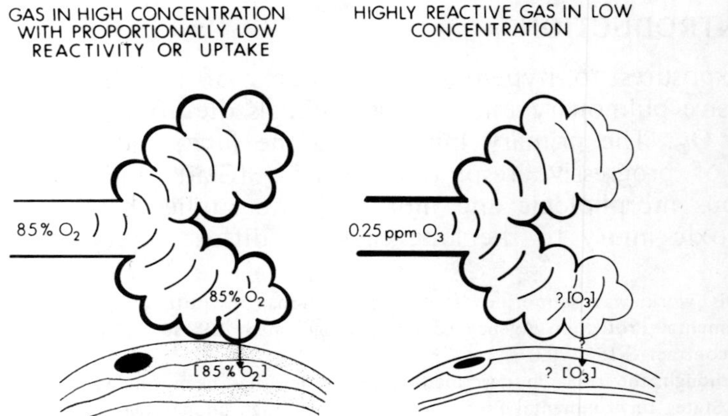

FIGURE 1. Schematic of distribution of lesions caused by hyperoxia and by ozone.

1976). Inflammatory cell infiltration and interstitial edema can occur as part of this pattern of injury. The pattern of injury by these oxidant gases is thought to be highly influenced by the distribution of the gases (Fig. 1). The more proximal portions of the lung will experience a higher level of exposure than in the distal lung parenchyma because as the reactive gas is consumed in tissue reactions proximally less of the oxidant is available for diffusion to distal regions of the lung. Thus significant stratified inhomogeneity will exist for these gases (Crapo et al., 1978a).

The following studies review work involving lung injury in rats caused by exposures to hyperoxia and to relatively low concentrations of ozone and NO_2. Similar morphometric and morphologic analyses permit the patterns of lung injury caused by these highly different oxidant gases to be compared and contrasted.

METHODS

Animals and Exposures

For oxygen exposures, specific-pathogen-free Charles Rivers CD strain male rats weighing 300–350 g each were used. Half of these animals were exposed continously to 85% oxygen for periods of 3–14 d using polystyrene chambers and following procedures previously described (Crapo et al., 1980). The O_2 concentration varied less than 2%, and CO_2 was maintained below 0.5% by providing 7 to 8 complete gas changes per hour. Matched animals were kept in air under identical exposure conditions to be used as controls. A detailed analysis of morphologic and morphometric studies of these animals has been reported previously (Crapo et al., 1978b, 1980).

For ozone and NO_2 exposures, male Fisher-344 rats were kept in 0.323-m^2 stainless-steel chambers during exposures to air, O_3, or NO_2. The O_3 and NO_2 exposures were not done concurrently, and therefore separate control groups of animals were used for each oxidant exposure. Ozone was generated from 100% oxygen with an Orec model 03V5-0 Ozonizer (Ozone Research and Equipment Corp., Phoenix, Ariz.) mixed with dry, filtered air and directed into the top of the chambers. Flow through the chambers was 323 l/min. The concentration of ozone in the chambers was continuously monitored with a chemiluminescence ozone analyzer, EPA reference method RFOA-0176-007 (Bendix Corporation Model 8002, Ronceverte, West Virginia). The NO_2 exposures were achieved by adding 1% NO_2 in nitrogen (from a cylinder) to the air flowing through the chamber which was maintained at a constant flow of 323 l/min. The concentration of NO_2 in the chambers was continously monitored using a chemiluminescence analyzer, EPA reference method RFNA-0777-022 (Bendix Corporation Model 8101-C). Computer-controlled feedback circuiting was used to maintain the chamber NO_2 concentrations at the desired levels. The Fisher-344 rats were 6 wk old (200–250 g) at the beginning of the exposures and were exposed for 6 wk, continuously to filtered air to 0.25 ± 0.01 ppm ozone for 12 h/d, 7 d/wk, or to 2.0 ± 0.02 ppm NO_2 for 23 h/d with twice daily spikes of $\frac{1}{2}$ h each to 6.0 ± 0.05 ppm NO_2. The animals were weight matched at the beginning of the exposures. All animals received food and water *ad libitum.*

Tissue Preparation

Following the exposures, each rat was anesthetized with intraperitoneal pentobarbital. In rapid succession the abdomen was opened, both hemidiaphragms were punctured to allow collapse of the lungs, the trachea was cannulated, and fixative was infused down the trachea at a constant pressure of 20 cm of fixative above the midpoint of the chest wall. The initial flow of fixative was adjusted to provide an infusion rate rapid enough to inflate the lungs to 70–80% of TLC as determined by gas dilution methods (Hayatdavoudi et al., 1980). The fixative was 2% glutaraldehyde in cacodylate buffer at pH 7.4 and having a total osmolality of 350 (Weibel et al., 1979). After a minimum of 30 min of fixation inside the chest wall, the lungs were dissected out and the volume of the fixed lungs determined by volume displacement (Scherle, 1970).

Morphologic Analyses of the Distal Lung Parenchyma

From each animal in the oxygen or ozone exposure groups tissue samples were taken from four peripheral sites in the left lung. Those sites were upper anterior, upper posterior, lower anterior, and lower posterior. These tissue samples were kept in 2% glutaraldehyde for a minimum of 1 h and then washed in cacodylate buffer and postfixed with 2% OsO_4. They were then dehydrated in a graded series of alcohol, transferred to propylene oxide, and embedded in Epon. Thin sections were cut and stained with uranyl acetate and lead citrate prior to viewing. A total of 60 random micrographs were taken for each animal and printed at 8000X-8500X on 11 X 14-inch photographic paper.

Procedures for carrying out the morphometric analyses of peripheral lung tissue have been well described by Weibel (1979). The proportion of the whole lung volume that is the parenchymal alveolar region was determined by using a light microscope and serial step-cut paraffin sections through the entire right cardiac lobe of each animal. Approximately 1000 uniformly distributed points on the right cardiac lobe were sampled for each animal studied. The percentages falling on the alveolar and nonalveolar regions were determined. Methods for determining the volume, surface, and numerical densities of tissues and harmonic and arithmetic mean thicknesses were those described by Weibel (1979). Determination of the numerical density of lung cells was done as previously described (Crapo et al., 1978b). This requires determining the number of nuclear profiles occurring per unit area of tissue section (these counts were done using ultrathin sections and an electron microscope) and determining the mean caliper diameter $\bar{D}$ of the major types of lung cell nuclei. The method for determining the mean caliper diameter has been reported previously (Woody et al., 1979).

Morphologic and Morphometric Analyses of the Proximal Alveolar Regions

In order to isolate the alveolar region adjacent to terminal bronchioles, the left lobe of the lung was transversely cut into a series of slices. Using a dissecting microscope and a sharpened 20-gauge Becton Dickenson luer stub cannula, terminal bronchioles that could be identified as being perpendicular to the cut surface of the lobe were punched out. Terminal bronchioles were assumed to be perpendicular to the surface of the lobe only if their cut surface was completely circular. These tissue blocks were then washed in cacodylate buffer, postfixed in OsO_4, dehydrated in a graded series of alcohols and propylene oxide, and embedded in Epon 812. During embedment the blocks were oriented in a BEEM capsule so that the cut surface of the terminal bronchioles was at the leading edge (Fig. 2). Using a Sorvall MT 5000 ultramictrotome the blocks were serially sectioned through the terminal bronchiole until the first alveolar duct bifurcation was reached.

Excess tissue was then trimmed and thin sections containing the proximal alveolar region were cut and placed on 200-mesh grids. Electron micrographs were taken in each of the four corners of each of 10–15 grid

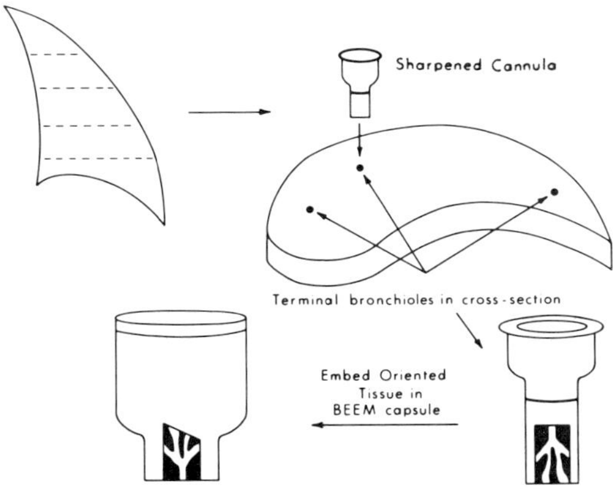

FIGURE 2. Schematic of methods used to punch out and orient cylinders of tissue containing the proximal alveolar region. The terminal bronchiole ends up at the bottom of the BEEM capsule oriented perpendicular to the angle at which the block will be sectioned.

squares and enlarged to X8500 on 11 X 14-inch photographic paper. Morphometric analysis of these electromicrographs was then carried out using techniques as described for the distal alveolar region. Because values for the volume of "proximal alveolar region" in a normal rat lung are not known, the data calculated for volume, surface, and numerical densities of various tissue components cannot be converted to absolute volumes, surfaces, and numbers, as they were for the distal alveolar region. The data from the proximal alveolar region were normalized by dividing each of the volume, surface, and numerical density parameters by the surface density of alveolar epithelial basement membrane in the same sections. Bachofen and Weibel (1977) used this technique to express morphometric data from human lungs after acute lung injury. This method reduces variance caused by differences in the degree of inflation of fixed lungs, since the densities of tissue and epithelial basement membrane surface area increase or decrease proportionately as the inflation volume changes. The basement membrane is a collagenous structure that is not thought to stretch at physiological levels of inflation and can therefore serve as a stable reference point unless extensive destructive lung injury is present.

Statistics

All comparisons of statistical significance are comparisons of control to treated animals using a two-tailed Student's *t*-test; $p < 0.05$ was considered to be statistically significant.

RESULTS

Distal Alveolar Region after Exposure to Hyperoxia

Exposure to hyperoxia was found to cause an acute reaction involving all major tissue components of the lung. The time course of changes in volumes of the major tissue components in the alveolar region is shown in Fig. 3. After 3 d of exposure to 85% O_2, the lungs appeared fundamentally normal except for a moderate increase in the volume of type II epithelial cells and in the volume of interstitial cells. None of these changes were statistical significant. At 5 and 7 d of exposure, dramatic and statistical significant increases of the volume of type II epithelium and of interstitial cells occurred. At the same time a statistically significant increase in the volume of interstitial matrix occurred. The increase in interstitial matrix was predominantly due to development of interstitial edema, although moderate increases in collagen could be qualitatively identified (Crapo et al., 1980). The total volume of type I epithelial cells and the capillary endothelium did not change during the 14 d of hyperoxic exposure. The pattern of cellular changes occurring during exposure to 85% O_2 are illustrated in Fig. 4. No change in cell number in any compartment occurred after 3 d exposure; however, after 5 d

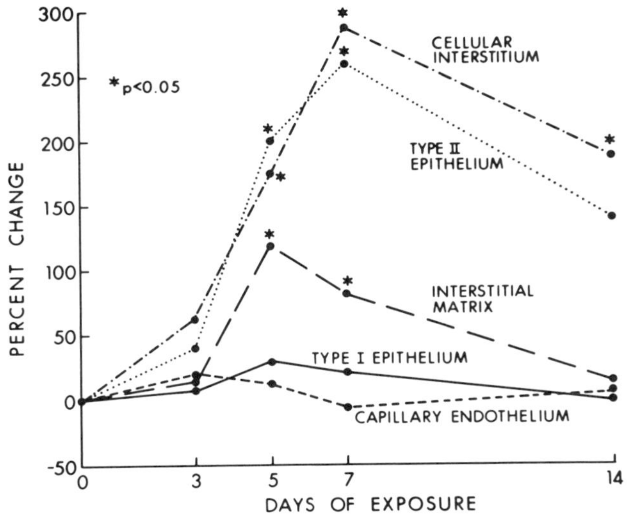

FIGURE 3. Changes in the volume of components of the alveolar septum during continuous exposure to 85% oxygen.

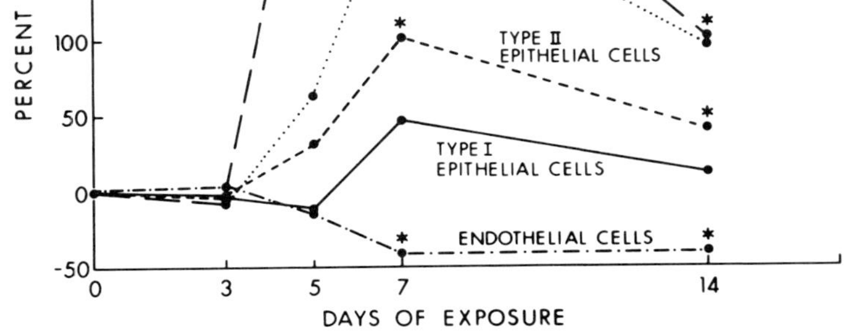

FIGURE 4. Changes in the numbers of cells in the alveolar region of the lung during continuous exposure to 85% oxygen.

exposure a two- to threefold increase in the number of interstitial cells occurred. This change in interstitial cell number involved a significant increase in inflammatory cells, which were predominantly polymorphonuclear leukocytes (Crapo et al., 1980). The number of fibroblasts also increased by 200–300% (Crapo et al., 1980). Alveolar macrophages showed a 200% increase after 7 d of 85% O_2 exposure. Although this increase was proportionally large, macrophages still represented only 5% of the total lung cells because they are normally present in relatively small numbers. At no time were intraalveolar polymorphonuclear leukocytes present. Type II epithelial cells doubled after 7 d exposure to 85% O_2 and type I cells showed a modest increase in number that did not achieve statistical significance.

It is of particular importance that the number of endothelial cells showed a steady decline during the first 7 d of 85% O_2 exposure, with loss of approximately half of the original complement of capillary endothelial cells. After 3 d in 85% O_2 the pulmonary capillary endothelium appeared to be normal. Between 3 and 7 d in 85% O_2, 41% of the capillary endothelial cells disappeared from the lung. After this point destruction of the pulmonary capillary appeared to stop with no further loss of endothelial cells occurring between d 7 and 14. Pulmonary capillary surface area was lost during exposures to 85% O_2 in direct proportion to the number of cells destroyed (Crapo et al., 1980). Since the number of endothelial cells decreased during exposure to 85% O_2 (Fig. 4) while their total volume remained unchanged (Fig. 3), it can be calculated that the mean cell volume substantially increased. Mean cell volume in fact increased almost twofold (Crapo et al., 1980) and the arithmetic mean thickness of the endothelium increased from 0.36 μm to 0.65 μm after 7 d in 85% O_2.

The appearance of a normal alveolar septum from a rat is illustrated in Fig. 5. Note the vascularity of the septum and the thin alveolar capillary membrane. The interstitium is small in proportional volume and located predominantly on the thick side of the alveolar septum and in the intercapillary spaces. The acute phase of alveolitis caused by exposure to 85% O_2 for 5 d is illustrated in Fig. 6. The interstitium is remarkably expanded with both edema fluid and increased collagen present. Inflammatory cells, predominantly polymorphonuclear leukocytes, are present in the interstitial space. The alveolar epithelium remained intact across the entire air surface although morphometric data demonstrated an increase in the number of alveolar type II cells. The remaining capillary endothelium was hypertrophic (Fig. 6).

Distal Alveolar Region after Exposure to 0.25 ppm Ozone

Results of morphometric analysis of the distal lung parenchyma after exposure to 0.25 ppm ozone are given in Tables 1 and 2. Body weight was

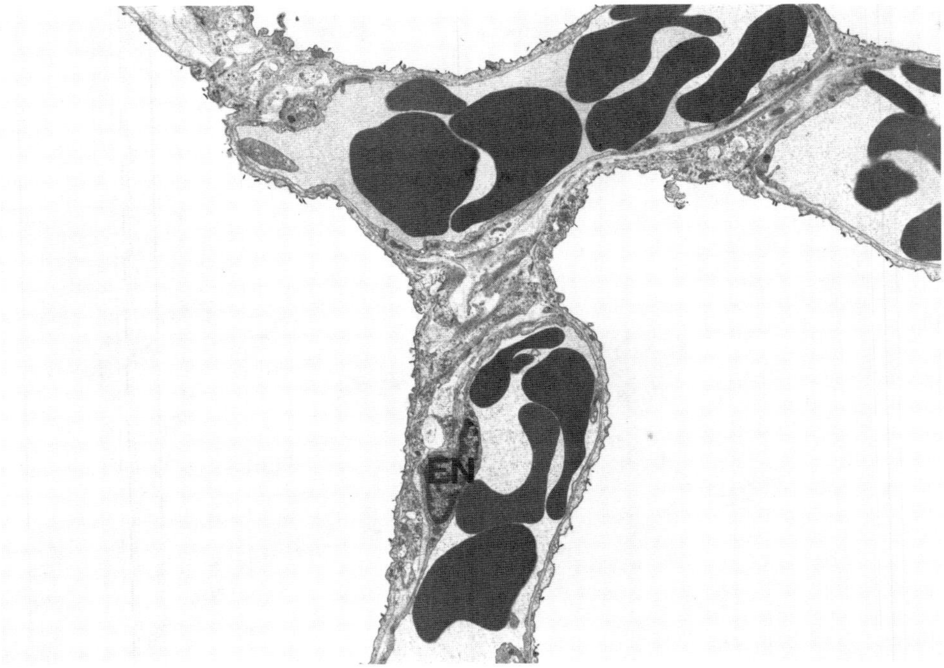

FIGURE 5. Electronmicrograph of the alveolar region of a normal rat lung. EN = capillary endothelial cell. X3600.

significantly lower in the animals exposed to ozone for 6 wk. Previous investigators have reported this to occur in rats exposed to 1 ppm or less of ozone (Stokinger et al., 1957; Schwartz et al., 1976). The arithmetic mean thickness of the epithelium and the endothelium increased about 10% ($p = 0.08$). No other statistically significant changes were found to occur in the volume of each of the major tissue compartments, the alveolar and capillary surface areas, the thickness of the various tissue compartments or in the characteristics of alveolar cells (Table 2).

The Effects of 0.25 ppm Ozone or 2 ppm NO_2 on the Proximal Alveolar Region

Results of the morphometric analysis of the proximal alveolar region of rats exposed to 0.25 ppm ozone or 2 ppm NO_2 for 6 wk are summarized in Tables 3 and 4. Some of the more significant changes are illustrated in Figs. 7 and 8. Although approximately a tenfold difference in dose and different exposure protocals were used for these exposures, approximately the same pattern of tissue reaction was found for both oxidants. Total tissue in the proximal alveolar region increased by approximately 20% (Fig. 7). This increase in tissue included statistically

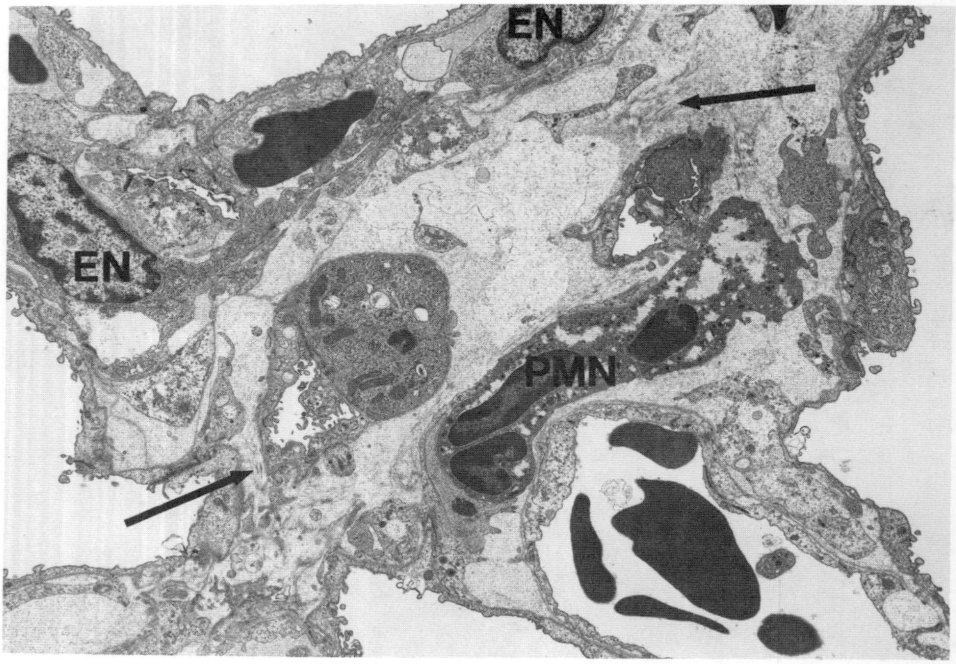

FIGURE 6. Alveolar septum of a rat after exposure to 85% O_2 for 5 d. The interstitium is markedly enlarged, predominantly with edema; however, increased collagen is present (arrows). A polymorphonuclear leukocyte (PMN) is present in the interstitium. Capillary endothelial cells (EN) show diffuse evidence of injury and hypertrophy. ×3600.

significant increases in the volume of type I epithelium, type II epithelium, interstitial cells, capillary endothelium, and alveolar macrophages in animals exposed to 0.25 ppm ozone. Exposure to NO_2 caused significant changes only in the volumes of alveolar type I epithelium, alveolar type II epithelium, and cellular interstitium. The magnitude of the changes in the volume of epithelium and interstitum were approximately the same (Fig. 7). The fraction of the alveolar surface that was covered by type II epithelium was about 2% in the controls and increased to over 3% after exposure to 2 ppm NO_2. Thickness of the alveolar epithelium was significantly increased after exposure to both oxidants (Table 3).

The total number of epithelial cells/mm^2 of basement membrane surface increased from 165 in the control animals to 423 after exposure to 0.25 ppm ozone and increased from 311 in the controls to 403 in the animals exposed to 2 ppm NO_2. The number of type I epithelial cells almost doubled after exposure to ozone, whereas they increased only about 25% after exposure to 2 ppm NO_2 (Fig. 8). After exposure to both oxidants the alveolar type II epithelial cells were smaller in average volume and had spread out to cover a substantially greater amount of basement membrane surface area per cell (Table 4). Although the majority of the epithelium appeared normal to quantitative examination, there was evidence of type I cell injury. Small areas of denuded alveolar basement membrane were seen in the animals exposed to 0.25 ppm O_3 and the plasma membranes of some type I cells were ruptured and/or vesiculated. Areas of increased electron lucency were sometimes found in the cytoplasm of these epithelial cells.

Figure 9 illustrates both the increased number and increased thickness of alveolar type I epithelium morphometrically determined to be present after exposure to 0.25 ppm O_3. A similar response occurred after

TABLE 1. Absolute Volumes, Surface Areas, and Diffusion Capacity for Parenchymal Portion of the Lungs of Rats Exposed to 0.25 ppm Ozone for 6 wk: Distal Alveolar Region[a]

Parameter	Control	0.25 ppm Ozone	*p* Value
n	8	8	
Body weight (g)	268 ± 4	249 ± 8	0.05
Lung volume (ml)	7.85 ± 0.17	7.97 ± 0.26	0.72
Total volumes (cm^3/both lungs)			
Air	5.60 ± 0.13	5.18 ± 0.21	0.64
Capillary lumen	0.44 ± 0.04	0.41 ± 0.04	0.59
Tissue	0.32 ± 0.01	0.32 ± 0.02	0.77
Type I epithelium	0.07 ± 0.01	0.08 ± 0.01	0.55
Type II epithelium	0.02 ± 0.00	0.02 ± 0.00	0.64
Cellular interstitium	0.06 ± 0.01	0.05 ± 0.01	0.29
Noncellular interstitium	0.07 ± 0.01	0.07 ± 0.01	0.80
Endothelium	0.09 ± 0.01	0.09 ± 0.01	0.43
Macrophages	0.01 ± 0.00	0.01 ± 0.00	0.78
Surface area (cm^2/both lungs)[b]			
Alveolar epithelium	3165 ± 160	2967 ± 181	0.43
Type I	3064 ± 170	2851 ± 171	0.39
Type II	101 ± 17	117 ± 27	0.63
Capillary endothelium	3376 ± 119	3154 ± 270	0.48
D_T (ml O_2/min·mm Hg)	2.969 ± 0.01	2.604 ± 0.17	0.16
Tissue thickness (μm)			
Harmonic mean (ari to plasma)	0.367 ± 0.01	0.388 ± 0.01	0.21
Arithmetic mean			
Epithelium	0.298 ± 0.01	0.327 ± 0.02	0.22
Type I	0.231 ± 0.01	0.267 ± 0.02	0.08
Type II	2.569 ± 0.35	2.185 ± 0.45	0.51
Interstitium	0.409 ± 0.02	0.412 ± 0.04	0.61
Endothelium	0.261 ± 0.01	0.303 ± 0.02	0.08

[a]All data are mean ± SE.

[b]Type I SA is the SA of basement membrane under type I cells; type II SA is the air surface of type II cells, excluding the extra SA contributed by microvilli; endothelial SA is the luminal surface of the endothelial cells.

TABLE 2. Morphometrically Determined Characteristics of the Major Parenchymal Lung Cells in Rats Exposed to 0.25 ppm Ozone for 6 wk: Distal Alveolar Region[a]

Parameter	Control	0.25 ppm Ozone	*p* Value
n	8	8	
Total number of cells			
per both lungs (×10⁶)	533 ± 28	520 ± 17	0.68
Alveolar type I cells			
Total number (×10⁶)	39 ± 3	42 ± 3	0.56
Average volume (μm^3)	1860 ± 138	1920 ± 237	0.81
Average surface area (μm^2)	9412 ± 849	8469 ± 958	0.47
Average basement			
membrane SA/cell (μm^2)	8189 ± 720	7203 ± 795	0.37
% Total lung cells	7	8	
% Alveolar surface covered	97	96	
Alveolar type II cells			
Total number (×10⁶)	59 ± 3	60 ± 3	0.82
Average volume (μm^3)	401 ± 47	340 ± 59	0.43
Average surface area (μm^2)			
With microvilli	179 ± 39	219 ± 47	0.51
Without microvilli	177 ± 32	195 ± 42	0.73
% Total lung cells	11	12	
% Alveolar surface covered	3	4	
Endothelial cells			
Total number (×10⁶)	296 ± 14	294 ± 11	0.74
Average volume (μm^3)	298 ± 18	327 ± 31	0.43
Average surface area (μm^2)			
% Total lung cells	55	57	0.43
Interstitial cells			
Total number (×10⁶)	131 ± 9	116 ± 6	0.19
Average volume (μm^3)	484 ± 56	475 ± 56	0.76
% Total lung cells	25	22	
Macrophages			
Total number (×10⁶)	9 ± 1	9 ± 1	0.81
Average volume (μm^3)	1044 ± 396	959 ± 299	
% Total lung cells	2	2	

[a]All data are mean ± SE.

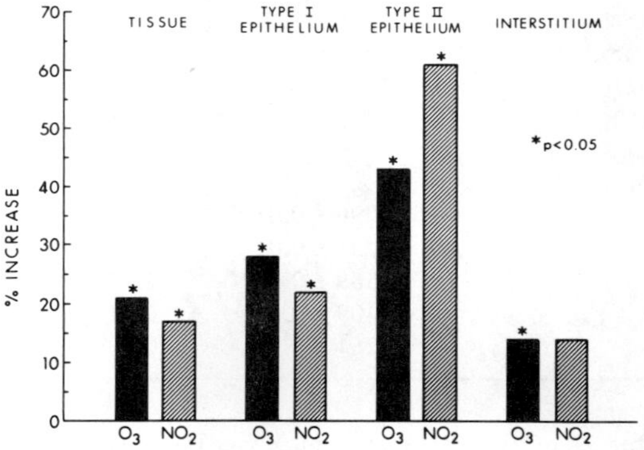

FIGURE 7. Percent change in the volume of components of alveolar septa from the proximal alveolar region after exposure to low doses of O_3 or NO_2 for 6 wk.

TABLE 3. Absolute Volumes, Surface Areas and Diffusion Capacity for Parenchymal Portion of the Lungs of Rats Exposed to 0.25 ppm Ozone or 2 ppm NO_2 for 6 wk: Proximal Alveolar Region[a]

Parameter	Control	0.25 ppm Ozone	Control	2 ppm NO_2
n	8	8	4	4
Total lung volume (cm^3)	7.85 ± 0.17	7.97 ± 0.26	8.2 ± 0.32	8.47 ± 0.40
Basement membrane surface density (cm^2/cm^3)	487 ± 16	514 ± 27	360 ± 18	403 ± 35
Volumes/S_{BM} (μm)[c]				
Air	18.377 ± 0.66	17.424 ± 1.07	25.033 ± 1.21	22.351 ± 2.17
Capillary lumen	1.363 ± 0.06	1.239 ± 0.08	1.719 ± 0.10	1.684 ± 0.06
Tissue	0.972 ± 0.03	1.174 ± 0.03[b]	1.172 ± 0.04	1.366 ± 0.06[b]
Type I epithelium	0.186 ± 0.01	0.238 ± 0.01[b]	0.259 ± 0.02	0.316 ± 0.01[b]
Type II epithelium	0.069 ± 0.01	0.099 ± 0.01[b]	0.071 ± 0.01	0.114 ± 0.01[b]
Cellular interstitium	0.192 ± 0.01	0.226 ± 0.01[b]	0.209 ± 0.01	0.238 ± 0.01
Noncellular interstitium	0.263 ± 0.01	0.291 ± 0.02	0.279 ± 0.02	0.320 ± 0.02
Endothelium	0.242 ± 0.01	0.273 ± 0.01[b]	0.311 ± 0.01	0.323 ± 0.01
Macrophages	0.020 ± 0.00	0.047 ± 0.01[b]	0.044 ± 0.01	0.054 ± 0.01
Surface area				
% Type I	98.3 ± 0.3	97.6 ± 0.2	97.9 ± 0.1	96.7 ± 0.5[b]
% Type II	1.7 ± 0.3	2.4 ± 0.2	2.0 ± 0.1	3.3 ± 0.4[b]
Capillary endothelium	97.3 ± 0.9	97.9 ± 0.2	118.4 ± 0.5	127.1 ± 4.6
D_T (μl O_2/min·mm Hg·cm^2)	0.8 ± 0.02	0.7 ± 0.02[b]	1.04 ± 0.02	0.991 ± 0.50
Tissue thickness (μm)				
Harmonic mean (air to plasma)	0.385 ± 0.01	0.426 ± 0.01[b]	0.345 ± 0.01	0.38 ± 0.02
Arithmetic mean				
Epithelium	0.255 ± 0.01	0.337 ± 0.01[b]	0.330 ± 0.02	0.431 ± 0.02[b]
Type I	0.190 ± 0.01	0.244 ± 0.01[b]	0.264 ± 0.02	0.327 ± 0.01[b]
Type II	4.327 ± 0.48	4.263 ± 0.37	3.681 ± 0.35	3.508 ± 0.24
Interstitium	0.461 ± 0.02	0.522 ± 0.02[b]	0.446 ± 0.01	0.491 ± 0.17
Endothelium	0.249 ± 0.01	0.279 ± 0.01[b]	0.263 ± 0.01	0.254 ± 0.01

[a]All data are mean ± SE.
[b]$p < 0.05$ for comparison with the respective control group.
[c]Volume density/epithelial basement membrane surface density.

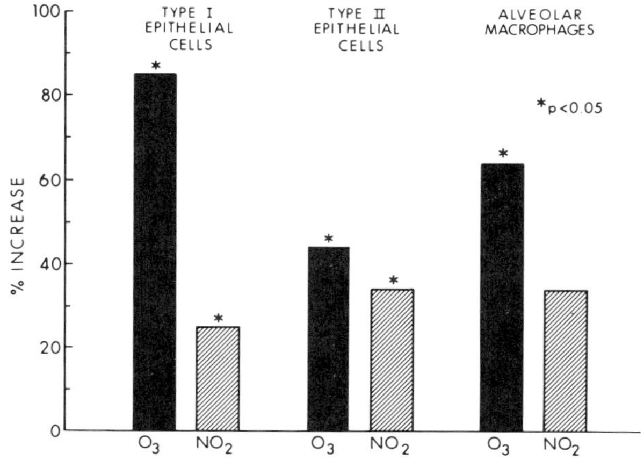

FIGURE 8. Percent change in the numbers of cells in alveolar septa from the proximal alveolar region after exposure to low doses of O_3 or NO_2 for 6 wk.

TABLE 4. Morphometrically Determined Characteristics of the Major Parenchymal Lung Cells in Rats Exposed to 0.25 ppm Ozone or 2 ppm NO_2 for 6 wk: Proximal Alveolar Region[a]

Parameter	Control	0.25 ppm Ozone	Control	2 ppm NO_2
n	8	8	4	4
Total number of cells/mm²	1529 ± 68	1778 ± 101	1477 ± 67	1602 ± 53
Alveolar type I cells				
Number/mm²	100 ± 6	185 ± 15[b]	142 ± 7	177 ± 10[b]
Average volume (μm^3)	1937 ± 190	1374 ± 158[b]	1820 ± 54	1803 ± 109
Average surface area (μm^2)	10646 ± 949	5744 ± 556[b]	8402 ± 401	6895 ± 433[b]
Average basement membrane SA/cell (μm^2)	10206 ± 959	5559 ± 507	6943 ± 343	5529 ± 351[b]
% Total lung cells	7	10	10	11
% Alveolar surface covered	98	98	97	95
Alveolar type II cells				
Number/mm²	165 ± 14	238 ± 26[b]	169 ± 15	226 ± 9[b]
Average volume (μm^3)	421 ± 25	420 ± 22	436 ± 62	507 ± 62
With microvilli	191 ± 28	131 ± 25	245 ± 40	295 ± 55
Without microvilli	105 ± 13	105 ± 12	118 ± 10	146 ± 19
% Total lung cells	11	13	11	14
% Alveolar surface covered	2	2	3	5
Endothelial cells				
Number/mm²	851 ± 28	855 ± 43	765 ± 36	757 ± 34
Average volume (μm^3)	286 ± 12	322 ± 12[b]	408 ± 14	427 ± 7
Average surface area (μm^2)	1152 ± 38	1160 ± 49	1556 ± 65	1688 ± 83
% Total lung cells	56	48	52	47
Interstitial cells				
Number/mm²	385 ± 41	454 ± 35	368 ± 20	399 ± 20
Average volume (μm^3)	534 ± 54	520 ± 44	568 ± 200	598 ± 277
% Total lung cells	25	26	25	25
Macrophages				
Number/mm²	28 ± 4	46 ± 6[b]	32 ± 3	43 ± 3
Average volume (μm^3)	755 ± 117	1009 ± 98	1372 ± 460	1265 ± 232
% Total lung cells	2	3	2	3

[a] All data are mean ± SE.
[b] $p < 0.05$.

exposure to 2 ppm NO_2 and is illustrated in Fig. 10. These two electronmicrographs also illustrate the small increase in the interstitium that was found to occur. A 10–15% increase was found to occur in both interstitial cells and interstitial matrix after exposure to either 0.25 ppm O_3 or 2 ppm NO_2. Of these changes only the increase in interstitial cells after exposure to O_3 reached statistical significance. Qualitative examination of the tissue sections, however, readily identified changes in the interstitial matrix. Small electron-lucent areas were found in increased numbers in pericapillary regions after exposure to both oxidants (Figs. 9–11). These changes are consistent with mild, early interstitial edema.

The number of alveolar type II epithelial cells increased by about 40% after exposure to either ozone or NO_2. Their mean cell volume, surface area, and thickness were not statistically different from controls (Table 4). The total amount of epithelial surface area covered by type II epithelial cells did increase, particularly in the animals exposed to NO_2. No qualitative changes in the morphology of alveolar type II cells were found as illustrated in Fig. 11.

DISCUSSION

The use of morphometric techniques to evaluate patterns of tissue and cell reactions after exposure to oxidants permits a more rigorous comparison of both the extent and distribution of effects caused by different agents. Oxygen represents a gas of relatively low toxicity which is present in high concentrations. As illustrated in Fig. 1, its relative uniform distribution throughout the alveolar region of the lung combined with its rapid diffusion across tissues leads to a uniform distribution of the O_2-mediated lung injury. Morphometric techniques have previously been used to study monkeys (Kapanci et al., 1969) and rats (Kistler et al., 1967; Crapo et al., 1980) exposed to pure oxygen atmospheres. Histologic

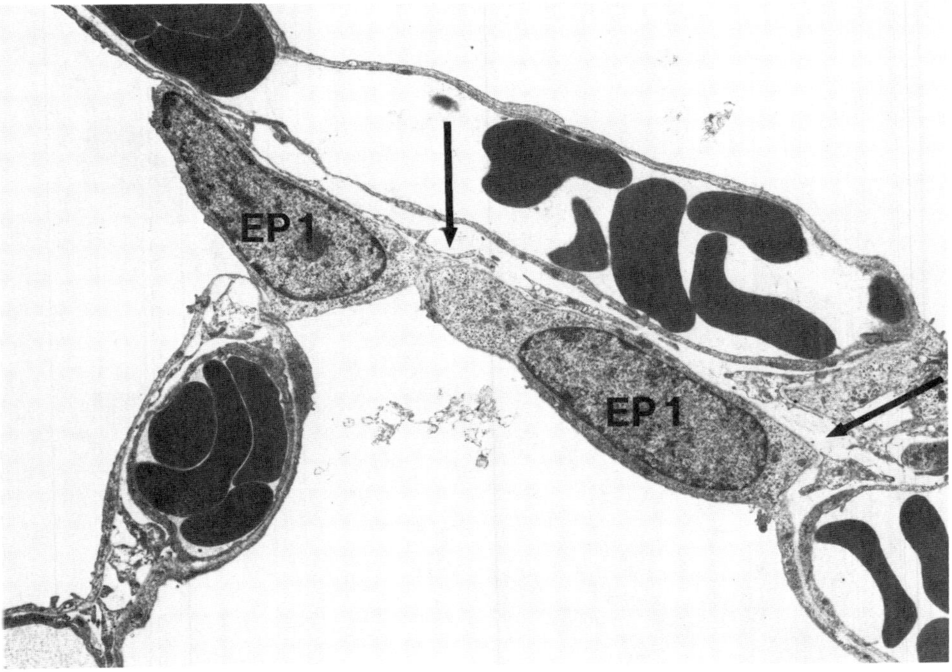

FIGURE 9. Electronmicrograph of an alveolar septa from the proximal alveolar region after exposure to 0.25 ppm O_3 for 6 wk. Two type I epithelial cells (EPI) are shown, illustrating the increased thickness and decreased mean surface area occurring in these cells. Electron lucent areas (arrows) in the pericapillary region suggest early interstitial edema. X3600.

damage has been found by many investigators to occur in all major compartments of the alveolar septum but to be particularly important in the capillary endothelium (Kistler et al., 1978). These cells have been postulated to be the primary site of injury that is related to animal death. The mechanism of oxygen injury to cells is probably related to the production of toxic metabolites of oxygen rather than to direct toxic effects of molecular oxygen (Freeman and Crapo, 1982). The production of toxic species such as H_2O_2, O_2^-, and $OH^\bullet$ by intracellular metabolism or by activated inflammatory cells is thought to be a major factor in oxygen-mediated lung injury. Thus the site of the injury is likely to be primarily influenced by which cells produce the highest rates of toxic oxygen metabolites and/or which cells have the least adequate cell defense mechanisms against these partially reduced species of oxygen.

Ozone and NO_2 represent oxidants for which the pattern of lung injury is likely to be highly influenced by the distribution of the gases. Both are capable of rapid reactions with common tissue constituents. Ozone has two unpaired electrons in its outer shell, giving it the potential to oxidize many cell components (Mustafa and Tierney, 1978), possibly through free radical formation (Menzel, 1976). Lipids in plasma mem-

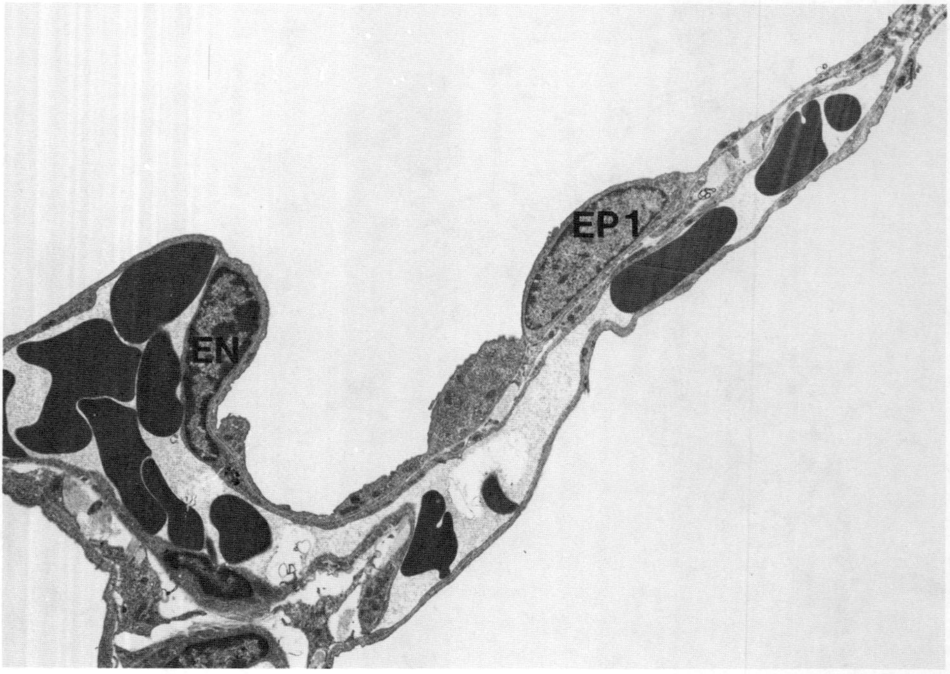

FIGURE 10. Electronmicrograph of an alveolar septa from the proximal alveolar region after exposure to 2.0 ppm NO_2 for 6 wk. Increased thickening of the type I epithelium (EP1) is present. ×3600.

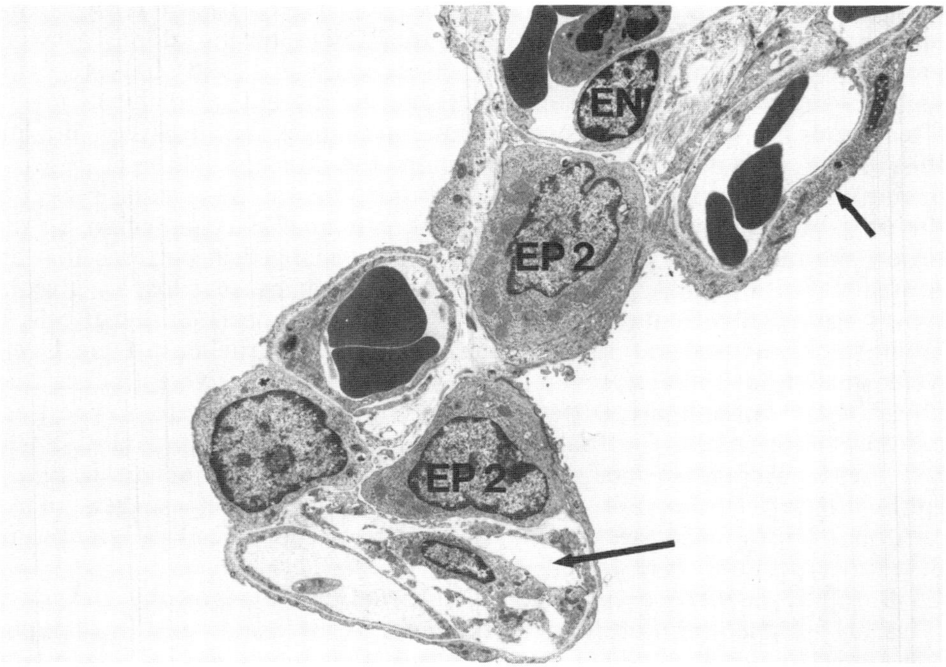

FIGURE 11. Alveolar septa from the proximal alveolar region after exposure to 0.25 ppm ozone for 6 wk. Increased number of type II epithelial cells (EP2) as well as thickening of the alveolar type I epithelium (short arrow) are illustrated. An increased incidence of areas of electron lucency in the interstitium (long arrow) suggest early interstitial edema. ×3600.

branes can undergo peroxidation or ozonolysis after exposure to ozone. Similar reactions with proteins and amino acids can lead to altered enzymatic activities and toxic effects on the cell (Mudd and Freeman, 1977). Nitrogen dioxide may react predominantly by formation of nitric and nitrous acid upon hydration, either in airways or upon contacting the surface of the alveolar septum. All of the above mechanisms suggest that these two oxidant gases may react rapidly with tissues they contact in crossing the alveolar septum. The concentration of the lesions caused by these oxidant gases in the proximal portions of the acinus may be due to the fact that the mass movement of fresh gas during each inspiration ends in the proximal acinus and movement of the gas into the distal alveoli is dependent on diffusion (Farhi, 1969). The reaction of the oxidants with tissue components in the proximal portions of the acinus is presumed to rapidly reduce the concentrations of the oxidants as they flow toward distal alveoli (Fig. 1). Thus the toxic response would tend to occur most intensely in the proximal airways and proximal portions of the acinus and would also tend to involve tissues closest to the gas-tissue interface.

In evaluating the reaction to oxidants present in low concentrations, morphometry offers a sensitive, quantitative tool to detect changes that may be difficult to substantiate by qualitative means alone. Previous investigators (Plopper et al., 1979; Schwartz et al., 1976) have reported slight changes in terminal bronchioles and in the adjacent alveoli in rats after exposure to 0.2 ppm ozone, but no changes in the alveoli beyond this region were found. Our data support the concept that low doses of ozone will not cause injury to the distal portions of the alveolar region of the lung but that injury does occur in the proximal alveolar region. In this region the numbers of type I and type II cells changed in patterns that are consistent with injury to the alveolar epithelium causing loss of alveolar type I epithelial cells and proliferation of alveolar type II epithelial cells in order to regenerate and replace damaged type I epithelium. Type I cells were smaller in volume, covered less surface area and were thicker than those of control animals. Differentiation of a type II cell into a type I cell requires more than a fourfold increase in mean cell volume and a 48-fold increase in average surface area. The changes found to occur in type I epithelium after exposure to low doses of O_3 or NO_2 suggest that a portion of these cells are at an intermediate stage of cell differentiation between type I and type II cells.

Both the qualitative evidence of epithelial cell hypertrophy and injury (Figs. 9–11) and the morphometric changes in the characteristics of type I and type II epithelial cells (Table 4) suggest that inhalation of 2 ppm NO_2 or 0.25 ppm ozone will cause type I epithelial cell injury and an increase in the rate of epithelial cell proliferation. This conclusion is supported by the work of Evans et al. (1976), who found increased mitotic activity in alveolar type II epithelial cells in the proximal alveolar region of rats exposed to less than 1 ppm of ozone. Castleman et al. (1973a,b) used histochemical methods to demonstrate an increase in the amount of acid phosphatase and β-glucuronidase present in epithelial cells of terminal bronchioles and their adjacent alveoli after exposure of rats to ozone. They postulated that ozone may have damaged the plasma membrane of lysosomes and allowed leakage of lysosomal enzymes into the cell cytoplasm. Thus the results of morphometric, autoradiographic, and histochemical studies all suggest that low doses of O_3 and NO_2 can cause injury to type I epithelial cells and can increase rates of epithelial cell turnover in the proximal alveolar region of the lung.

Other changes noted by morphometry were that macrophages increased by 50–100% and there were small increases in the volume of both cellular and noncellular components of the interstitium (Tables 3 and 4). Hussain et al. (1976) have demonstrated an increased concentration of prolyl hydroxylase and hydroxyproline in rat lung homogenates after exposure of animals to 0.8 ppm ozone for 7 d. We qualitatively identified evidence of increased collagen in the proximal alveolar region (Figs. 9 and 11). Histologic evidence of fibrotic lesions caused by higher doses of ozone has been reported (Freeman et al., 1974) and ozone has been used to produce pulmonary fibrosis in rats (Hesterberg and Last, 1981). The small changes in the volume of the interstitium demonstrated in Table 3 are likely to be associated with a mild but significant chronic fibrotic reaction.

Evidence of significant injury to the capillary endothelium after exposure to either ozone or NO_2 was not found. The number of endothelial cells and their average volume and average surface area did not change, other than there being a small increase in average volume of endothelial cells after exposure to 0.25 ppm ozone. The most likely reason for the difference in degree of injury occurring in alveolar epithelial cells and endothelial cells is that an oxidant gas reacts with tissue components as it diffuses across the alveolar septum. This would give a relatively large dose to the epithelium while reducing the dose to which the endothelial cells are exposed (Fig. 1).

The following conclusions are supported by the studies reported here:

1. The pattern of lung injury caused by oxidant gases is influenced by the concentration of the gas, reactivity of the gas with tissue components, the distribution of the gas throughout the acinus and the sensitivity of various cells to specific oxidants.

2. Oxygen is a gas of low relative toxicity that, when present in high concentrations, will cause a diffuse injury to all components of the alveolar septum. Injury to the pulmonary capillary endothelium is a physiologically important component of the response.

3. Ozone and NO_2 are highly reactive gases that occur in relative low concentrations and that cause a pattern of injury which is most extensive in the proximal portions of the pulmonary acinus.

4. The pattern of injury caused by ozone and NO_2 in low concentrations are similar in that both involve predominantly an epithelial injury. This pattern of injury involves a proliferation of alveolar type II epithelium. recruitment of macrophages, and an interstitial fibrotic response in the alveolar region.

5. The toxicity of ozone is roughly tenfold greater than that of NO_2. This is estimated by single points on a dose-response curve and involves substantially different exposure protocols. A full dose-response curve needs to be done to evaluate whether or not this relationship holds true for all concentrations of the gases.

REFERENCES

Bachofen, M., and Weibel, E. R. 1977. Alterations of the gas exchange apparatus in adult respiratory insufficiency associated with septicemia. *Am. Rev. Respir. Dis.* 116:589–615.

Clark, J. M., and Lambertson, C. J. 1971. Pulmonary oxygen toxicity: A review. *Pharmacol. Rev.* 23:37–133.

Castleman, W. L., Dungworth, D. L., and Tyler, W. S. 1973a. Cytochemically detected alterations of lung acid phosphates reactivity following ozone exposure. *Lab. Invest.* 29:310–319.

Castleman, W. L., Dungworth, D. L., and Tyler, W. S. 1973b. Histochemically detected enzymatic alterations in rat lungs exposed to ozone. *Exp. Mol. Pathol.* 19:102–121.

Crapo, J. D., Marsh-Salin, J., Pratt, P., and Ingram, P. 1978a. Studies of tolerance and cross-tolerance using nitrogen dioxide and oxygen. II. Morphology and morphometry. *J. Appl. Physiol.: Respirat. Environ. Exercise Physiol.* 44:370–379.

Crapo, J. D., Peters-Golden, M., Marsh-Salin, J., and Shelburne, J. S. 1978b. Pathologic changes in the lungs of oxygen treated rats—A morphometric analysis. *Lab. Invest.* 39:640–653.

Crapo, J. D., Barry, B. E., Foscue, H. A., and Shelburne, J. 1980. Structural and biochemical changes in rat lungs occurring during exposures to lethal and adaptive doses of oxygen. *Am. Rev. Respir. Dis.* 122:123–145.

Evans, M. J., Johnson, L. V., Stephens, R. J., and Freeman, G. 1976. Cell renewal in the lungs of rats exposed to low levels of ozone. *Exp. Mol. Pathol.* 24:70–83.

Farhi, L. E. 1969. Diffusive and convective movement of gas in the lung. In *Circulatory and Respiratory Mass Transport,* eds. G. E. W. Wolstenholme and J. Knight, pp. 277–293. New York: Little, Brown.

Freeman, B. A., and Crapo, J. D. 1981. Free radicals and tissue injury. *Lab. Invest.* 47:412–426.

Freeman, G., Crane, S. C., Stephens, R. J., and Furiosi, N. J. 1968a. Pathogenesis of the nitrogen dioxide-induced lesion in the rat lung. A review and presentation of new observations. *Am. Rev. Respir. Dis.* 98:429–443.

Freeman, G., Stephens, R. J., Crane, S. C., and Furiosi, N. J. 1968b. Lesion of the lung in rats continuously exposed to two parts per million of nitrogen dioxide. *Arch. Environ. Health* 17:181–192.

Freeman, G., Juhos, L. T., Furiosi, N. J., Mussenden, R., Stephens, R. J., and Evans, M. J. 1974. Pathology of pulmonary disease from exposure to interdependent ambient gases (nitrogen dioxide and ozone). *Arch. Environ. Health* 29:203–210.

Hayatdavoudi, G., Crapo, J. D., Miller, F. J., and O'Neil, J. J. 1980. Factors determining degree of inflation in intratracheally fixed rat lungs. *J. Appl. Physiol.* 48:389–393.

Hesterberg, T. W., and Last, J. A. 1981. Ozone-induced acute pulmonary fibrosis in rats. Prevention of increased rates of collagen synthesis by methylprednisolone. *Am. Rev. Respir. Dis.* 123:47–152.

Hussain, M. Z., Cross, C. E., Mustafa, M. G., and Bhatnagar, R. S. 1976. Hydroxyproline contents and prolyl hydroxylase activities in the lungs of rats exposed to low levels of ozone. *Life Sci.* 18:897–904.

Kapanci, Y., Weibel, E. R., Kaplan, H. P., and Robinson, F. S. 1969. Pathogenesis and reversibility of the pulmonary lesions of oxygen toxicity in monkeys. II. Ultrastructural and morphometric studies. *Lab. Invest.* 20:101–118.

Kistler, G. S., Caldwell, P. R. B., and Weibel, E. R. 1967. Development of fine structural damage to alveolar and capillary lining cells in oxygen-poisoned rat lungs. *J. Cell Biol.* 33:605–628.

LaForce, R. C., and Lewis, B. N. 1970. Diffusional transport in the human lung. *J. Appl. Physiol.* 28:291–298.

Menzel, D. B. 1976. Toxicity of nitrogen oxides and ozone. In *Free Radicals In Biology,* Vol. II, ed. W. A. Pryor, pp. 181–202. New York: Academic.

Mudd, J. B., and Freeman, B. A. 1977. Reaction of ozone with biological membranes. In *Biochemical Effects of Environmental Pollutants,* ed. S. D. Lee, pp. 97–133. Ann Arbor, Mich.: Ann Arbor Science.

Mustafa, M. F., and Tierney, D. F. 1978. Biochemical and metabolic changes in the lung with oxygen, ozone and nitrogen dioxide toxicity. *Am. Rev. Respir. Dis.* 118:1061–1090.

Plopper, C. G., Dungworth, D. L., Tyler, W. S., and Chow, C. K. 1979. Pulmonary alterations in rats exposed to 0.2 and 0.1 ppm ozone: A correlated morphological and biochemical study. *Arch. Environ. Health* 34:390–395.

Scherle, W. 1970. A simple method for volumetry of organs in quantitative stereology. *Mikroskopie* 26:57–60.

Schwartz, L. W., Dungworth, D. L., Mustafa, M. G., Tarkington, B. K., and Tyler, W. S. 1976. Pulmonary responses of rats to ambient levels of ozone. Effects of 7-day intermittent or continuous exposure. *Lab. Invest.* 34:565–578.

Stokinger, H. E., Wagner, W. D., and Dobrogorski, O. H. 1957. Ozone toxicity studies. III. Chronic injury to lungs of animals follow exposure at a low level. *Arch. Ind. Health* 16:514–522.

Weibel, E. R. 1979. *Stereological Methods. I. Practical Methods for Biological Morphometry.* London: Academic.

Woody, D. M., Woody, E. Z., and Crapo, J. D. 1979. Determination of the mean caliper diameter of lung nuclei by a method which is independent of shape assumptions. *J. Microscopy* 118:421–427.

THE PATHOBIOLOGY AND EPIDEMIOLOGY OF HUMAN EMPHYSEMA

William M. Thurlbeck

Department of Pathology, University of British Columbia, Vancouver, British Columbia, Canada

Emphysema is defined in anatomical terms as enlargement of the gas-exchanging part of the lung (the acinus) accompanied by destruction of respiratory tissue. Emphysema is classified by the way that the acinus is dominantly involved. In proximal acinar emphysema, the proximal part of the acinus—respiratory bronchioles—is dominantly involved and two forms of proximal acinar emphysema are usually recognized: centrilobular emphysema and simple pneumoconiosis of coalworkers. The acinus is more or less uniformly involved in panacinar emphysema, and several clinical associations have been described with this lesion. In distal acinar emphysema, alveolar ducts and sacs are particularly involved, and spontaneous pneumothorax of young adults is associated with this form of emphysema. Scarring is usually associated with irregular involvement of the acinus (irregular emphysema) and is usually asymptomatic.

No uniform agreement exists as to the application of this classification and there is widespread discrepancy of classification of emphysematous lungs between experts, especially when emphysema is severe. The precise definition of destruction of respiratory tissue in emphysema has not been agreed on, and this had led to wide variations in the assessment of prevalence of emphysema in autopsy series. Tobacco smoking is the most important cause of emphysema and is thought to bring it about by imbalance between the protease-antiproteinase mechanisms in the lung.

Increasing severity of emphysema is accompanied by increasing frequency of symptoms, but a substantial proportion of subjects with severe emphysema will be apparently free from symptoms. The major functional characteristics of severe emphysema are reduction in expiratory flow, increase in lung volumes, and diminished diffusing capacity. Diminished expiratory flow in emphysema is determined in part by loss of elastic recoil and in part by associated airway disease. Loss of recoil in emphysematous lungs may be brought about by functional changes in the apparently normal intervening lung between the emphysematous spaces.

INTRODUCTION

I propose to review the pathology of pulmonary emphysema in humans on the assumption that many readers are not fully familiar with human emphysema and are more concerned with experimental models of lung injury and repair. I will first define emphysema, classify it, discuss its pathogenesis and etiology, and finally describe its clinicopathologic correlations. I will also discuss the epidemiology of emphysema but, for reasons that will become obvious, will not be able to discuss the natural history of emphysema. I will also emphasize problems in the various sections, since these are not always obvious to those not working in the field.

DEFINITION

Emphysema is defined as abnormal enlargement of the airspaces of the lung, accompanied by destruction of respiratory tissue (American Thoracie Society, 1962; World Health Organization, 1961). Emphysema thus is a disorder of anatomy, and it is essential to know the appropriate normal anatomy if one is to understand the pathology of emphysema. The lung consists of a series of conducting airways—bronchi and bronchioles—that lead to the gas-exchanging part of the lung. Bronchi are distinguished from bronchioles in that they contain cartilage in their walls, and the transition from bronchi to bronchioles occurs in airways about 2 mm in internal diameter in humans. The last purely conducting structure is the terminal bronchiole; the structures distal to it comprise the acinus. The acinus is a highly complex structure, but a simplified model will suffice for our purposes (Fig. 1). The structures immediately beyond to the terminal bronchiole are the respiratory bronchioles, which have both alveoli and nonalveolated epithelium forming their walls; thus respiratory bronchioles both conduct and exchange gas. Proceeding distally, progressively more alveoli appear in their walls. There are characteristically three orders of respiratory bronchioles. Respiratory bronchioles are succeeded by alveolar ducts whose walls are entirely alveolated, and these in turn lead into alveolar sacs, the terminal respiratory structures, which are likewise completely alveolated.

CLASSIFICATION

Emphysema is classified by the way it involves the acinus, and four forms of emphysema are usually recognized (Thurlbeck, 1976).

Proximal Acinar Emphysema

In proximal acinar emphysema, respiratory bronchioles are selectively or dominantly involved. The respiratory bronchioles are enlarged and destroyed both in series and parallel. The emphysematous spaces lie approximately at the center of secondary lobules (the smallest portions of the lung surrounded by connective tissue septa), and this was the source of the original term, centrilobular emphysema. Proximal acinar emphysema occurs under two distinct circumstances, and it is important to recognize this distinction.

Proximal acinar emphysema is the common form of nonindustrial emphysema and is almost invariably associated with cigarette smoking and inflammation of distal airways and the walls of the emphysematous spaces. This form of emphysema is usually referred to as centrilobular emphysema

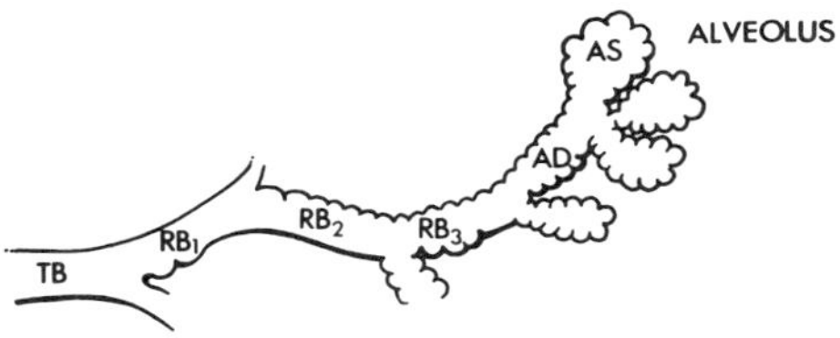

FIGURE 1. A diagram simplifying the anatomy of the acinus (the gas exchanging part of the lung, beyond a terminal bronchiole, TB) shows that it comprises: respiratory bronchioles (RB) with increasing numbers of alveoli in their walls (RB_1 to RB_3), but part of the walls are still formed by nonalveolated epithelium; alveolar ducts (AD), which are entirely alveolated; and alveolar sacs (AS), the terminal structures.

(Figs. 2–4). The emphysematous spaces are found more frequently in the upper zones of the lung, and centrilobular emphysema is usually more severe in the upper zones of the lungs (Thurlbeck, 1963). Involvement of the lung is characteristically quite uneven, with some respiratory bronchioles being spared or slightly involved whereas nearby ones may be severely involved producing large emphysematous spaces. Centrilobular emphysema is frequently associated with chronic bronchitis and is the form of emphysema most commonly encountered in patients with symptomatic chronic airflow obstruction.

Simple pneumoconiosis of coalworkers or focal emphysema also involves the proximal part of the acinus. It differs from centrilobular emphysema in that there is always a heavy deposit of coal around the emphysematous spaces, the enlargement of respiratory bronchioles is modest and emphysema less obvious, and the process is more uniform through the lung. Simple penumoconiosis is associated with mild impairment of function only, producing minor abnormalities of gas exchange (Morgan and Seaton, 1975).

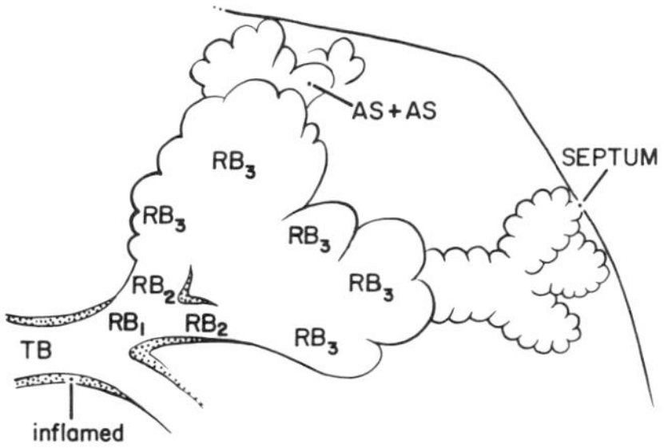

FIGURE 2. In centrilobular emphysema, respiratory bronchioles are enlarged and destroyed producing lesions in the proximal acinus. The proximal bronchiole and the walls of the emphysematous spaces usually show evidence of inflammation. Reproduced courtesy W. B. Saunders & Co. (Thurlbeck, 1976).

Panacinar Emphysema

In panacinar emphysema, there is more or less uniform involvement of the acinus (Figs. 5 and 6). Controversy exists concerning the distinction between centrilobular emphysema and panacinar emphysema—some believe them to be distinctly different diseases (Anderson and Foraker, 1973); others believe them to have the same associations (Mitchell et al., 1970). The reason for the disagreement will be discussed below. There are at least four different associations of panacinar emphysema, each with distinct clinico-pathologic associations (Thurlbeck, 1976), but this view is not shared by all.

The most classical associationsof panacinar emphysema is with $alpha_1$

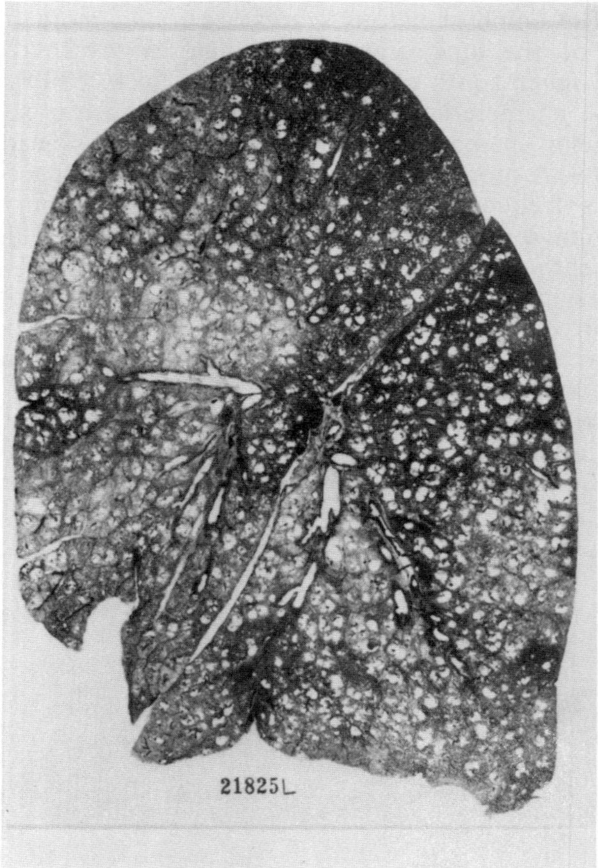

FIGURE 3. A paper-mounted whole lung section illustrates centrilobular emphysema that is fairly uniformly distributed through the lung. Reproduced courtesy of the editor, *Thorax* (Thurlbeck, 1967b).

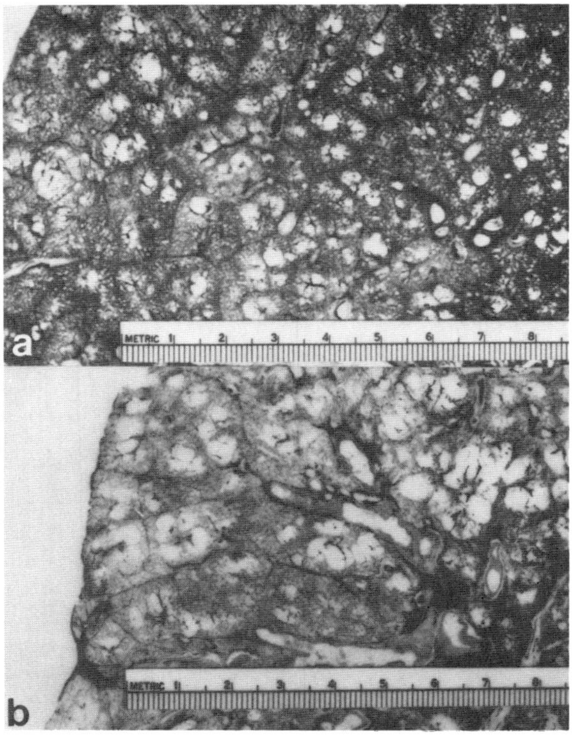

FIGURE 4. (*a*) A close-up of Fig. 3 shows the centrilobular emphysematous spaces and the normal distal part of the acini which separate the spaces. (*b*) Centrilobular emphysema is now more severe and extends almost to the lobular septa. Some observers would dispute that this is centrilobular emphysema, but the similarity between (*a*) and (*b*) seems obvious. Reproduced courtesy of the editor, *Hum. Pathol.* (Thurlbeck et al., 1970).

antitrypsin deficiency (Eriksson, 1965), most commonly with the PiZ phenotype. It is assumed, but unproven, that other forms of Pi-associated emphysema (PiSZ, for example) are also panacinar in type. Familial emphysema unassociated with Pi insufficiently has been recorded as being panacinar in type (Martelli et al., 1974). The emphysema is characteristically worse in the lower zones of the lungs. Since cigarette smoking greatly enhances the susceptibility to develop emphysema in Pi deficiency, it is not uncommon to find some centrilobular emphysema in the upper zones of the lung, but this is usually mild and the remaining lung is extensively involved by severe panacinar emphysema. Severe, pure panacinar emphysema is uncommon, and this finding should suggest that the emphysema is familial in type.

Localized mild panacinar emphysema is found fairly frequently at autopsy. It is found more commonly in older subjects and is not

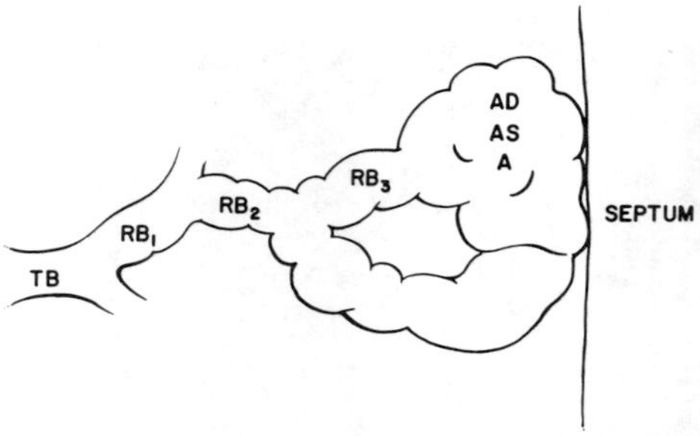

FIGURE 5. In panacinar (panlobular) emphysema, the acinus is more or less uniformly involved. Reproduced courtesy W. B. Saunders & Co. (Thurlbeck, 1976).

FIGURE 6. A paper-mounted whole lung section shows panacinar emphysema which is worse anteriorly and inferiorly. A close-up view shows uniform destruction of acini and lobules. Compare with Figs. 3 and 4. Reproduced courtesy W. B. Saunders & Co. (Thurlbeck, 1976).

associated with clinical evidence of chronic airflow obstruction in these subjects. Under these circumstances, it is more frequent in the lower and anterior parts of the lung. The lung undergoes a characteristic pattern of change with increasing age (see below), and it may be appropriate to think of panacinar emphysema associated with aging as being a focal exaggeration of the aging process.

Bronchial and bronchiolar obliteration may be associated with panacinar emphysema. Occasionally this is a localized, incidental finding, but more commonly it is associated with the Swyer–James (Swyer and James, 1953) or McLeod's (1954) syndrome of unilateral pulmonary transradiancy. In this condition, one lung or major portions of a lung are unduly transradiant and the involved regions of the lung characteristically trap air on expiration so that the mediastinum moves to the unaffected side during this manuever. The syndrome is usually due to severe acute bronchitis and bronchiolitis in childhood leading to obliteration of airways. Perhaps more frequently than producing this syndrome, airway obliteration results in collapse of lung parenchyma and more proximal airway dilatation—postinfective bronchiectasis. In unilateral pulmonary transradiancy, the pattern of airway obliteration is such that parenchymal inflation is maintained, presumably by collateral ventilation via unobliterated airways. The proximal airways may be dilated, and some have referred to this syndrome as "bronchiectasis without atelectasis." It is important to recognize that the air trapping is due to associated airway disease and that emphysema is presumably a consequence of the air trapping. A detailed study of the lung parenchyma in cases of unilateral pulmonary transradiancy is not available, but it seems likely that emphysema may not be present in the affected lung tissue. However, when there is emphysema it is panacinar in type.

Panacinar emphysema may occur in the lower zones of the lung in patients with upper zonal centrilobular emphysema. These patients have all the association of centrilobular emphysema—cigarette smoking, chronic bronchitis, and chronic airflow obstruction. The combination of the two forms of emphysema is probably the most classical finding in patients with chronic airflow obstruction, and it is also one source of much of the controversy mentioned above. Transitions, real or imagined, may be seen in the morphology between the upper zonal centrilobular emphysematous spaces and lower zonal panlobular emphysema. Those that believe that transitions are real maintain that centrilobular emphysema has progressed to panacinar emphysema and these lungs should be classified as examples of centrilobular emphysema. It is clear, however, that the progression has to be assumed, and it is probably preferable to take the pragmatic view that since the emphysema in the lower zones of the lung involves the acinus more or less uniformly, it is panacinar in type. Those who take the former view provide evidence to support the notion that centrilobular and panacinar emphysema are different processes with different clinicopathologic correlations, and those who take the latter view find evidence that the associations are similar. An additional problem resulting from this form of mixed centrilobular and panacinar emphysema is that some may regard one type of emphysema as dominant whereas others may regard the other type as being dominant. Thus, even two observers who believe that they are different types of emphysema may classify the same lung differently.

The distal part of the acinus, alveolar ducts, and sacs may be dominantly involved with emphysema, and there is often a good deal of associated fibrosis when this distal acinar emphysema occurs (Figs. 1, 7, 8). The distal part of the acinus abuts on bronchi, blood vessels, pleura, and lobular septa. The occurrence along the last structures led to the term "paraseptal emphysema" (Heard, 1958) and the occurrence of a rim under the posterior pleura to the term "mantel emphysema" or "superficial emphysema" (McLean, 1956). It is the subpleural location that leads to the characteristic associations of distal acinar emphysema—spontaneous pneumothorax of young adults (Edge et al., 1966). While the classical finding at surgery is the presence of blebs in the pleura (see below), these blebs result from dissection of air into the pleura from paraseptal emphysema caused by alveolar rupture. It has also been claimed that paraseptal emphysema is the type of emphysema associated with the curious syndrome of "Bullous disease of the lung" (Edge et al., 1966; Laurenzi et al., 1962). In this syndrome, usually encountered in subjects in the third and fourth decade, multiple bullae are seen radiologically and the patients show minimal evidence of chronic airflow obstruction until they develop bronchitis. In my own experience, scattered distal acinar emphysema is quite common in patients with symptomatic chronic airflow obstruction with centrilobular and/or panacinar emphysema and occasionally it is the dominant form of emphysema in patients dying with or of chronic airflow obstruction (Thurlbeck, 1976).

In irregular emphysema the acinus is irregularly involved (Figs. 9 and 10). It is for practical purposes always associated with scarring. It may be the commonest form of emphysema, since nearly all lungs on close

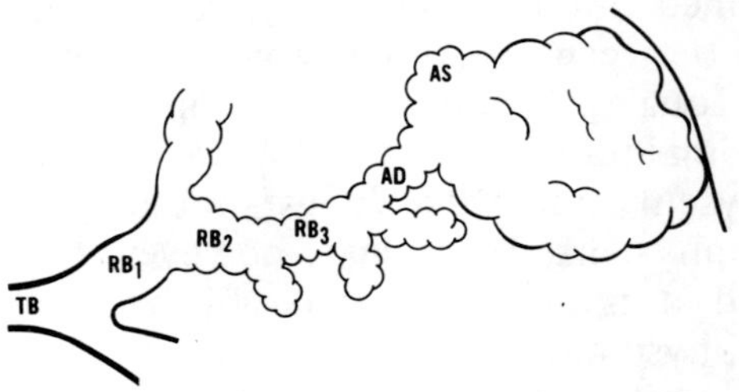

FIGURE 7. In paraseptal (distal acinar) emphysema, alveolar ducts and sacs are dominantly involved. Reproduced courtesy W. B. Saunders & Co. (Thurlbeck, 1976).

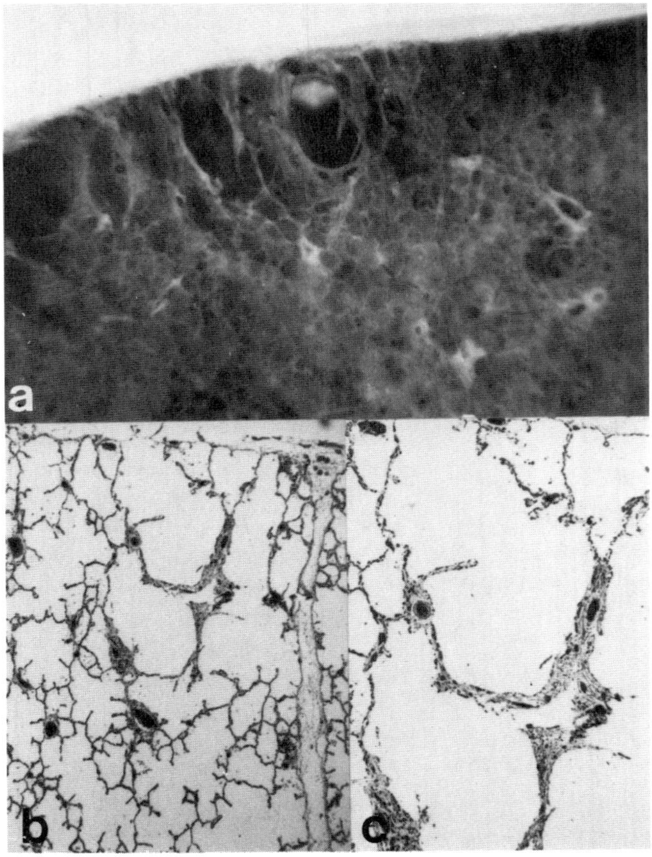

FIGURE 8. (*a*) Distal acinar emphysema is seen just deep to the pleura. (*b*) Photomicrograph of (*a*) shows that the emphysema involves the distal acinus; the pleura is at the top margin of the picture. (*c*) A closer view of (*b*) shows the associated fibrosis. Reproduced courtesy W. B. Saunders & Co. (Thurlbeck, 1976).

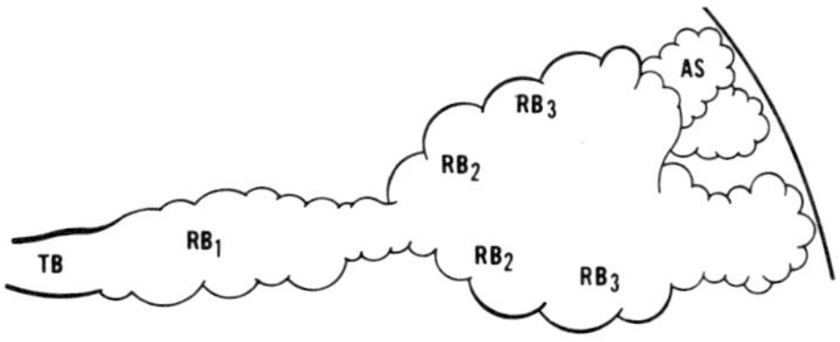

FIGURE 9. In irregular emphysema there is irregular enlargement and destruction of air spaces. Reproduced courtesy W. B. Saunders & Co. (Thurlbeck, 1976).

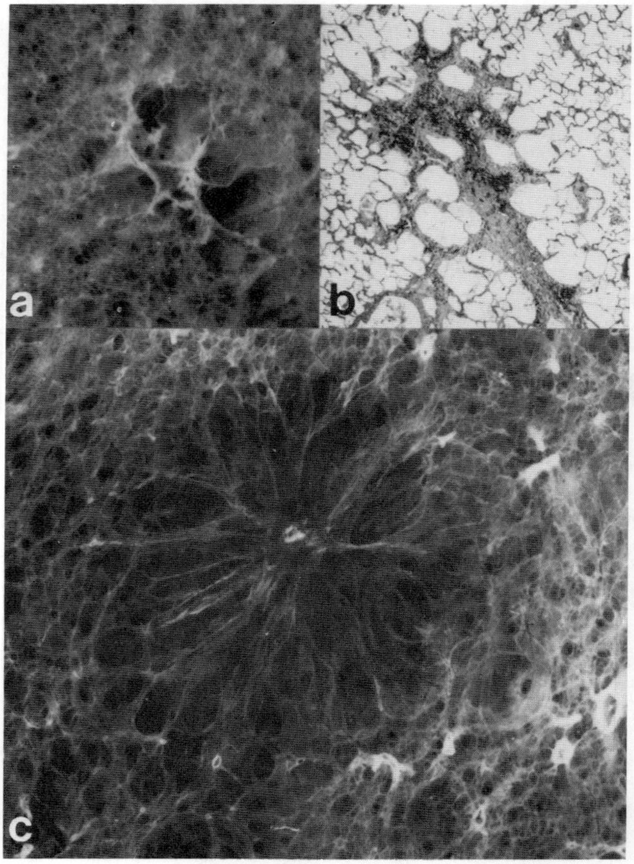

FIGURE 10. (*a*) A scar in the lung has associated emphysema. (*b*) A photomicrograph of the lesion in (*a*) illustrates the scarring better, showing that the acinus is irregularly involved. (*c*) A granuloma in the center of the field has caused enlargement and destruction of adjacent respiratory tissue. Reproduced courtesy W. B. Saunders & Co. (Thurlbeck, 1976).

examination will disclose a scar with associated emphysema. The great majority of examples of irregular emphysema are unassociated with symptoms, but occasionally many scattered granulomas throughout the lung may have associated emphysema producing symptoms. Since there is, almost by definition, extensive fibrosis, increase in lung volume—one of the characteristic features of emphysema—often does not occur, and such patients may have a mixed restrictive and obstructive pattern of pulmonary function.

PROBLEMS WITH CLASSIFICATION

The apparently neat and orderly classification just described and the classical examples of emphysema illustrated in textbooks and monographs (and this contribution) should not obscure the fact that there is no agreement between expert observers on the classification of severely emphysematous lungs (Thurlbeck et al., 1968). Once emphysema gets severe, it becomes atypical and usually more than one form of emphysema is present. The situation is analogous to classification of cirrhosis of the liver and the kidney in chronic renal failure. Perhaps a more rational approach would be to speak of "end-stage emphysema." The problem of a "transition" from centrilobular to panacinar emphysema has been touched upon. In severe upper zonal emphysema, frequently most of the lobule is destroyed, but some portions of the peripheral acinus may be preserved along lobular margins and the degree of preservation is variable (Fig. 4). If one defines centrilobular emphysema as being emphysema selectively and *only* involving respiratory bronchioles, one cannot classify lesions which enroach on the peripheral acinus as centrilobular emphysema. This has led to the notion that centrilobular emphysema is not associated with chronic airflow obstruction (Reid, 1967)—since mild emphysema is usually not associated with chronic airflow obstruction (see below), and only mild centrilobular emphysema fits the definition, it follows that centrilobular emphysema, thus defined, will rarely be associated with symptomatic chronic airflow obstruction. A further extension of this argument is that if emphysema does not fit into the definition of centrilobular emphysema,

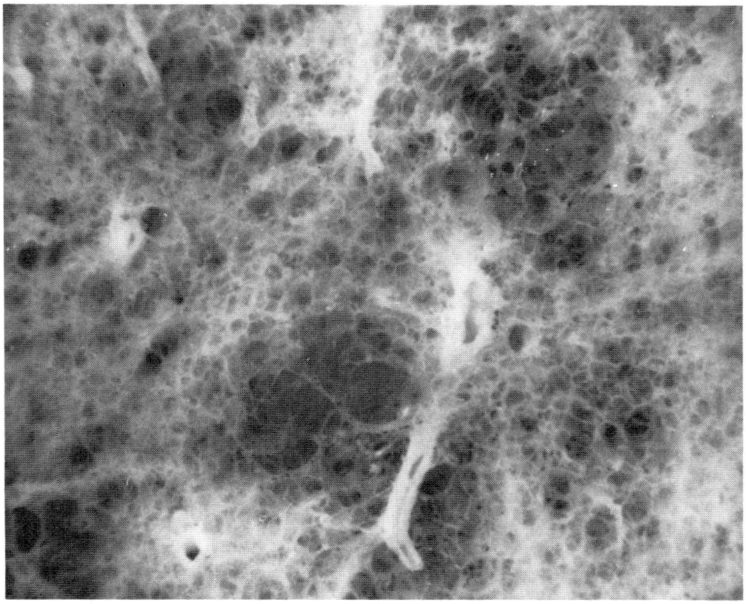

FIGURE 11. Mild centrilobular emphysematous spaces (thick arrows and elsewhere) are separated from each other by apparently normal respiratory tissue (thin arrows). Reproduced courtesy W. B. Saunders & Co. (Thurlbeck, 1976).

distal acinar emphysema, and irregular emphysema, it must be panacinar emphysema. It will then follow that panacinar emphysema is the form of emphysema associated with chronic airflow obstruction. That is to say, as emphysema gets worse it gets more difficult to classify and then falls into this particular usage of the term panacinar emphysema. Since severe emphysema is more likely to produce symptoms, panacinar emphysema will be found to be the type of emphysema usually associated with symptoms when this approach is taken.

CONDITIONS THAT ARE NOT EMPHYSEMA

As indicated above, a bleb is a collection of air in the pleura (Ciba Guest Symposium Report, 1959). An extension of this notion is that a bleb is a collection of air anywhere in the interstitial tissue of the lung. Thus interstitial emphysema can be regarded as multiple pulmonary blebs and is, of course, not true emphysema at all.

With age, the lung undergoes a predictable set of changes (Thurlbeck, 1976). Grossly the lung changes shape (but not volume) as the antero-posterior diameter increases and the cephalocaudad dimension diminishes (Anderson et al., 1964). Lungs may lose weight with age (Howell, 1970), the proportion that alveolar walls form of total lung volume decreases (Thurlbeck and Angus, 1975), alveolar surface area decreases (Thurlbeck, 1967a), and the internal geometry of the lung changes so that the proportion of the lung formed by alveoli decreases and the volume proportion of alveolar ducts (the core of air internal to the mouths of alveoli in alveolar ducts, alveolar sacs and respiratory bronchioles) increases (Thurlbeck and Angus, 1975). The mean linear intercept (average distance between alveolar walls) increases so that there is enlargement of air spaces (Thurlbeck, 1967a; Butler, 1976; Hasleton, 1972; Niewoehner et al., 1975). Since there may also be destruction of respiratory tissue (reduction of volume proportion of alveolar walls, above), this change can be regarded, and is regarded by some, as emphysema. I have been opposed to this notion, preferring to call this the "aging lung" rather than "senile emphysema." The reasons for this point of view are (1) that this appears to be a normal process (rather than being an abnormal one), since it effects nearly all subjects and thus does not fulfill all the criteria of the definition of emphysema, and (2) that there appear to be no serious functional consequences of aging of the lung. The former reason, which is the logical one for not using the term emphysema, may not be well founded. We have recently shown that BALB/cNiaa mice brought up under specific-pathogen-free conditions and in an environment almost free of environmental insults do not show the morphologic pulmonary changes considered so characteristic of age (Kawakami et al., in press). Thus, it may be that the "aging" changes in the human lung represent a response to environment rather than age per se. Nonetheless, the changes remain "normal" in that most people respond to them.

Following pneumonectomy or atelectasis, the remaining lung increases in dimension, as do the individual airspaces, and thus there is abnormal enlargement of airspaces. However, this change is not emphysema since there is no destruction of lung tissue, and this should be referred to as compensatory overinflation. The most dramatic example of overinflation is found in congenital lobar overinflation (Murray, 1967; Lincoln et al., 1971). In this condition, there is rapid overinflation of a lobe or part of a lobe, most commonly the left upper lobe in infants. The overinflation is often sufficiently great to compress the normal lung and may produce severe respiratory distress. The overinflation is most commonly due to absent or diminished amounts of bronchial cartilage. The precise change in these lungs has not been described, but it is likely mostly overdistension of the acinus.

PROBLEMS WITH DEFINITION

The application of the definition of emphysema as described above will have profound effects on the epidemiology of emphysema. The three essential components of the definition are (1) abnormal, (2) enlargement of airspaces, and (3) accompanied by destruction of respiratory tissue. Item 1 has been addressed, above. Item 2 requires a precise knowledge of variations in airspace size in "normal" lungs, and item 3 requires a precise description of destruction. The information concerning item 2 is not available, and item 3 has not been agreed upon. We can avoid the problem about dimensions of airspaces if we assume that the presence of destruction of respiratory tissue is more important and precedes or accompanies airspace enlargement. It has been suggested that the best definition of destruction is the presence of fenestrae or holes in the alveolar walls greater than 20 μm in longest diameter (Boren, 1962; Pump, 1976). (Holes smaller than this are thought to be normal structures, the pores of Kohn.) This concept has never been adequately tested, so that a more empirical approach is usually taken. Mild emphysema is characteristically patchy and can be recognized by contrasting it to apparently normal lung tissue that is adjacent to it (Fig. 11). The apparently normal lung has an orderly structure and may or may not have increased airspace dimensions and/or number of holes in the alveolar walls. This definition—obvious destruction in contrast to the apparently normal adjacent lung—can be applied reasonably easily, and when this is done, then almost all examples of mild emphysema, say the least affected 50% of emphysematous lungs, will have no symptoms attributable to emphysema. Put another way, by using a much cruder criterion of emphysema, most emphysema will not be associated with symptoms. The frequency of asymptomatic emphysema might be increased if a more subtle definition of emphysema is used, but this would have little clinical impact. A much more difficult decision exists when there is no apparently normal tissue with which to contrast the lesion. Figure 12 illustrates a spectrum of changes. Nearly all observers would regard Fig. 12*a* as normal and Fig. 12*f* as an example of panacinar emphysema. The point of transition between normal and abnormal in this series of pictures is debatable. I regard *b* as a normal aged lung, but *c* to *f* as abnormal. This distinction is not made on the basis of abnormal fenestrae (above), but because the normal orderly organization of lung into alveoli and ducts is not apparent from *c* onwards.

THE EPIDEMIOLOGY OF EMPHYSEMA

No true epidemiologist would ever be satisfied by a description of the epidemiology of emphysema. Since emphysema is defined anatomically, epidemiologic studies are almost exclusively confined to autopsy studies, and these are always cross-sectional rather than longitudinal in character. There is almost no such thing as an unbiased sample of autopsy cases, although in certain circumstances the autopsy rate may approach 100% (Thurlbeck et al., 1974). However, even then selective admission to certain hospitals is inevitable. The current low autopsy rate and studies limited to teaching hospitals must produce dubious data. The threshold of recognition of emphysema by observers is highly variable (Fig. 12) and is likely to affect apparent prevalence rates almost as much as true differences between populations. I have reviewed the 20 studies of the prevalence of emphysema and it is apparent that widespread differences in prevalence exist (Thurlbeck, 1976). Within these studies, it is also apparent that emphysema is more frequent and severe in smokers than nonsmokers, and it is frequent in all industrial countries, but relatively infrequent in nonindustrial countries. It is more common in coalminers than in noncoalminers (Cockcroft et al., 1982). Another approach has been to consider the effect of age, sex, smoking, and geographical location on average *severity* of emphysema (Thurlbeck et al., 1974). This has the advantage that while absolute assessments of severity in subjectively judged systems vary considerably from observer to observer, rank order correlations are quite good between observers (Thurlbeck et al., 1968). Some examples of average severity are shown in Figs. 13–16. Emphysema is commonly found at autopsy with an average score rising to about 20 in males in the seventh decade and to about 6 in females (Fig. 13). Smoking exerts a major effect, with heavy smokers having an average emphysema scoring about seven times higher than non-smokers. In subjects matched for smoking or nonsmoking, males have increased severity of emphysema (Figs. 15 and 16). With increasing age the average severity of emphysema increases, and this is true of both smokers and nonsmokers. In both sexes,

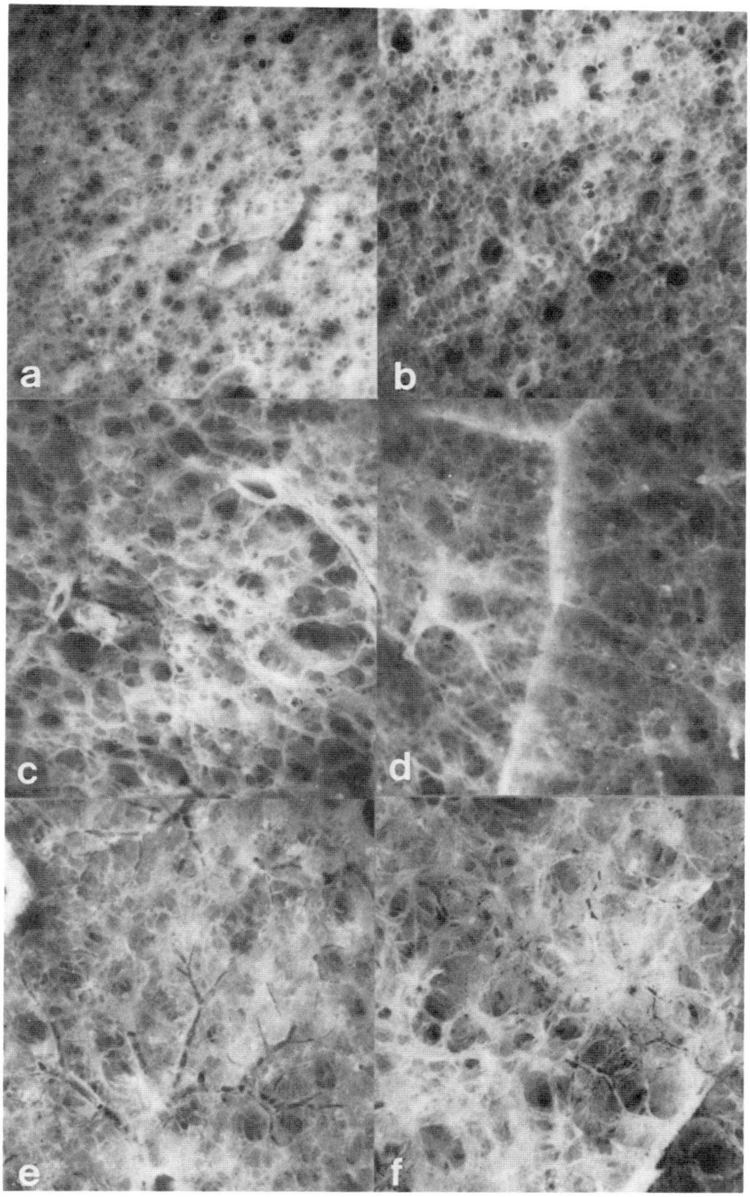

FIGURE 12. This panel shows an apparent progression of changes from (*a*) "normal" lung parenchyma to (*f*) obvious panacinar emphysema. The borderline of change is arbitrary, and I feel that (*b*) is a normal lung from an old subject, whereas (*c–f*) are panacinar emphysema (see text). Reproduced courtesy W. B. Saunders & Co. (Thurlbeck, 1976).

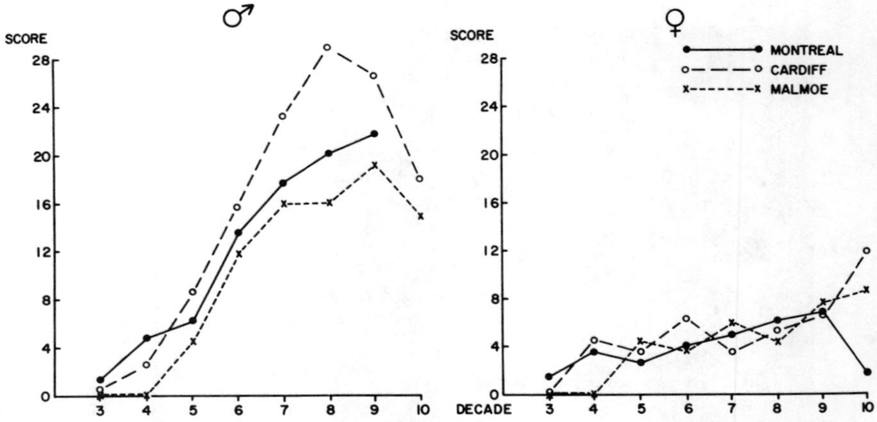

	Male 3	4	5	6	7	8	9	10		Female 3	4	5	6	7	8	9	10
	27	31	47	44	50	26	7	-	MONTREAL	14	12	26	26	28	22	6	3
	24	19	59	120	161	127	49	6	CARDIFF	11	13	33	45	92	103	50	8
	2	1	13	19	67	79	31	9	MALMOE	1	1	4	16	30	73	86	21

FIGURE 13. The average emphysema score is shown by decade for three geographic locations in males and females separately numbered. The average emphysema score using the panel grading method (33) is about fives times higher in males beyond the fourth decade. The average severity is higher in Cardiff males than the other centers. For comparison, the lung shown in Fig. 6 would be scored at about 40. The number of cases in each decade is shown at the bottom of the figure. Reproduced courtesy of the editor, *Am. Rev. Respir. Dis.* (Thurlbeck et al., 1974).

the average frequency declines in the oldest age groups, suggesting the earlier demise of subjects with severe emphysema. There may be some geographical differences, although these are dwarfed by the other influences.

PATHOBIOLOGY OF EMPHYSEMA

I am not sure what the pretentious term pathobiology means. I will assume that it includes the pathogenesis of emphysema and the effects of emphysema on lung function. As indicated, emphysema is causally associated with smoking. It is now generally thought that this is due to an imbalance between protease-antiprotease sysstems in the lung. Tobacco smoke leads to inflammation in the lung as manifested by respiratory bronchiolitis (Niewoehner et al., 1974) and increased numbers of macrophages and polymorphonuclear leucocytes in the lung. The macrophages are metabolically more active in smokers (Harris et al., 1970), and both macrophages and polymorphonuclear leucocytes contain elastolytic proteases, notably elastase. Tobacco smoke also has other effects—it decreases the antiproteolytic effect of $alpha_1$ antitrypsin (Janoff et al., 1979) and also diminishes the cross-linking of elastin (Laurent et al., 1983). Thus elastin synthesis is decreased, its destruction increased, and its protective mechanism impaired in smokers.

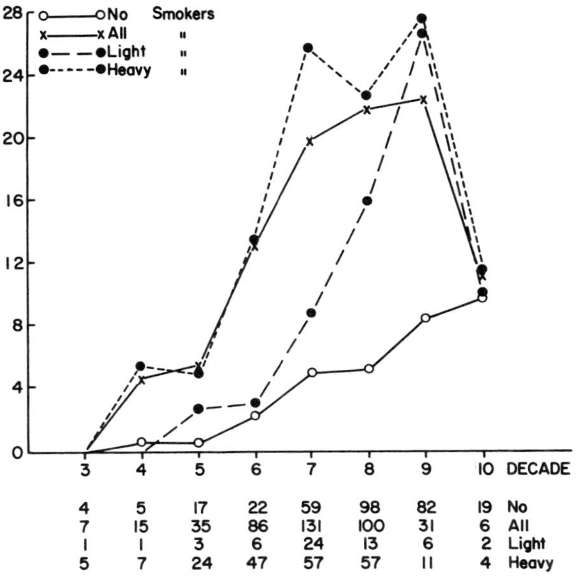

FIGURE 14. Various smoking categories from the cases shown in Fig. 13 are compared. Heavy smoking (20 or more cigarettes/d) increases the average severity of emphysema about sevenfold compared to nonsmokers. There is a rapid increase in the average severity of emphysema after the sixth decade in light smokers (less than 20 cigarettes/d). Reproduced courtesy of the editor, *Am. Rev. Respir. Dis.* (Thurlbeck et al., 1974).

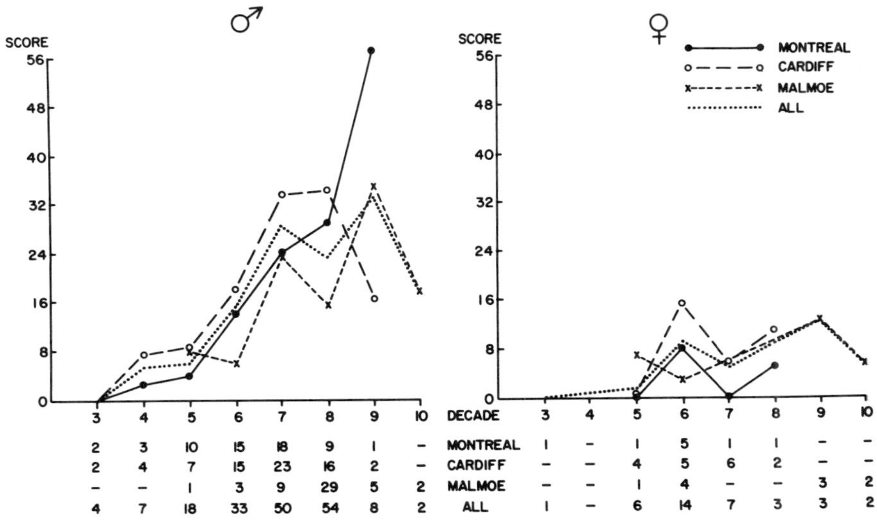

FIGURE 15. Female and male heavy smokers (20 or more cigarettes/d) are compared. "All" is the average of three geographic locations. The numbers are small, but the average severity of emphysema is about three times as great in males. Reproduced courtesy of the editor, *Am. Rev. Respir. Dis.* (Thurlbeck et al., 1974).

As emphysema gets worse, clinical evidence of airflow obstruction increases in frequency (Fig. 17). There are also numerous studies that show that expiratory flow rates, lung volumes, diffusing capacity, and lung elasticity all get more abnormal with increasing amounts of emphysema (Thurlbeck, 1976). However, the correlations are not very close and many exceptions are apparent. For example, about 40% of patients with the severest grade of emphysema were not clinically recognized as having emphysema by retrospective analysis of the clinical records (Fig. 17). Pulmonary function tests are more sensitive, but even then many exceptions are present with pulmonary function tests being relatively normal in the presence of moderate emphysema. It is appropriate to consider how emphysema may cause abnormalities of pulmonary function. The most significant alteration is flow limitation as assessed, for example, by the volume of air expired in the first second during a forced expiratory maneuver (FEV_1). This test is generally regarded as the most useful assessment of flow limitation. In theory, flow can be limited in only two ways—by decrease of the force applied to the lungs, which is elastic recoil of the lung, or by narrowing of the airways. In emphysema, elastic recoil is lost and thus will be associated with reduced expiratory flow. The relationship between recoil and diminished flow is not very strong, suggesting that other mechanisms are involved. It has been documented that flow is diminished more in emphysema than predicted by loss of recoil (Berend et al., 1979; Berend and Thurlbeck, 1982). This in turn suggests that there is associated airway disease. While it is relatively easy

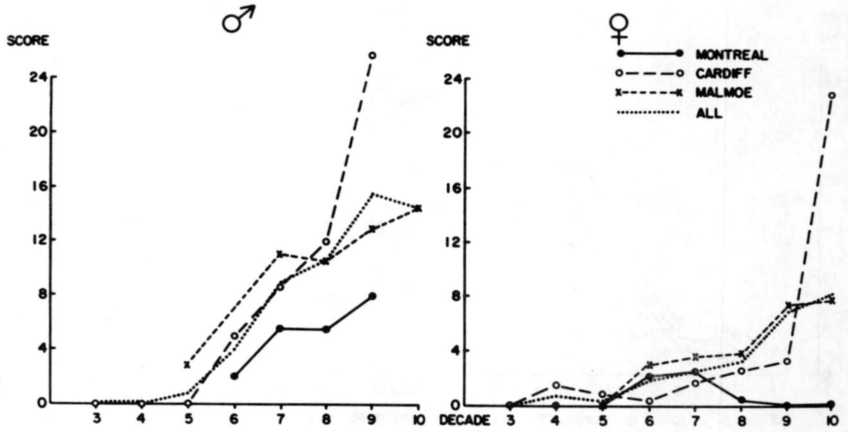

	♂ 3	4	5	6	7	8	9	10	♀ 3	4	5	6	7	8	9	10
MONTREAL	-	-	-	1	5	4	1	-	2	2	6	6	9	4	2	1
CARDIFF	1	1	3	2	8	7	3	-	1	2	6	6	16	20	6	1
MALMOE	-	-	1	-	9	14	9	4	-	-	1	7	12	49	61	13
ALL	1	1	4	3	22	25	13	4	3	4	13	19	37	73	69	15

FIGURE 16. A significant average emphysema score is seen in nonsmokers and increases with age. The average emphysema score is about three times greater in males than females. Reproduced courtesy of the editor, *Am. Rev. Respir. Dis.* (Thurlbeck et al., 1974).

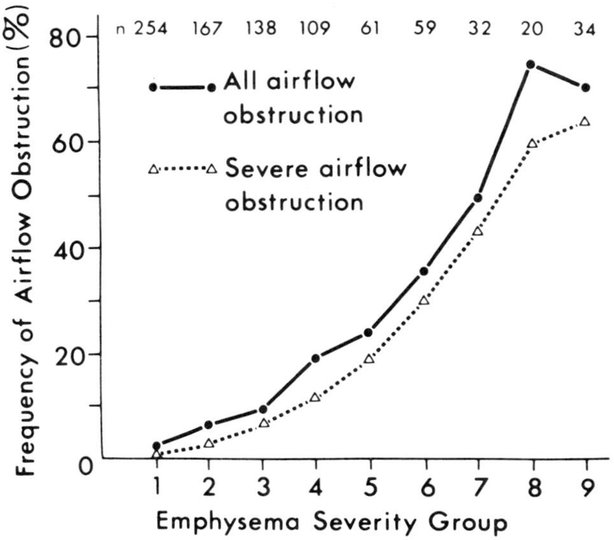

FIGURE 17. The frequency of airflow obstruction as determined by retrospective analysis of the patients' charts in groups showing increasing severity of emphysema is illustrated. Emphysema severity group 1 are patients without emphysema; emphysema severity group 9 are patients with severe emphysema. The lung illustrated in Fig. 6 would be an emphysema severity group 6. For details, see Thurlbeck (1977). Reproduced courtesy of the editor, *Chest* (Thurlbeck, 1977).

to show airway abnormalities in emphysematous lungs, the relationship between emphysema severity and airway lesions is not close. Perhaps the best explanation lies in the tortuous and irregularly narrowed airways that are seen so characteristically in emphysema (Linhartova et al., 1977). It has long been suggested that irregularly supported airways may collapse in expiration, compounding the effect of loss of recoil (Anderson and Foraker, 1962). The most characteristic functional finding in emphysema is loss of diffusing capacity, or transfer of gas from alveoli to blood. This is probably brought about by loss of capillary bed and impaired gas mixing brought about by the varying time constants in the lung and in series (stratified) inhomogeneity (Thurlbeck, 1984).

Finally, as has been pointed out, the relationship between emphysema and loss of elastic recoil is not very close. This may be because loss of recoil in emphysematous lungs occurs in the intervening parts of relatively normal looking lung between the obvious emphysema. The evidence for this has been reviewed in detail elsewhere (Thurlbeck, 1983), and it includes the observation that an increase in lung volume and loss of elastic recoil occur with trivial emphysema when most of the lung appears normal, that compliance of the lung between centrilobular emphysematous spaces is increased whereas the spaces themselves do not have increased compliance, and loss of recoil in a given lung is greater in the lobe less affected by emphysema than in the lobe more affected. If this is true, it suggests very subtle changes in the apparently normal intervening lung parenchyma. These may be at the molecular level and may not be visible even ultrastructurally.

REFERENCES

American Thoracic Society. 1962. Chronic bronchitis, asthma and pulmonary emphysema. A statement by the committee on diagnostic standards for nontuberculous respiratory diseases. *Am. Rev. Respir. Dis.* 85:762–768.

Anderson, A. E., and Foraker, A. G. 1962. Relative dimensions of bronchioles and parenchymal spaces in lungs from normal subjects and emphysematous patients. *Am. J. Med.* 32:218–226.

Anderson, A. E., Jr., and Foraker, A. G. 1973. Centrilobular emphysema and panlobular emphysema: Two different diseases. *Thorax* 28:547–550.

Anderson, W. F., Anderson, A. E., Jr., Hernandez, J. A., and Foraker, A. G. 1964. Topography of aging and emphysematous lungs. *Am. Rev. Respir. Dis.* 90:411–423.

Berend, N., and Thurlbeck, W. M. 1982. Correlations of maximum expiratory flow with small airway dimensions and pathology. *J. Appl. Physiol: Respirat. Environ. Exercise Physiol.* 52:346–335.

Berend, N., Woolcock, A. J., and Marlin, G. E. 1979. Correlation between the function and structure of the lungs in smokers. *Am. Rev. Respir. Dis.* 119:695–705.

Boren, H. G. 1962. Alveolar fenestrae. Relationship to the pathology and pathogenesis of pulmonary emphysema. *Am. Rev. Respir. Dis.* 85:328–344.

Butler, C. 1976. Lung surface area in various morphologic forms of human emphysema. *Am. Rev. Respir. Dis.* 114:347–352.

Ciba Guest Symposium Report. 1959. Terminology, definitions and classification of chronic pulmonary emphysema and related conditions. *Thorax* 14:286–299.

Cockcroft, A., Wagner, J. C., Ryder, Seal, R. M. E., Lyons, J. P., and Anderson, N. 1982. Post-mortem study of emphysema in coalworkers and non-coalworkers. *Lancet* 2:600–603.

Edge, J., Simon, G., and Reid, L. 1966. Peri-acinar (paraseptal) emphysema: Its clinical, radiological and physiological features. *Br. J. Dis. Chest* 60:10–18.

Eriksson, S. 1965. Studies in alpha-1-antitrypsin deficiency. *Acta Med. Scand.* 177(Suppl.):1–85.

Harris, J. O., Swenson, E. W., and Johnson, J. E. III. 1970. Human alveolar macrophages: Comparison of phagocytic ability, glucose utilization and ultrastructure in smokers and nonsmokers. *J. Clin. Invest.* 49:2086–2096.

Hasleton, P. S. 1972. The internal surface area of the adult human lung. *J. Anat.* 112:391–400.

Heard, B. E. 1958. Macroscopic anatomy of pulmonary emphysema. *Aust. Ann. Med.* 5:73–88.

Howell, T. H. 1970. *A Student's Guide to Geriatrics.* London: Staples Press.

Janoff, A., Carp, H., and Lee, D. K. 1979. Cigarette smoke inhalation decreases alpha-1-antitrypsin activity in rat lung. *Science* 206:1313–1314.

Kawakami, M., Paul, J. L., and Thurlbeck, W. M. 1984. The effect of age on lung structure in male BALB/cNNia inbred mice. *Am. J. Anat.* in press.

Laurent, P., Janoff, A., and Kagan, H. N. 1983. Cigarette smoke blocks cross-linking of elastin *in vitro. Am. Rev. Respir. Dis.* 127:189–192.

Laurenzi, G. A., Turina, G. M., and Fishman, A. P. 1962. Bullous disease of the lung. *Am. J. Med.* 32:361–378.

Lincoln, J. C. R., Stark, J., Subramainian, S., Aberdeen, E., Bonham-Carter, R. E., Berry, C. L., and Waterston, D. J. 1971. Congenital lobar emphysema. *Ann. Surg.* 173:55–62.

Linhartova, A., Anderson, A. E. Jr., and Foraker, A. G. 1977. Further observations on luminal deformity and stenosis of nonrespiratory bronchioles in pulmonary emphysema. *Thorax* 32:53–59.

MacLeod, W. M. 1954. Abnormal transradiancy of one lung. *Thorax* 9:147–153.

Martelli, N. A., Goldman, E., and Roncoroni, A. J. 1974. Lower zone emphysema in young patients without α_1-antitripsin deficiency. *Thorax* 29:237–244.

McLean, K. H. 1956. Macroscopic anatomy of pulmonary emphysema. *Aust. Ann. Med.* 5:73–88.

Mitchell, R. S., Silvers, G. W., Goodman, N., Dart, D., and Maisel, J. C. 1970. Are centrilobular emphysema and panlobular emphysema two different diseases? *Hum. Pathol.* 1:433–441.

Morgan, W. K. C., and Seaton, A. 1975. *Occupational Lung Disease.* Philadelphia: W. B. Saunders.

Murray, G. F. 1967. Congenital lobar emphysema. *Surg. Gynecol. Obstet.* 124:611–625.

Niewoehner, D. E., Kleinerman, J., and Rice, D. B. 1974. Pathologic changes in the peripheral airways of young cigarette smokers. *N. Engl. J. Med.* 291:755–758.

Niewoehner, D. E., Kleinerman, J., and Liotta, L. 1975. Elastic behaviour of postmortem human lungs: Effects of aging in mild emphysema. *J. Appl. Physiol.* 39(6):943–949.

Pump, K. K. 1976. Emphysema and its relation to age. *Am. Rev. Respir. Dis.* 114:5–13.

Reid, L. 1967. *The Pathology of Emphysema.* Chicago: Year Book Medical Publishers.

Swyer, P. R., and James, G. C. W. 1952. A case of unilateral pulmonary emphysema. *Thorax* 8:133–136.

Thurlbeck, W. M. 1963. The incidence of pulmonary emphysema. With observations on the relative incidence and spatial distribution of various types of emphysema. *Am. Rev. Respir. Dis.* 87:206–215.

Thurlbeck, W. M. 1967a. The internal surface area of non-emphysematous lungs. *Am. Rev. Respir. Dis.* 95:765–773.

Thurlbeck, W. M. 1967b. Internal surface area and other measurements in emphysema. *Thorax* 22:483–496.

Thurlbeck, W. M. 1976. *Chronic Airflow Obstruction in Lung Disease,* vol. 5, pp. 1–456. Philadelphia: W. B. Saunders.

Thurlbeck, W. M. 1977. Aspects of chronic airflow obstruction. *Chest* 72:341–349.

Thurlbeck, W. M. 1983. Overview of the pathology of pulmonary emphysema in the human. *Clin. Chest Med.* 4:337–350.

Thurlbeck, W. M. 1984. Morphology and clinico-morphologic correlations: State of the art. *Chest* in press.

Thurlbeck, W. M., and Angus, G. E. 1975. Growth and aging of the normal lung. *Chest* 67(Suppl.):3s–7s.

Thurlbeck, W. M., Anderson, A. E., Janis, M., Mitchell, R. S., Pratt, P., Restrapo, G., Ryan, S. F., and Vincent, T. 1968. A co-operative study of certain measurements of emphysema. *Am. Rev. Respir. Dis.* 98:217–228.

Thurlbeck, W. M., Dunnill, M. S., Hartung, W., Heard, B. E., Heppleston, A. G., and Ryder, R. C. 1970. A comparison of three methods of measuring emphysema. *Hum. Pathol.* 1:215–226.

Thurlbeck, W. M., Ryder, R. C., and Sternby, N. 1974. A comparative study of the severity of emphysema in necropsy populations in three different countries. *Am. Rev. Respir. Dis.* 109:239–248.

World Health Organizaton Report. 1961. Chronic cor pulmonale: Report of an expert committee. *WHO Tech. Rep. Ser.* 213:15.

RESPIRATORY FUNCTION RESPONSES OF ANIMALS AND MAN TO OXIDANT GASES AND TO PULMONARY EMPHYSEMA

Joe L. Mauderly

Lovelace Inhalation Toxicology Research Institute,
Lovelace Biomedical and Environmental Research
Institute, Albuquerque, New Mexico

Data on the respiratory functional responses of animals and humans to inhaled oxidant gases and to pulmonary emphysema were reviewed and compared. Comparisons included responses to short-term inhalation of ozone, nitrogen dioxide, and oxygen and the functional manifestations of chronic emphysema. The comparisons illustrated that animals and humans have qualitatively similar functional responses to the irritant, bronchoconstrictive, and sensitizing effects of acutely inhaled ozone and nitrogen dioxide. Animals and humans responded similarly to the inflammatory and edematous effects of inhaled oxygen. Similar changes in maximal expiratory flow–volume curves, pressure–volume curves, lung volumes, and alveolar–capillary gas exchange occurred in animals and humans with emphysema. These results suggest that similar respiratory functional changes occur in both animals and humans when similar morphological changes result from lung injury. This observation lends confidence to the use of laboratory animals in studies to predict the effects of long-term exposure of humans to inhaled oxidant gases.

INTRODUCTION

There continues to be considerable concern for the adverse health effects of human inhalation exposure to oxidant gases, particularly for ozone (O_3), formed in polluted atmospheres by photochemical reactions, and nitrogen dioxide (NO_2), formed by oxidation of atmospheric nitrogen during combustion. Air quality controls have resulted in lower ambient concentrations of these gases, but the effects of long-term exposures of humans to existing low levels remain unknown. Epidemiological data relating air pollution to reduced respiratory function and increased symptoms have been collected, but the dose-effect relationships for specific pollutants remain uncertain. Toxicological studies using animals have demonstrated relationships between oxidant gases and lung disease, but the exposure concentrations have typically been much higher than the environmental levels encountered by humans.

Review performed under U.S. Department of Energy contract DE-AC04-76EV01013.

The author thanks Dr. W. I. Christensen of the Lovelace Medical Center and Suzanne Likens of this institute for previously unpublished data used in this report. The author also thanks the members of the Inhalation Toxicology Research Institute staff who critically reviewed this report and those who provided word processing and illustrative services.

Measurements of respiratory function have been a primary tool in epidemiologic studies of the effects of air pollutants on human health. The utility of these assays lies in their ability to provide, by rapid and noninvasive means, quantitative data that relate to underlying structural changes in the lung. There is a strong association between measured functional abnormalities and subjective functional impairment in humans, and the tests are useful for determining the extent of disability in relation to the requirements of an individual's occupation (Morgan, 1982).

The development of methods for measuring the respiratory function of animals has roughly paralleled progress in human lung function testing (Likens and Mauderly, 1979). Considerable development has occurred since the mid-1970s in respiratory function testing of small laboratory animals, and essentially all tests applied to humans can now be applied in toxicological studies. Functional manifestations of various lung injuries have been demonstrated to be similar in animals and man (Mauderly et al., 1983b). It is expected that respiratory function will be an important endpoint for extrapolating information from future animal studies to assess the risk to humans from inhaled oxidant gases.

The purpose of the present report is to summarize the existing information allowing direct comparisons of the respiratory functional effects of inhaled oxidant gases in animals and humans. The goal underlying the comparisons is to evaluate the confidence with which we can predict functional effects in humans from those observed in animals. There are no data specific to long-term exposure of humans to oxidant gases; thus, direct interspecies comparisons will only be made for acute exposures. However, much of the present concern for chronic, oxidant-induced lung disease is for the induction of pulmonary emphysema. Pulmonary emphysema is a common clinical finding in man and part of the chronic obstructive lung disease complex thought to result, at least in part, from long-term exposure to airborne environmental pollutants. In addition, emphysema has been observed in lungs of animals exposed chronically to high concentrations of inhaled oxidant gases. It is expected that a major goal of future animal studies will be to determine whether chronic inhalation of low levels of oxidant gases causes emphysema. Therefore, comparisons of the known functional manifestations of pulmonary emphysema in animals and humans are also included in this report.

The comparisons made in this report demonstrate that qualitatively similar functional changes occur in animals and humans in response to acute oxidant gas exposures and to uncomplicated pulmonary emphysema. In many cases, the responses are also quantitatively similar. It is concluded that measurements of respiratory function of animals in toxicological studies can give information useful in assessing human health risk from environmental oxidant gas exposures.

METHODS FOR COMPARISONS

This report does not review all available data on the functional effects of oxidant gas inhalation or emphysema, but is intended as a summary of data allowing direct comparisons between animals and humans. Published reports of similar physiological measurements in animals and humans exposed to known concentrations of oxidant gases were reviewed, and only those presenting actual data in figure or tabular form are included. For the emphysema comparisons, certain recent, previously unpublished data from this institute are also included. Data were extracted graphically from published figures or calculated from published tables and replotted using similar relative scales to facilitate interspecies comparisons. Most data are expressed either as changes from preexposure states or as differences from predicted normal or measured control values. The exposures or the measured functional parameters often differed, but similar physiological events (e.g., airway constriction) were demonstrated by the comparisons. Several of the studies of humans included exercise during exposure, while the animals were all exposed at rest. The human subjects were all conscious during the physiological measurements, while the animals in several studies were anesthetized for testing. Because of these constraints, the comparisons can be considered as only semiquantitative.

The comparisons are grouped according to the following categories: acute responses to O_3; acute responses to NO_2; acute responses to oxygen; and functional manifestations of pulmonary emphysema.

RESULTS AND DISCUSSION

Acute Responses to Ozone

Acute inhalation exposure to O_3 alters the breathing patterns of both animals and humans. Changes in the respiratory frequencies and tidal volumes of humans and guinea pigs during and after a 2-h exposure are shown in Fig. 1. Bates and Hazucha (1973) exposed 10 healthy, nonsmoking humans to 0.75 parts per million (ppm) O_3 for 2 h, with the subjects alternately resting or exercising lightly during 15-min intervals. Respiratory function was measured at 0.5-h intervals during and for 4 h after the exposure. Murphy et al. (1964) exposed 10 conscious, young adult, male guinea pigs to 0.68 ppm O_3 via a nose cone for 2 h and measured respiration during and for 3.5 h after exposure.

In both guinea pigs and humans, the respiratory frequency increased (Fig. 1*a*) and the tidal volume decreased (Fig. 1*b*) during exposure to O_3 and returned toward the original values during the first 3 h after exposure. The similarities between the curves of the two species are striking. The magnitudes of the maximum relative changes in tidal volume were nearly identical

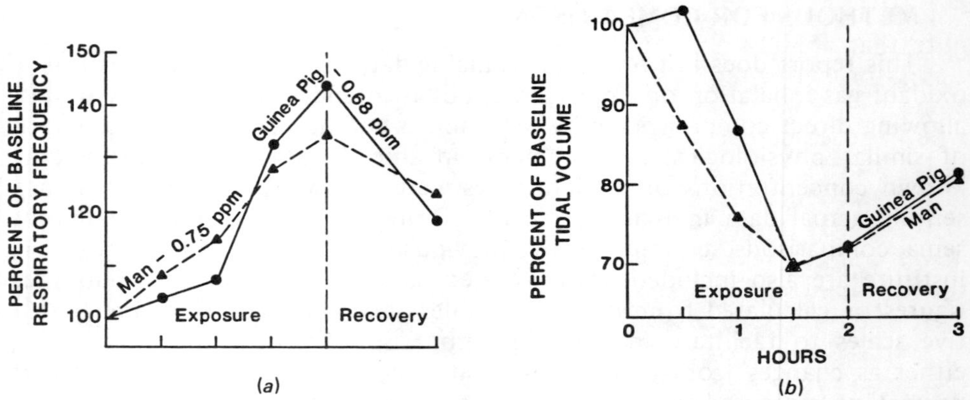

FIGURE 1. The effect of 2-h exposures to O_3 on (*a*) respiratory frequency and (*b*) tidal volume of man (dashed line) (Bates and Hazucha, 1973) and guinea pigs (solid line) (Murphy et al., 1964). Respiratory frequency was increased and tidal volume decreased in both species during exposure, and both were returning toward preexposure values by 3 h after exposure.

and both occurred at 1.5 h of exposure. This comparison demonstrates that similar concentrations of O_3 (approximately 0.7 ppm) inhaled for the same length of time cause nearly identical changes in the breathing patterns of guinea pigs and humans.

Acute exposure to O_3 has been demonstrated to cause airway constriction and airflow limitation in both animals and humans. The relative magnitudes of this effect during 2 h exposures of humans, guinea pigs, and cats are compared in Fig. 2. Bedi et al. (1979) exposed 9 normal, young adult males to 0.4 ppm O_3. Bates and Hazucha (1973) exposed the 10 subjects used in the preceding comparison to 0.75 ppm O_3. In both studies, the subjects alternately rested and exercised during exposure, and in both studies, airflow limitation was assessed by measuring the mean mid-expiratory flowrate during forced expiration. Amdur et al. (1978) exposed 10 conscious, young adult guinea pigs to 0.4 ppm O_3 and measured dynamic lung compliance and total pulmonary resistance. Dynamic compliance was used in the present comparison because it is a more sensitive indicator of lower airway constriction in guinea pigs than total pulmonary resistance (Silbaugh et al., 1981a). Watanabe et al. (1973) exposed 5 anesthetized, paralyzed, and mechanically ventilated cats to 0.5 ppm O_3 and measured total pulmonary resistance. Conductance, the reciprocal of resistance, was plotted in the present comparison to obtain the same directional change as occurred in the other species.

Figure 2 illustrates the progressive decreases in the measured values of all species during exposure, reflecting the similar occurrence of airflow limitation resulting from airway constriction. Although the two curves for humans suggest a dose-response relationship, the methodological variables involved preclude quantitative interspecies comparisons.

Although no published data are available for animals, it is interesting to note that animals and humans may develop similar tolerances to the acute respiratory effects of O_3. The phenomenon of tolerance in humans has become well known and is illustrated by the study of Horvath et al. (1981), who exposed subjects repeatedly to 0.5 ppm O_3 and measured airflow limitation during forced exhalation. After 2–5 daily acute exposures, the subjects had reduced responses, and this tolerance lasted for 7–20 d. Dungworth et al. (1975) observed that rhesus monkeys exposed for 8 h/d for 7 d to 0.5 or 0.8 ppm O_3 exhibited the maximum increase in respiratory frequency on the first exposure day and diminished frequencies thereafter until the fourth day, when the frequencies were normal.

Inhaled O_3 has also been shown to temporarily increase the reactivities of the airways of humans and animals to aerosolized bronchoactive agents, causing a greater change in total pulmonary resistance during challenge (Fig. 3). Golden et al. (1978) challenged 8 normal young adults with histamine before and after exposure to 0.6 ppm O_3 for 2 h. The subjects were seated and wore nose clips during O_3 exposure. Abraham et al. (1980a) challenged 8 conscious, female sheep with carbachol before and after exposure to 0.5 ppm O_3 for 2 h via a nasotracheal catheter. Lee et al. (1977) reported data from the challenge of 1 anesthetized, mechanically ventilated mongrel dog with histamine before and after exposure to 0.7 ppm O_3 for 2 h via an orotracheal catheter.

Figure 3 illustrates that O_3 exposure increased the responses to bronchoconstrictors of all three species; however, the time courses of the responses differed. The airway reactivity of man was increased immediately after O_3 exposure and had returned toward the preexposure level by 24 h after exposure. The airway reactivity of the sheep was slightly increased immediately after O_3 exposure, but was increased even more at 24 h. The response was

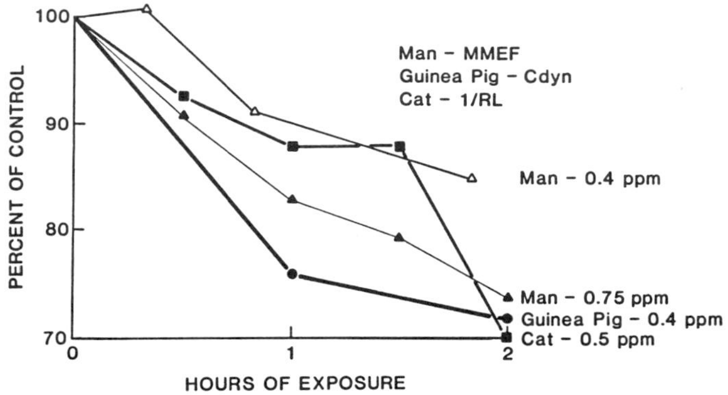

FIGURE 2. The effect of 2-h exposures to O_3 on airway constriction in man [0.4 ppm = Bedi et al. (1979); 0.75 ppm = Bates and Hazucha (1973)], guinea pigs (Amdur et al., 1978), and cats (Watanabe et al., 1973). MMEF = mean midexpiratory flowrate, Cdyn = dynamic lung compliance, and RL = total pulmonary resistance. Measured indices of airflow limitation were altered similarly in all species.

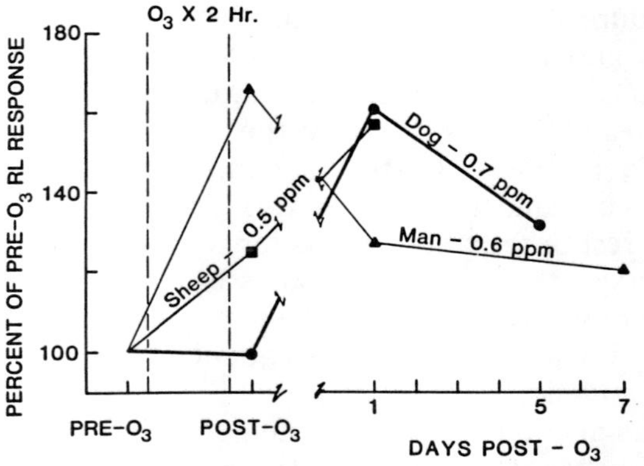

FIGURE 3. The effect of 2-h exposure to O_3 on airway reactivity to inhaled bronchoconstrictor aerosols in man (Golden et al., 1978), sheep (Abraham et al., 1980a), and dogs (Lee et al., 1977). RL = total pulmonary resistance, measured during bronchoprovocation challenge before and after O_3 exposure. The RL response was increased, indicating increased reactivity, in all species after O_3 exposure. The maximum response occurred immediately after exposure in man, but at 1 d after exposure in sheep and dogs.

not measured later; thus, its duraction was unknown. Airway reactivity of the dog was not increased until 24 h after O_3 exposure, and was diminished at 5 d after exposure. It is not known whether the different times of peak response resulted from true species differences or because the upper airways of the animals were not exposed to O_3 as in the mouth-breathing human subjects.

Acute Responses to Nitrogen Dioxide

Acute inhalation exposure to NO_2 results in respiratory changes similar to those elicited by O_3. Responses of humans, sheep, and guinea pigs to 4-h exposures to NO_2 are compared in Fig. 4. Stacey et al. (1983) exposed 10 normal, young adult male humans to 0.5 ppm. Specific airway resistance was measured before and after the exposure, which included two 15 min exercise periods. Abraham et al. (1980b) measured the total pulmonary resistances of 6 conscious sheep before and after exposure to 15 ppm via a head chamber. Murphy et al. (1964) measured the respiratory frequencies of 10 conscious guinea pigs before and after exposure to 5.2 ppm via nose cone.

The values of the sheep and guinea pigs were increased at the end of the 4-h NO_2 exposures, and were returning toward the preexposure levels after 6–24 h of recovery. The specific airway resistance of the human subjects was slightly increased at the end of exposure and remained elevated at 24 h. Although different parameters were measured (breathing pattern in guinea pigs and airflow limitation in humans and sheep), the comparison demonstrates that acute exposure to NO_2 affected the respiration of all three species. The data suggest a dose-response relationship, since the greatest response was observed in the sheep and the least in humans; however, such a relationship is highly speculative.

Inhaled NO_2 has been shown to temporarily increase the airway sensitivity of the guinea pig to inhaled aerosolized histamine in the same manner as described for O_3 above. Silbaugh et al. (1981b) reported that the cumulative dose of inhaled histamine causing a 50% reduction in dynamic lung compliance was reduced to 33% of the preexposure value at 10 min after a 1-h exposure to 56 ppm NO_2, and that considerable recovery occurred by 19 h after exposure. Although no data have been published demonstrating a similar response to NO_2 in humans, it seems reasonable to predict that NO_2 would increase the airway reactivity of humans as does O_3, since both O_3 and NO_2 cause the response in guinea pigs.

Inhalation of extremely high concentrations of NO_2 (tens to hundreds of ppm) also apparently cause similar responses in lungs of humans and animals, although accurate quantitative data are lacking for the accidental exposures of man. The clinical syndrome in nonfatal human cases lasts several hours to days and consists of breathing pattern changes, productive cough, and leukocytosis (Jones, 1980). The most prominent functional changes are an increased respiratory frequency and reduced CO diffusing capacity, resulting from alveolar epithelial damage, edema, and hemorrhage. Johnson et al. (1982) reported the same structural and functional responses in dogs exposed for 6 h to 69 ppm NO_2, including productive cough, increased respiratory frequency, and reduced CO diffus-

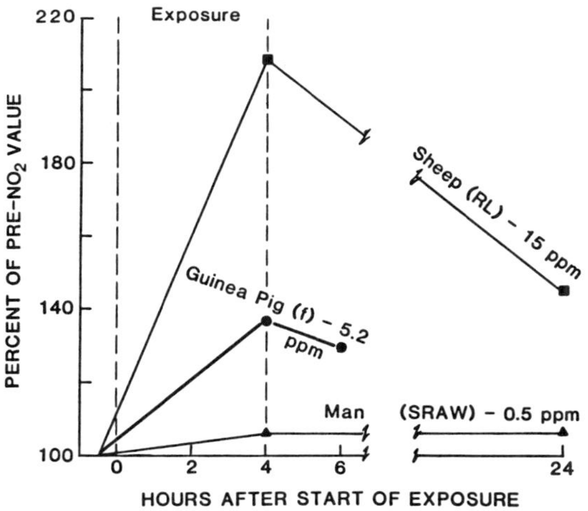

FIGURE 4. The respiratory effects of 4-h exposure to NO_2 in man (Stacey et al., 1983), sheep (Abraham et al., 1980b), and guinea pigs (Murphy et al., 1964). RL = total pulmonary resistance, f = respiratory frequency, and SRAW = specific airway resistance. These indicators of airway irritation increased in all species during NO_2 exposure and those of sheep and guinea pigs were returning toward preexposure values by 6–24 h after exposure.

ing capacity as the primary effects. The nature and time course of these changes in dogs were shown (Mauderly et al., 1983b) to be quantitatively similar to those of a human subject exposed to oxyacetylene welding fumes thought to include high levels of NO_2 (Jones et al., 1973). These results suggest that animals and humans respond similarly to high as well as low concentrations of inhaled oxidant gases.

Acute Responses to Oxygen

The concern for the toxic effects of high concentrations of (O_2) are related primarily to its clinical use in ventilation therapy. Oxygen toxicity has been studied in humans and animals, and similar effects have been observed in all species investigated (Clark and Lambertsen, 1971). The major effects after exposure for several hours are edema, congestion, and inflammation, changes more similar to those induced by high concentrations of NO_2 than to those caused by low levels of O_3 or NO_2. The functional effects of experimental exposures of a human subject and rats to approximately 100% O_2 at 1 atm pressure for 48 h are compared in Fig. 5. Caldwell et al. (1966) exposed a normal, young adult male human to 98% O_2 for 48 h. The subject did not exercise during exposure, but took 5 deep breaths and coughed once every 2 h. Harkema et al. (1982) exposed 10 young adult female rats to 100% O_2 for 48 h. Similar respiratory function tests of the anesthetized rats and the conscious human subject were performed before and after exposure.

The major functional changes in both human and rats were reductions in CO diffusing capacity and lung compliance, measured as dynamic compliance in the man and quasistatic compliance in the rats. These changes reflected the gas exchange and mechanical abnormalities resulting

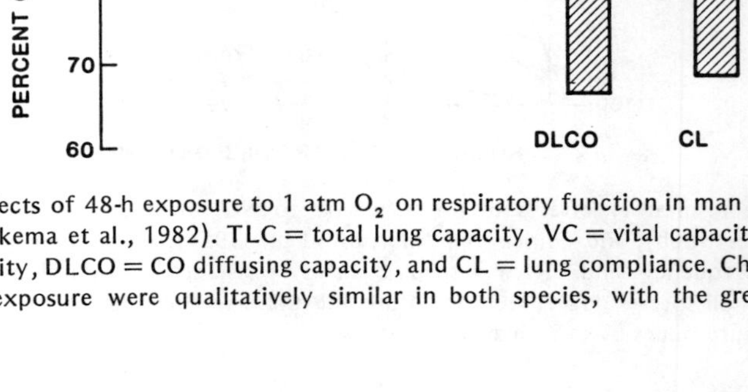

FIGURE 5. The effects of 48-h exposure to 1 atm O_2 on respiratory function in man (Caldwell et al., 1966) and rats (Harkema et al., 1982). TLC = total lung capacity, VC = vital capacity, FRC = functional residual capacity, DLCO = CO diffusing capacity, and CL = lung compliance. Changes measured immediately after exposure were qualitatively similar in both species, with the greatest effect on DLCO and CL.

from edema and congestion at the alveolar level. In addition, total lung capacity and vital capacity were slightly reduced and functional residual capacity was slightly increased in both species. Thus, the functional changes in rats and human were qualitatively identical and quantitatively similar, with the magnitudes of the changes slightly greater in rats.

It is interesting to note that the study of Harkema et al. (1982) demonstrated that rats with preexisting, experimentally induced pulmonary emphysema were not more sensitive to O_2 toxicity than normal rats and that similar results have been reported for Syrian hamsters (Kovnat et al., 1975). Although good quantitative data are lacking, it is recognized clinically that chronic lung disease does not predispose humans to the toxic effects of O_2 (Butler and Rhodes, 1980). These findings further support the existence of similar structural and functional responses to O_2 toxicity in animals and humans.

Functional Manifestations of Pulmonary Emphysema

Pulmonary emphysema in animals, induced experimentally by intratracheal instillation of proteinases, has become a common model for pathophysiological studies (Karlinsky and Snider, 1978). The lesion produced is chronic, largely uncomplicated by bronchitis, alveolitis, or fibrosis, and most closely resembles the panlobular form of human emphysema caused by a genetically induced antiproteinase deficiency. The following comparisons are drawn from reports of enzyme-induced emphysema in animals and reports of human subjects in which emphysema was the sole or primary finding. Four functional manifestations of emphysema are compared: flow limitation during forced expiration, static or quasistatic lung pressure–volume relationships, subdivisions of lung volume, and alveolar-capillary gas exchange measured as CO diffusing capacity.

The indices of lung function most commonly used in clinical, occupational and epidemiological settings are the flow rates or volume–time relationships during forced expiration. Changes in forced expiratory parameters have also been reported to follow exposure of rats to 0.2–2.0 ppm O_3 for 3 mo (Costa et al., 1983) and exposure of Syrian hamsters to 20 ppm NO_2 for 12–15 mo (Kleinerman and Niewoehner, 1973). One of the most physiologically descriptive methods of displaying these events is the maximal expiratory flow–volume (MEFV) curve, obtained by plotting expiratory flow rate against exhaled volume. Although often reported for animals, published MEFV curves of humans with emphysema are rare. The MEFV curves from normal and emphysematous humans, dogs, rabbits, and rats are compared in Fig. 6.

Gelb et al. (1973) reported the MEFV curve of a 79-yr-old man in whom panlobular emphysema was later confirmed in tissue taken during a lung tumor resection, and compared it to a mean curve from age-matched, normal nonsmokers. Klassen et al. (1981) excised lungs from 17 mongrel dogs and instilled 3–5 ml of 0.5–5.0% papain solution into one lung from

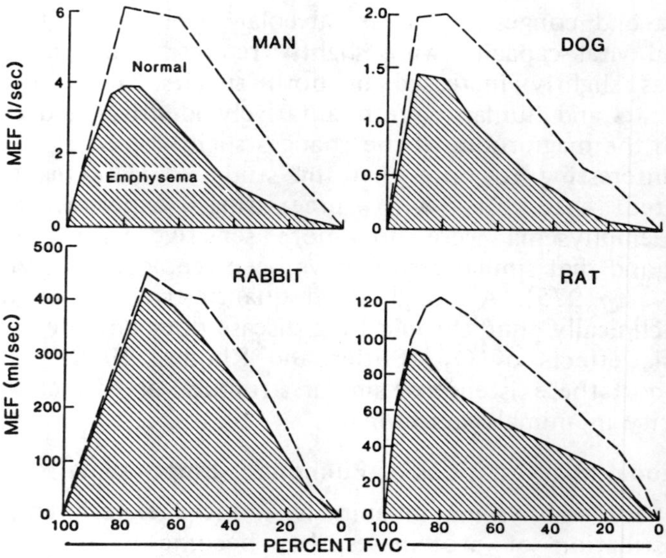

FIGURE 6. The effects of pulmonary emphysema on the maximal expiratory flow-volume curves of man (Gelb et al., 1973), dogs (Klassen et al., 1981), rabbits (Caldwell, 1971), and rats (J. L. Mauderly, Inhalation Toxicology Research Institute). Maximal expiratory flowrates (MEF) during forced expiration are plotted against the expired forced vital capacity (FVC). Flowrates of all species were similarly reduced by emphysema, with the greatest effect in the mid-volume range. The lesser effect in rabbits was thought to be related to concurrent lung infection.

each dog. They reported flow–volume data from the treated lungs at 60–90 min after instillation and control data from the contralateral, untreated lungs. Caldwell (1971) instilled 1 mg papain on alternate weeks for a total of 10 instillations into the lungs of 22 white rabbits and reported flow–volume data measured in the intact, anesthetized rabbits. The effect of emphysema on the MEFV curves of rats has been documented by several studies (Damon et al., 1982; Harkema et al., 1982; Likens and Mauderly, 1982; Mauderly et al., 1983a). The MEFV curves of single emphysematous and normal young adult, male, Fischer-344 rats recently measured at this institute (Mauderly, previously unpublished) illustrate typical changes. The emphysematous rat was intratracheally instilled 3 wk previously with 0.5 IU elastase/g body weight in 1.0 ml saline, and the normal rat was similarly instilled with saline alone. Forced expiratory tests were performed on the intact, anesthetized rats by methods previously reported (Harkema et al., 1982).

The MEFV curves from all species show the characteristic reduction of flow rates throughout the exhaled volume, with the greatest effect in the middle and lower volume ranges. The curves of both normal and emphysematous humans, dogs, and rats are strikingly similar in appearance, while those of the rabbits demonstrate a lesser degree of flow limitation with emphysema. The flow rates of the emphysematous human, dog, rabbit, and rat MEFV curves at 40% of the forced vital capacity were 32, 42, 80, and 57%, respectively, of those of the corresponding normal curves. Caldwell (1971) reported that microscopic lesions suggestive of infection were found in 30% of the control and 45% of the treated rabbits. These lesions consisted of foci of inflammation and early fibrosis, and were probably responsible for the relatively small degree of apparent flow limitation. Emphysema-related flow limitation is thought to be primarily caused by a loss of elastic recoil and parenchymal support of small airways (Mead et al., 1967), and fibrosis would act to increase elastic recoil. A similar lack of significant flow limitation in the presence of mixed emphysematous, inflammatory, and fibrotic lesions has been observed at this institute in rats exposed chronically to diesel engine exhaust (Mauderly et al., 1982).

Lung pressure-volume (*P–V*) curves demonstrate the loss of tissue elastic recoil caused by destruction of alveolar walls in emphysema. The *P–V* curves of normal and emphysematous man, dogs, rats and Syrian hamsters are shown in Fig. 7. Zamel et al. (1976) reported *P–V* curves from 4 male and 1 female subject, 42–60 yr old, with emphysema

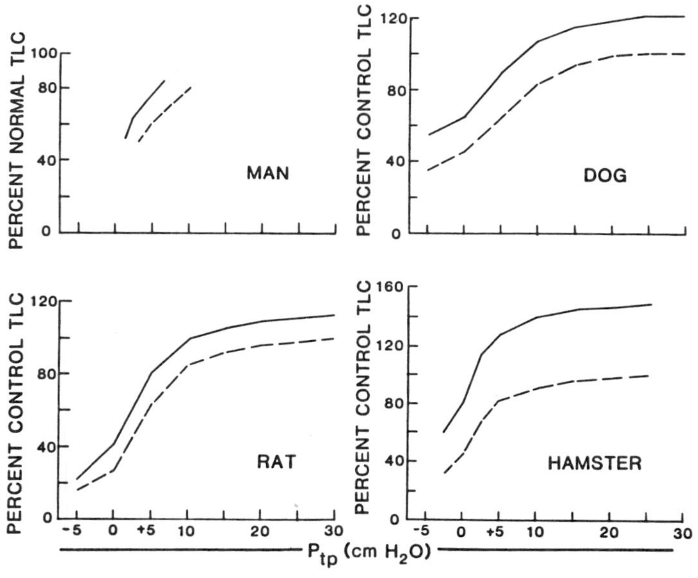

FIGURE 7. The effects of pulmonary emphysema on the lung pressure–volume curves of normal (dased line) and emphysematous (solid line) men (Zamel et al., 1976), dogs (Pushpakom et al., 1970), rats (Mauderly et al., 1983a), and Syrian hamsters (Snider et al., 1977). Volumes, expressed as percentages of total lung capacity (TLC), are plotted against transpulmonary pressure (Ptp). The pressure–volume curves of all species were shifted upward and to the left, indicating a loss of lung elastic recoil. The effect at high lung volumes was greatest in the hamsters, but the effects in all species were similar in the range of tidal breathing (i.e., 0 to +10 cm H_2O).

confirmed histologically in tissue taken during lung tumor resection, and compared them to curves from 15 normal, nonsmoking controls. Only those portions of the curves in the range of tidal breathing were reported. Pushpakom et al. (1970) measured the *P–V* curves of 8 anesthetized, mongrel dogs before and 7–28 d after 1–4 instillations of a total of 48–192 mg papain. Mauderly et al. (1983a) measured the *P–V* curves of 11 anesthetized, young adult, male rats at 3 wk after instillation of 0.5 IU elastase/g body weight and compared them to curves from 10 saline-instilled controls. Snider et al. (1977) measured the *P–V* curves of 8 anesthetized hamsters 18–21 after instillation of 0.2 mg elastase/100 g body weight and compared them to curves from 15 untreated controls.

The *P–V* curves of all four species were similarly affected by emphysema. The curves of the emphysematous subjects were shifted upward and to the left, reflecting a loss of elastic recoil over the entire lung volume range. The effect at high lung volumes was greatest in the hamsters, but the magnitude of the effect in the tidal breathing range (0–10 cm H_2O) was nearly identical to all species.

Emphysema-induced changes in lung elastic recoil and airway collapse during exhalation cause characteristic changes in the relative magnitudes of the various lung volume subdivisions. No published data allowing calculation of all physiological subdivisions of lung volume in emphysematous man were found; however, unpublished data from 2 men, 43 and 51 yr old, with genetic antiproteinase deficiencies and considered to have emphysema by clinical criteria were provided by Dr. W. I. Christensen of the Lovelace Medical Center, Albuquerque, N. Mex. The mean lung volumes of those subjects, plotted as percentages of total lung capacity, are compared to normal volumes predicted on the basis of sex, age, and height in Fig. 8. Likens and Mauderly (1982) measured the lung volumes of 15 anesthetized, young adult, male rats 3 wk after instillation with 1.0 IU elastase/g body weight and those of 15 untreated controls.

Portions of the predicted normal human total lung capacity contributed by functional residual capacity and residual volume were larger than those of the normal, anesthetized rats. In both humans and rats, emphysema resulted in increases in these volumes at the cost of decreases in inspiratory capacity and vital capacity. The changes were proportionately greater in the humans; however, a survey of reported values for vital capacity of emphysematous humans suggests that the examples used in the present comparison may represent extreme changes.

Information on emphysema-induced changes in lung volumes were also reported for dogs by Pushpakom et al. (1970), for rabbits by Caldwell (1971), and for hamsters by Snider et al. (1977) (all reports used in preceeding comparisons). The effects of emphysema on selected lung volumes of all these species can be compared by describing the volumes as percentages of the predicted or measured normal values. Total lung capacities of emphysematous men, dogs, rabbits, rats, and hamsters were

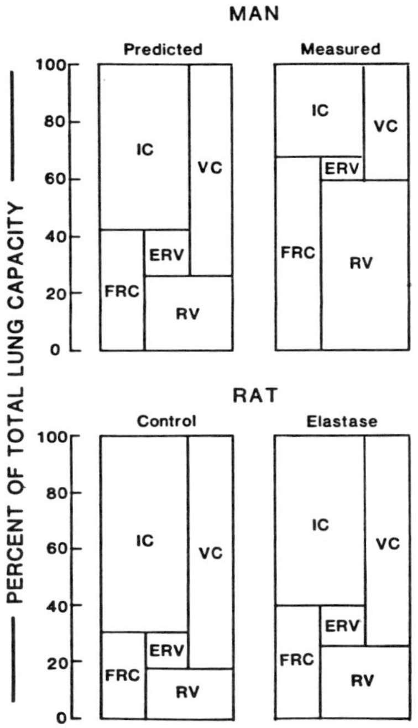

FIGURE 8. The effects of pulmonary emphysema on the lung volumes of man (W. I. Christensen, Lovelace Med. Center, Albuquerque, N. Mex.) and rats (Likens and Mauderly, 1982). IC = Inspiratory capacity, VC = Vital capacity, FRC = functional residual capacity, ERV = expiratory reserve volume, and RV = residual volume. The FRC and RV of normal, anesthetized rats were proportionately smaller than those of normal, conscious man. Emphysema increased the FRC and RV of both species and reduced IC and VC. The effect was greater in the human subjects.

107, 115, 98, 120, and 147% of normal, respectively. Functional residual capacities of emphysematous men, dogs, rabbits, rats, and hamsters were 174, 121, 109, 161, and 170% of normal, respectively. Residual volumes of emphysematous men, dogs, rabbits, rats, and hamsters were 252, 122, 81, 170, and 163% of normal, respectively. Thus, functional residual capacity was increased in all species. Total lung capacity and residual volume were increased in all species except rabbits, in which they were slightly decreased. The different response of rabbits was probably due to the inflammatory and fibrotic complications, as described above, rather than to a species difference in the responses to emphysema. It must be kept in mind that total lung capacity and residual volume are set in humans by voluntary inspiration and expiration, and in animals by inflation and deflation to preselected pressures. Regardless of differences in methodology, similar qualitative changes in lung volumes appear to occur in emphysematous humans and animals.

An emphysema-induced decrease in the efficiency of alveolar–

capillary gas exchange, as reflected by reduced CO diffusing capacity (DLCO), is thought to result primarily from the loss of alveolar surface area caused by destruction of alveolar walls. Four reports of reduced DLCO in human subjects with emphysema were included in the present comparison. Macklem et al. (1965) reported a mean DLCO of 6 males, 47–66 yr old, that was 61% of the predicted value. Thurlbeck et al. (1965) reported a mean DLCO of 6 subjects (sex and age not listed) that was 60% of predicted. In both reports, the subjects were described as emphysematous by clinical criteria. In the following two reports (both used in preceeding comparisons), emphysema was confirmed histologically. Gelb et al. (1973) reported a mean DLCO of 7 subjects that was 50% of predicted. Zamel et al. (1976) reported a mean DLCO of 5 subjects that was 49% of predicted. Thus, the 24 human subjects in these four reports had a mean DLCO that was 55% of the predicted normal value. Pushpakom et al. (1970) measured the DLCO of the 8 dogs described above, before and after treatment with papain. Damon et al. (1982) measured the DLCO of 9 anesthetized rats (males and females) 3 wk after instillation with 1.0 IU elastase/g body weight and that of 20 saline-treated controls. Harkema et al. (1982) measured the DLCO of 10 anesthetized, female rats 3 wk after instillation with 0.3 IU elastase/g body weight and that of 10 saline-treated controls. Raub et al. (1982) measured the DLCO of groups of anesthetized, male Syrian hamsters at 4 wk after treatment with elastase and that of buffer-instilled controls. Sixteen hamsters treated with 3 IU elastase were compared to 16 controls and 15 treated with 6 IU were compared to 10 controls. The effects of emphysema on the DLCO of all the above subjects were compared by plotting the mean values as percentages of normal in Fig. 9.

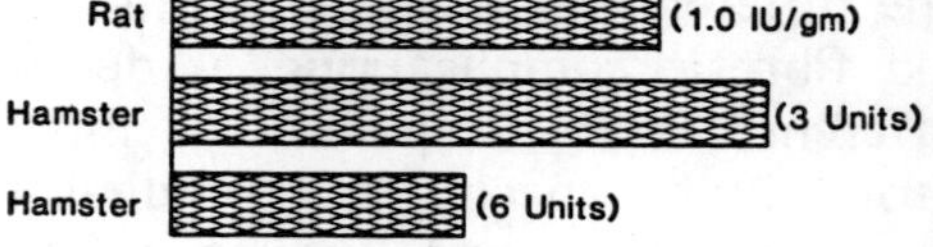

FIGURE 9. The effect of pulmonary emphysema on CO diffusing capacity of man [composite of 24 subjects from Gelb et al. (1973); Macklem et al. (1965); Thurlbeck et al. (1965); and Zamel et al. (1976)], dogs (Pushpakom et al., 1970), rats [0.3 IU elastase/g = Harkema et al. (1982); 1.0 IU elastase/g = Damon et al. (1982)], and Syrian hamsters (Raub et al., 1982: two doses of elastase). Emphysema reduced the CO diffusing capacities of all species. A dose-response relationship was demonstrated by the data from two treatment groups of rats and hamsters.

Emphysema caused a reduction of DLCO in all species. The inclusion of two group of rats and hamsters in the comparison was intended to demonstrate the relationship between the severity of emphysema and DLCO. Neither of the reductions of DLCO of the lower dose rats or hamsters were statistically significant; however, the changes in the higher dose groups were. This illustrates that the lack of effect found in some studies of experimental emphysema in animals is related to the enzyme dose used and does not indicate a species-related difference in the response to emphysema.

SUMMARY AND CONCLUSIONS

These comparisons illustrate many similarities between the respiratory functional responses of animals and humans to inhaled oxidant gases and pulmonary emphysema. Inferences that can be drawn from these comparisons are limited by the small number of reports providing comparable data, the different treatments and evaluations employed, and the small number of species for which data are available. It seems safe to say, however, that the currently available data provide no reason to doubt that qualitatively similar responses occur in humans and laboratory animals to acutely inhaled oxidant gases and to pulmonary emphysema.

These comparisons provide little information on the quantitative similarities in responses among species. Some differences in sensitivity to inhaled materials are well known to exist among animals, but direct interspecies comparisons employing identical treatments and functional evaluations are generally lacking.

The similarities of acute responses to oxidant gases and the similar functional manifestations of emphysema lend confidence to the use of laboratory animals in studies directed toward the prediction of adverse pulmonary effects in humans exposed chronically to low levels of oxidant gases. There are differences in lung structure among animals, and no known animal model exactly duplicates the human lung structure. Studies are needed to better understand how these differences influence the morphological manifestations of oxidant gas exposure and how information from various species can be used in the extrapolation to humans. It is reasonable to expect that different morphological responses, if they occur, would result in different functional expressions of the injury. Conversely, it is reasonable to expect that if similar morphological changes occur in lungs of humans and animals, then similar functional changes will also occur.

REFERENCES

Abraham, W. M., Januszkiewicz, A. J., Mingle, Welker, M., Wanner, A., and Sackner, M. A. 1980a. Sensitivity of bronchoprovocation and tracheal mucous velocity in detecting airway responses to O_3. *J. Appl. Physiol.: Respirat. Environ. Exercise Physiol.* 48:789–793.

Abraham, W. M., Welker, M., Oliver, W., Mingle, M., Januszkiewicz, A. J., Wanner, A., and Sackner, M. A. 1980b. Cardiopulmonary effects of short-term nitrogen dioxide exposure in conscious sheep. *Environ. Res.* 22:61–72.

Amdur, M. O., Ugro, V., and Underhill, D. W. 1978. Respiratory response of guinea pigs to ozone alone and with sulfur dioxide. *Am. Ind. Hyg. Assoc. J.* 39:958–961.

Bates, D. V., and Hazucha, M. 1973. The short-term effects of ozone on the human lung. *Proc. Conf. on Health Effects of Air Pollutants.* Nat. Acad. Sci., Nat. Res. Council. Oct. 3–5, 1973, pp. 507–540.

Bedi, J. F., Folinsbee, L. J., Horvath, S. M., and Ebenstein, R. S. 1979. Human exposure to sulfur dioxide and ozone: Absence of a synergistic effect. *Arch. Environ. Health* 34:233–239.

Butler, J. A., and Rhodes, M. L. 1980. Oxygen toxicity. *J. Indiana St. Med. Assn.* 73:587–591.

Caldwell, E. J. 1971. Physiologic and anatomic effect of papain on the rabbit lung. *J. Appl. Physiol.* 31:458–465.

Caldwell, P. R. B., Lee, W. L., Schildkraut, H. S., and Archibald, E. R. 1966. Changes in lung volume, diffusing capacity and blood gases in men breathing oxygen. *J. Appl. Physiol.* 21:1477–1483.

Clark, J. M., and Lambertsen, C. J. 1971. Pulmonary oxygen toxicity: A review. *Pharmacol. Rev.* 25:38–133.

Costa, D. L., Kutzman, R. S., Lehmann, J. R., Popenoe, E. A., and Drew, R. T. 1983. A subchronic multidose ozone study in rats. In *Advances in Modern Environmental Toxicology,* vol. V, eds. S. D. Lee, M. G. Mustafa, and M. A. Muhlman, pp. 369–393. Princeton, N.J.: Princeton Science.

Damon, E. G., Mauderly, J. L., and Jones, R. K. 1982. Early effects of intratracheal instillation of elastase on mortality, respiratory function and pulmonary morphometry of F-344 rats. *Toxicol. Appl. Pharmacol.* 64:465–475.

Dungworth, D. L., Castleman, W. L., Chow, C. K., Mellick, M. G., Tarkington, B., and Tyler, W. S. 1975. Effect of ambient levels of ozone on monkeys. *Fed. Proc.* 34:1670–1674.

Gelb, A. F., Gold, W. M., Wright, R. R., Bruch, H. R., and Nadel, J. A. 1973. Physiologic diagnosis of subclinical emphysema. *Am. Rev. Resp. Dis.* 107:50–63.

Golden, J. A., Nadel, J. A., and Boushey, H. A. 1978. Bronchial hyperirritability in healthy subjects after exposure to ozone. *Am. Rev. Respir. Dis.* 118:287–294.

Harkema, J. R., Mauderly, J. L., and Hahn, F. F. 1982. The effects of emphysema on oxygen toxicity in rat. *Am. Rev. Respir. Dis.* 126:1058–1065.

Horvath, S. M., Gliner, J. A., and Folinsbee, L. J. 1981. Adaptation to ozone: Duration of effect. *Am. Rev. Respir. Dis.* 123:496–499.

Johnson, W. K., Mauderly, J. L., Hahn, F. F., and Muggenburg, B. A. 1982. Lung function and morphology of dogs after sublethal exposure to nitrogen dioxide. *J. Toxicol. Environ. Health* 10:201–222.

Jones, G. R., Proudfoot, A. T., and Hall, J. I. 1973. Pulmonary effect of acute exposure to nitrous fumes. *Thorax* 28:61–65.

Jones, R. N. 1980. Acute and accidental exposures to irritant gases. In *Pulmonary Diseases and Disorders,* vol. I, ed. A. P. Fishman, pp. 793–799. New York: McGraw-Hill.

Karlinsky, J. B., and Snider, G. L. 1978. Animal models of emphysema. *Am. Rev. Respir. Dis.* 117:1109–1133.

Klassen, T., Thurlbeck, W. M., and Berend, N. 1981. Correlation between lung structure and function in a canine model of emphysema. *J. Appl. Physiol.: Respirat. Environ. Exercise Physiol.* 51:321–326.

Kleinerman, J., and Niewoehner, D. 1973. Physiological, pathological and morphometric studies of long-term nitrogen dioxide exposures and recovery in hamsters. *Am. Rev. Resp. Dis.* 107:1081 (abstr.).

Kovnat, D. M., Lory, P., Snider, G. L., and Brody, J. S. 1975. Susceptibility of the emphysematous lung to oxygen toxicity. *Chest* 68:417. (abstr.).

Lee, L. Y., Bleecker, E. R., and Nadel, J. A. 1977. Effect of ozone on bronchomotor response to inhaled histamine aerosol in dogs. *J. Appl. Physiol.: Respirat. Environ. Exercise Physiol.* 43:626–631.

Likens, S. A., and Mauderly, J. L. 1979. Respiratory measurements in small laboratory animals: A literature review. Inhalation Toxicology Research Institute (Albuquerque, N.M.), LF-68.

Likens, S. A., and Mauderly, J. L. 1982. Effect of elastase or histamine on single-breath N2 washouts in the rat. *J. Appl. Physiol.: Respirat. Environ. Exercise Physiol.* 52:141-146.

Macklem, P. T., Fraser, R. G., and Brown, W. G. 1965. Bronchial pressure measurements in emphysema and bronchitis. *J. Clin. Invest.* 44:897-905.

Mauderly, J. L., Benson, J. M., Bice, D. E., Henderson, R. F., Jones, R. K., McClellan, R. O., Mokler, B. V., Pickrell, J. A., Redman, H. C., and Wolff, R. K. 1982. Life span study of rodents inhaling diesel exhaust: Results through 18 months. Inhalation Toxicology Research Institute (Albuquerque, N.M.) Annual Report LMF-102.

Mauderly, J. L., Harkema, J. R., Gregory, R. E., and Pickrell, J. A. 1983a. Comparison of starvation and elastase emphysema models in rats. *Fed. Proc.* 42:797 (abstr.).

Mauderly, J. L., Silbaugh, S. A., Likens, S. A., Harkema, J. R., and Johnson, W. K. 1983b. Comparisons of the respiratory functional responses of animals and man. In *Proc. of the 13th Conf. on Environ. Toxicol.* USAF Report No. AFAMRL-TR-82-101, in press.

Mead, J., Turner, J. M., Macklem, P. T., and Little, J. B. 1967. Significance of the relationship between lung recoil and maximum expiratory flow. *J. Appl. Physiol.* 22:95-108.

Morgan, W. C. K. 1982. Pulmonary disability and impairment: Can't work? Won't work? In *Basics of Respiratory Disease,* vol. 10, no. 5, ed. R. G. Loudon, pp. 1-6. New York: American Thoracic Society.

Murphy, S. D., Ulrich, C. E., Frankowitz, S. H., and Xintaras, C. 1964. Altered function in animals inhaling low concentrations of ozone and nitrogen dioxide. *Am. Ind. Hyg. Assoc. J.* 25:246-253.

Pushpakom, R., Hogg, J. C., Woolcock, A. J., Angus, A. E., Macklem, P. T., and Thurlbeck, W. M. 1970. Experimental papain-induced emphysema in dogs. *Am. Rev. Respir. Dis.* 102:778-789.

Raub, J. A., Mercer, R. R., Miller, F. J., Graham, J. A., and O'Neil, J. J. 1982. Dose response of elastase-induced emphysema in hamsters. *Am. Rev. Respir. Dis.* 125:432-435.

Silbaugh, S. A., Wolff, R. K., Johnson, W. K., Mauderly, J. L., and Macken, C. A. 1981a. Effects of sulfuric acid aerosols on pulmonary function of guinea pigs. *J. Toxicol. Environ. Health* 7:339-352.

Silbaugh, S. A., Mauderly, J. L., and Macken, C. A. 1981b. Effects of sulfuric acid and nitrogen dioxide on airway responsiveness of the guinea pig. *J. Toxicol. Environ. Health* 8:31-45.

Snider, G. L., Sherter, C. B., Koo, K. W., Karlinsky, J. B., Hayes, J. A., and Franzblau, C. 1977. Respiratory mechanics in hamsters following treatment with endotracheal elastase or collagenase. *J. Appl. Physiol.: Respirat. Environ. Exercise Physiol.* 42:206-215.

Stacey, R. W., Seal, E., House, D. E., Green, J., Roger, L. J., and Raggio, L. 1983. A survey of effects of gaseous and aerosol pollutants on pulmonary function of normal males. *Arch. Environ. Health* 38:104-115.

Thurlbeck, W. M., Fraser, R. G., and Bates, D. V. 1965. The correlation between pulmonary structure and function in chronic bronchitis, emphysema and asthma. *Med. Thorac.* 22:295-303.

Watanabe, S., Frank, R., and Yokoyama, E. 1973. Acute effects of ozone on lungs of cats. *Am. Rev. Respir. Dis.* 108:1141-1151.

Zamel, N., Hogg, J., and Gelb, A. 1976. Mechanisms of maximal expiratory flow limitation in clinically unsuspected emphysema and obstruction of the peripheral airways. *Am. Rev. Respir. Dis.* 113:337-345.

Part Four

EXTRAPOLATION OF METABOLIC RESPONSES TO OXIDANT EXPOSURE IN ANIMALS AND HUMANS

INTRODUCTION

David J. P. Bassett

Department of Environmental Health Sciences,
School of Hygiene and Public Health, Johns
Hopkins University, Baltimore, Maryland

DISCUSSION

During the last few years the development of techniques to isolate viable preparations of different lung cell populations represented a major contribution to studies of lung metabolic responses resulting from oxidant exposures. The papers presented by Rasmussen (1984) and Autor et al. (1984) clearly illustrated how isolated cells might be exposed to different concentrations of ozone and nitrogen dioxide, and how they can be subsequently used to establish mechanisms by which oxidants might alter major metabolic functions of the lung. Block (1984) very nicely demonstrates how information obtained from isolated cells can be used to characterize, in an isolated perfused lung, the early effects of elevated oxygen on the ability of the lung to take up and metabolize circulating serotonin. As he indicates in his presentation, such functional assays might provide the basis for evaluating early alterations in human lung metabolism by relatively non-invasive methods.

The majority of animal studies that investigate the metabolic responses of the lung to toxic substances, have depended upon morphological evaluations and measurements of whole-lung enzyme levels. If we are going to be able to extrapolate such data to the mixed exposures and to the varying predisposing conditions of the human population, it will be necessary to establish whether observed changes in lung biochemistry and pathophysiology are reversible, or manifestations of normal defense mechanism, or adaptive responses, or representative of adverse reactions that cause or contribute to the development of irreversible diseases. A detailed understanding of the chemistry of inhaled substances is clearly a prerequisite to any predictions of their possible interactions with biological macromolecules. Extrapolation to the establishment of how such interactions might result in adverse health effects requires the use of increasingly complex models each serving as a passport to the next.

Studies using isolated tissue fractions have provided the basis of many hypotheses linking the chemistry of toxic substances to inactivation of key metabolic enzymes, to damage of cellular lipids, and to alterations of macromolecular structure. Oxidant exposures of subcellular organelles have also demonstrated the production of reactive species of oxygen and the production of such lipid breakdown products as malondialdehyde. Although these types of *in vitro* studies have indicated possible mechanisms and intracellular targets, they do not indicate whether such interactions occur within the intact cell when under the influence of normal, or perhaps modified cellular control and protective mechanisms. They therefore cannot be used alone to define those early biochemical events that lead to cytotoxicity, nor to adequately establish concentration-effect relationships. Although biochemical assays of microsomal, mitochondrial, and cytosolic enzymes in lungs taken from acute oxidant-exposed animals, have been useful to demonstrate metabolic impairments prior to any major pathological damage, similar evaluations in chronically exposed animals have not been so readily interpretable. Alterations in whole tissue enzyme activities under conditions of altered pathology can result from specific enzyme-protein induction, from proliferation to those cells rich in that particular enzyme, or from a change in cell population brought about by replacement of aged cells, or by inflammatory cell infiltration from the circulation. It should be noted that such *in vitro* assay procedures in some cases represents the only available technique to examine human material. As Crapo et al. (1984) note, interpretation of biochemical data requires a detailed knowledge of enzyme distributions amongst the different cell populations, combined with morphometric evaluations of exposed lung tissue.

Autor et al. demonstrate artificial generation of superoxide anion in isolated cell cultures can be used during oxidant stress to define early biochemical events, and to establish relationships between, lipid membrane damage, energy generation, and anti-oxidant defense mechanisms. When combined with the methods of exposing cells to gaseous pollutants, as described by Rasmussen, such metabolic studies will dramatically enhance our ability to predict cellular responses of the lung to oxidant exposures of defined concentration. The presently available isolated cell types, macrophages, type II epithelium, bronchiolar Clara cells, and capillary endothelium, all have specific metabolic functions that include phagocytosis, surfactant biosynthesis, xenobiotic metabolism, and amine uptake, respectively. Studies of how these metabolic processes are controlled and influenced by toxic substances are a prerequisite to interpreting similar investigations with whole-tissue systems.

Unfortunately, the present isolation procedures do not necessarily produce a true representation of young and old cells that occur in the intact organ. Studies that isolate cells following chronic exposures to toxic substances, might therefore be of limited use. A major disadvantage of using purified isolated-cell populations, whether freshly prepared or in culture, is that they are no longer under the influence of other cell types, or in association with components of the extracellular matrix. Understanding the role of the basement membrane in damage and repair processes will be aided by the increasing use of cells cultured on collagenated supports. Cultured organ explants have also provided methods for looking at cell development, differentiation and cell-cell interactions. They should also be considered as a suitable method of studying the effects of oxidants on inflammatory responses, since synthesis and release of proteases and chemotaxins can be readily measured in the absence of serum inactivators. It should be noted that both explant and primary cell cultures can lose some of their *in vivo* characteristics as a result of physical and enzyme isolation techniques.

Metabolic studies using isolated perfused organs obviously provide the best *in vitro* approximation to what occurs in the intact animal. Toxic substances can be applied by physiological routes, under conditions where ventilation and perfusion rates can be controlled. The heterogeneity of the lung can be overcome by studies of metabolic functions specific to particular cell types, as Block demonstrates on endothelial serotonin uptake and metabolism. As noted by him, the use of the isolated perfused lung to evaluate tissue taken from chronically exposed animals might be limited by possible changes in blood flow distribution, and morphological alterations in the capillary bed. Metabolic measurements using this preparation are also complicated by varying interstitial pool sizes, recycling of metabolic intermediates, and the length of time it can be maintained without physiological deterioration associated with edema fluid accumulation.

In conclusion, better extrapolation modeling will only be possible when we combine the relatively recent developments in the use of isolated cell populations with an expanded application of isolated perfused lungs from chronically exposed animals. More detailed understanding of lung biochemistry in normal and oxidant exposed animals will also be necessary in order to interpret pathophysiological alterations in routine animal studies and in the human population. A special note, as Bromberg (1984) emphasizes, concerns the need to understand how metabolic functions are altered by, or influence, those early pathological events at the onset of exposure, that result in further damage or subsequent repair. It is only by studying the whole spectrum of events, from initial interaction of an inhaled agent to the development of a definable adverse reaction, that we will be able to identify human populations at risk and so provide better scientific base from which public policy judgments can be made.

REFERENCES

Autor, A. P., et al. 1984. Toxicity of oxygen radicals in cultured pulmonary endothelial cells. *J. Toxicol. Environ. Health* 13:387–395.

Block, E. 1984. Early metabolic changes in response to lung injury: Extrapolation from animals to humans. *J. Toxicol. Environ. Health* 13:369–386.
Bromberg, P. A. 1984. Introduction. *J. Toxicol. Environ. Health* 13:297–299.
Crapo, J. D., et al. 1984. Alterations in lung structure caused by inhalation of oxidants. *J. Toxicol. Environ. Health* 13:301–321.
Rasmussen, R. E. 1984. *In vitro* systems for exposure of lung cells to NO_2 and O_3. *J. Toxicol. Environ. Health* 13:397–411.

EARLY METABOLIC CHANGES IN RESPONSE TO LUNG INJURY: EXTRAPOLATION FROM ANIMALS TO HUMANS

Edward R. Block

Medical Service, Veterans Administration Medical Center, and Department of Medicine, University of Florida, College of Medicine, Gainesville, Florida

The pulmonary capillary endothelium provides a nonthrombogenic, semipermeable barrier between pulmonary blood and tissues. In recent years, particular attention has been focused on the ability of these cells to metabolize a variety of circulating biologically active substances either by interiorizing the substance through specific membrane transport processes or by directly altering the substance by way of enzymatic activity at the plasma membrane. Serotonin, norepinephrine, and several prostaglandins are examples of biologically active substances that are removed from the circulation by the pulmonary endothelium by way of specific transmembrane transport processes. Concomitant with the increased interest in metabolic functions of the pulmonary endothelium, there has been a growing awareness of the central role of endothelial cell abnormalities in the pathogenesis of various lung injuries and disease states. During the past several years, considerable evidence has accumulated in support of the hypothesis that alterations in the metabolic functions of the lung provide a method of detecting lung injury in vivo, *and tests of the metabolic functions of the lung have progressed from* in vitro *systems to animal models to humans. This paper reviews some of the evidence responsible for this progression and discusses some of the limitations inherent in the extrapolation of lung metabolism studies from animal models to humans. In this discussion, particular emphasis is placed on the pulmonary uptake and metabolism of serotonin, norepinephrine, and prostaglandins E and F by mammalian lungs.*

INTRODUCTION

The lung, once viewed exclusively as an organ of gas exchange, has attracted increasing interest during the past decade as a metabolic organ (Fishman and Pietra, 1974; Said, 1982). Particular attention has been paid to the ability of the mammalian lung to metabolize a variety of circulating biologically active substances. A wide range of amines, peptides, lipids, and cyclic nucleotides are metabolized in the pulmonary vascular bed (Table 1). Although the lung will process substances from a number of chemical groups, within a given chemical class it is quite selective in distinguishing between substances that differ very slightly in their structure. For example, angiotensin I, the decapeptide prohormone of the renin-angiotensin system, is almost completely metabolized in a single passage through the pulmonary vascular bed, whereas its octapeptide lower

TABLE 1. Biologically Active Substances Handled in the Pulmonary Vascular Bed[a]

Cleared by the lung
Bradykinin
Adenine nucleotides
Norepinephrine
Serotonin (5-hydroxytryptamine)
Prostaglandins E and F
Activated by the lungs
Angiotensin I
Cyclic endoperoxides
Unaffected by the lungs
Epinephrine
Prostaglandins A and I_2
Angiotensin II
Vasopressin

[a]Modified from Fishman and Pietra (1974) with permission

homolog, angiotensin II, passes through the lung unaffected. Similarly, the lung removes norepinephrine and prostaglandins E and F while permitting transpulmonary passage of such close congeners as epinephrine and prostaglandins A and I_2 (prostacyclin).

Subsequent to the observations that passage through the lungs alters the concentrations of a variety of biologically active substances, the pulmonary capillary endothelial cell was identified as the cell type responsible for these functions (Ryan and Ryan, 1977). Metabolism may occur by two distinct pathways, both of which involve interaction with the plasma membrane of the endothelial cells (Table 2). Thus, pulmonary endothelial cells metabolize substances in the circulation either by interiorizing the substance through specific membrane transport processes or by directly altering the substance by way of enzymatic activity at the plasma membrane.

TABLE 2. Handling of Biologically Active Substances by the Pulmonary Endothelium[a]

Metabolized at the endothelial surface without uptake
Bradykinin
Angiotensin
Adenine nucleotides
Metabolized after uptake by the endothelial cell
Serotonin
Norepinephrine
Prostaglandins E and F

[a]Modified from Fishman and Pietra (1974) with permission.

Concomitant with the increased interest in metabolic functions of the pulmonary endothelium, there has been a growing awareness of the central role of endothelial-cell abnormalities in the pathogenesis of various lung injuries and disease states. For example, the pulmonary capillary endothelial cell is the critical or initial site of injury in oxygen toxicity (Crapo et al., 1980; Kistler et al., 1967), radiation-induced lung injury (Gross, 1977), a number of drug-induced lung injuries (Adamson and Bowden, 1974; Gould and Miller, 1975; Meyrick et al., 1972), and the many varieties of the adult respiratory distress syndrome (National Institutes of Health, 1977). Since damage to the pulmonary endothelium is an early manifestation of many lung injuries, it is reasonable to assume that one or more of the metabolic functions of these cells may be simultaneously altered. As such, several investigators have hypothesized that these metabolic functions may provide an early and sensitive marker of injury to the pulmonary endothelium (Block and Fisher, 1977; Catravas et al., 1981; Gillis et al., 1979).

During the past several years, considerable evidence has accumulated in support of the hypothesis that alterations in the metabolic functions of the lung provide a method of detecting lung injury *in vivo*, and we have seen tests of the metabolic functions of the lung progress from *in vitro* systems to animal models to human patients. This paper will review some of the evidence responsible for this progression and will discuss some of the limitations inherent in the extrapolation from animal models to humans. In doing so, I will focus on several representative examples of lung metabolic function, namely, the uptake and subsequent metabolism of biogenic amines and prostaglandins.

UPTAKE AND METABOLISM OF 5-HYDROXYTRYPTAMINE (SEROTONIN), NOREPINEPHRINE, AND PROSTAGLANDINS

5-Hydroxytryptamine (5-HT) and norepinephrine (NE) are biologically active amines found in a wide variety of mammalian cells (Fishman and Pietra, 1974). 5-HT, a potent platelet aggregatory agent, contributes to the functioning of the central nervous system; NE, the neurotransmitter of most sympathetic postganglionic fibers, is a major hormone of the adrenal medulla. In addition, 5-HT and NE are potent vasopressors for systemic and pulmonary vascular beds, and both are removed from the circulation by the lungs of many animal species, including mice, rats, rabbits, hamsters, guinea pigs, dogs, cats, and humans (Fishman and Pietra, 1974; Junod, 1975). The role of the lungs in removing circulating 5-HT was first demonstrated by Gaddum et al. (1953) and has been confirmed by subsequent investigators. Ginn and Vane (1968), the first to report that NE was removed from the pulmonary circulation, found that 40% of infused NE was removed during a single passage through the pulmonary circulation. Numerous studies have shown that the magnitude of uptake of

5-HT and NE by the lung is influenced by differences in species, perfusion flow rate and heterogeneity, pulmonary vascular surface area, and substrate concentration (Stalcup et al., 1982).

Removal (uptake) of 5-HT and NE by the lung involves transmembrane transport of these amines by the pulmonary endothelium, followed by intracellular metabolism and subsequent release of metabolic by-products (Fig. 1). Transmembrane transport of 5-HT and NE by endothelial cells requires energy and involves a sodium-dependent, carrier-mediated co-transport process that is linked to Na^+,K^+-ATPase (Junod, 1975; Steinberg et al., 1975). In the normal lung, transmembrane transport of 5-HT and NE has been shown to be the rate-limiting step in the uptake process, and this transport is independent of subsequent intracellular metabolism of 5-HT and NE. Although 5-HT and NE are taken up into lung endothelial cells by a similar transport process, pharmacologically distinct sites for the transmembrane transport of these two amines exist (Junod, 1975). The intracellular fates of these two amines also differ. 5-HT is metabolized to 5-hydroxyindoleacetic acid by the action of monoamine oxidase and aldehyde dehydrogenase, whereas monoamine oxidase and catechol-*o*-methyl-transferase are responsible for the metabolic degradation of NE by endothelial cells (Fishman and Pietra, 1974; Junod, 1975).

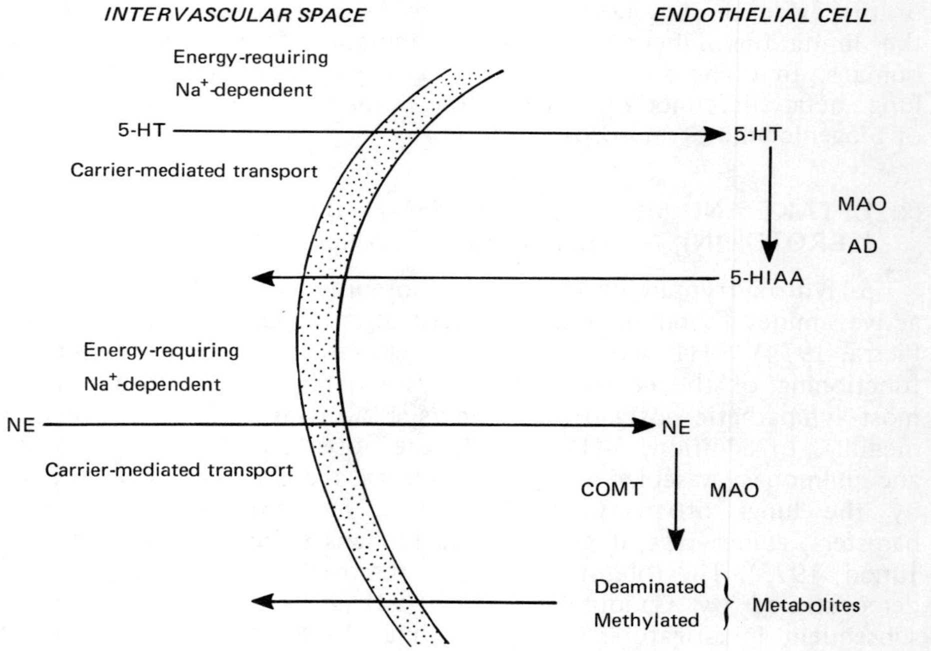

FIGURE 1. Handling of 5-hydroxytryptamine (5-HT) and norepinephrine (NE) by mammalian lungs. 5-HIAA = 5-hydroxyindoleacetic acid; MAO = monoamine oxidase; AD = aldehyde dehydrogenase; COMT = catechol-*o*-methyl-transferase.

Prostaglandins (PG) E_1, E_2, and $F_{2\alpha}$, like the biogenic amines 5-HT and NE, are taken up and metabolized during a single passage through the pulmonary circulation of experimental animals and humans (Anderson and Eling, 1976; Bito et al., 1977; Fishman and Pietra, 1974; Linehan and Dawson, 1979). Bito et al. (1977) and Anderson and Eling (1976) have shown that removal of the PGs by the lung involves a carrier-mediated, energy-requiring membrane-transport step, followed by enzymic inactivation by 15-OH-PG dehydrogenase. Therefore, for PGs, as for biogenic amines, a transmembrane transport process must precede metabolism, and many pharmacological agents that depress the pulmonary removal of PGs probably do so by inhibiting membrane transport of PG. Although the cellular site of PG uptake and metabolism is unknown, *in vitro* and *in vivo* evidence indicates that the process occurs in pulmonary endothelial cells or at some other site close to the vascular space (Hook and Gillis, 1975; Linehan and Dawson, 1979).

EFFECT OF EXPERIMENTAL LUNG INJURY ON UPTAKE AND METABOLISM OF BIOGENIC AMINES AND PROSTAGLANDINS

Although much is known about how the normal mammalian lung handles 5-HT, NE, and PGs, we are just beginning to understand how lung injury in general and endothelial damage in particular alter the handling of these agents. Recent studies using toxicants that indisputedly damage pulmonary capillary endothelium indicate that impaired removal of radiolabeled 5-HT, NE, and PG E_1 and E_2 by the lungs may be an early and sensitive index of endothelial cell injury. To date, the most widely studied example of the interaction between lung injury and lung metabolic function is the effect of normobaric oxygen (O_2) toxicity on the pulmonary removal of biogenic amines and PGs.

Prolonged exposure to high partial pressures of O_2 can alter the structure and function of many organs (Clark and Lambertsen, 1971). In adult animals breathing normobaric gas mixtures with high O_2 tensions, lung injury predominates, and the pulmonary endothelial cell is one of the critical target cells (Crapo et al., 1980; Kistler et al., 1967). Block and Fisher (1977) reported that removal of 5-HT by isolated, perfused lungs from rats exposed to 100% O_2 at 1 atm absolute (ATA) was significantly decreased. This depression in uptake occurred after an 18-h exposure to O_2, was potentiated by vitamin E deficiency, and antedated morphologic and physiologic alterations in the lung (Fig. 2). Hyperoxia has also been reported to depress uptake and metabolism of NE (Block and Cannon, 1978), PGE_2 (Klein et al., 1978), and $PFG_{2\alpha}$ (Chaudhari et al., 1979) *in vitro* by the lungs of rats and guinea pigs. Dobuler et al. (1982) extended these studies to conscious, chronically instrumented rabbits. Using multiple indicator dilution methods, these investigators measured the removal of

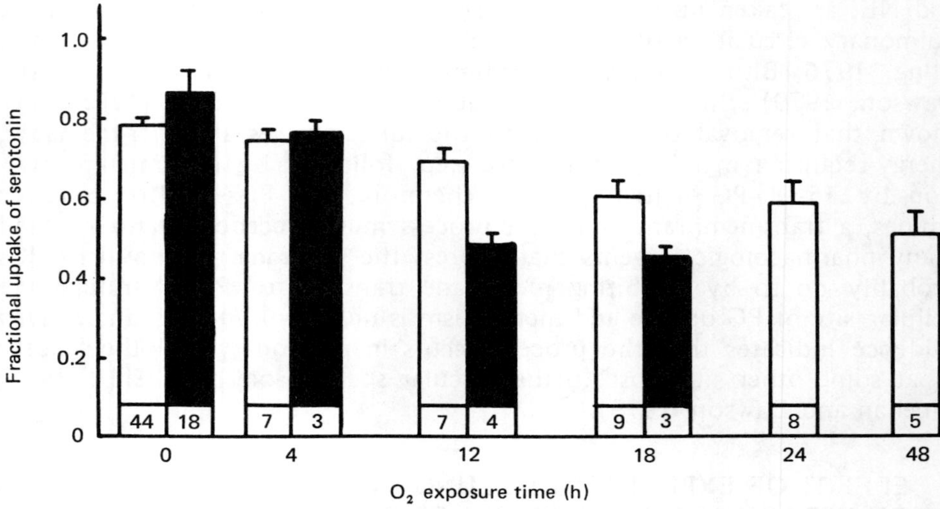

FIGURE 2. Effect of exposure to 100% O_2 at 1 ATA on serotonin (5-hydroxytryptamine) uptake in normal (open bars) and vitamin E-deficient (closed bars) rats. Number of animals studied is at the bottom of each bar. Since fractional uptake by lungs of air-exposed controls did not vary with duration of air exposure, results from all exposure times were combined and represented as 0 h of O_2 exposure. Difference from 0 h of O_2 exposure were statistically significant at 18 ($p < 0.01$), 24 ($p < 0.01$), and 48 h ($p < 0.001$) of exposure in normal rats and 12 ($p < 0.01$) and 18 h ($p < 0.001$) of exposure in vitamin E-deficient rats. Reprinted with permission from Block and Fisher (1977).

[^{14}C]5-HT during a single transpulmonary passage *in vivo*. Exposure to normobaric hyperoxia (96 ± 2% O_2) for 24 h resulted in a significant decrease in pulmonary [^{14}C]5-HT removal, while light and electron microscopy of the lungs of these rabbits revealed normal endothelial and type I cell ultrastructure (Fig. 3).

In addition to pulmonary O_2 toxicity, many other lung injuries that involve endothelial cell damage (e.g., drug, toxicant, vascular, or surgically induced injuries) have been reported to interfere with the metabolic functions of the pulmonary endothelium (Table 3). For example, the following have been reported to depress pulmonary uptake of 5-HT or NE or both in various animal species: monocrotaline, a pyrrolizidine alkaloid that produces swelling and proliferation of pulmonary capillary endothelial cells (Gillis et al., 1978; Huxtable et al., 1978); alpha-naphthylthiourea, a rodenticide that specifically injures pulmonary endothelium (Block and Schoen, 1981); and paraquat (Block and Schoen, 1982; Roth et al., 1979), an herbicide, and bleomycin (Catravas et al., 1981), an antineoplastic agent, both of which injure lung vascular endothelium. Similarly, pulmonary microembolization (Flink et al., 1982), pulmonary hypertension

(Flink et al., 1982; Gewitz et al., 1982; Sole et al., 1979), and cardiopulmonary bypass (Gillis et al., 1974; Pitt et al., 1982) have been shown to decrease pulmonary removal of 5-HT, NE, or PGE_1. In those instances when changes in lung metabolic function were correlated with morphologic changes in the lung, the alterations in the metabolic functions were documented to to occur before or shortly after evidence of structural injury developed.

In contrast to lung injuries in which the pulmonary capillary endothelial cell is a critical or early target, uptake of biogenic amines is preserved in the lungs of animals in which the primary or major injury is unrelated to the lung endothelium. For example, lungs from hamsters with elastase-induced emphysema and lungs from rats with pulmonary fibrosis have unimpaired ability to remove perfused 5-HT (Block and Schoen, 1982; Ryerson et al., 1981). Similarly, streptozotocin-induced diabetes (Watkins and Rannels, 1982) and moderate increases in lung water associated with pulmonary edema (Steinberg et al., 1975) do not significantly affect uptake of 5-HT.

On the basis of these observations, uptake and metabolism of 5-HT, and presumably of NE and PGs as well, appear to provide sensitive metabolic indices of injury to the pulmonary microvascular endothelium. Although the pulmonary endothelial cell is known to be the primary site of damage in many disease states and injuries of the lung, endothelial

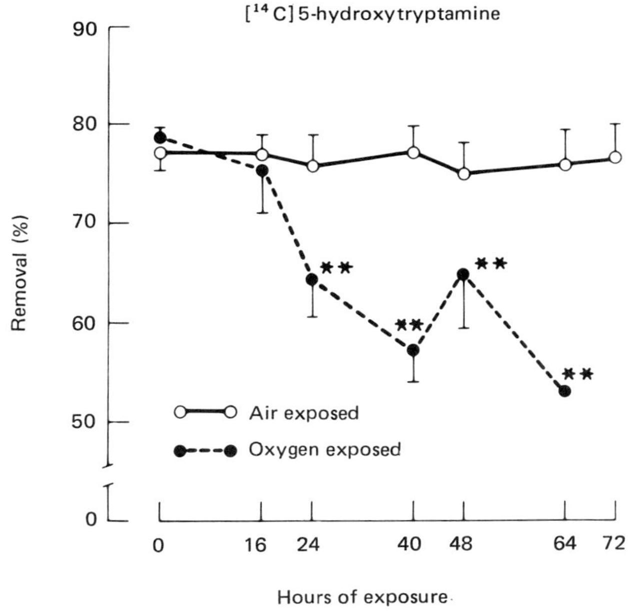

FIGURE 3. Removal of [^{14}C]5-hydroxytryptamine by the lungs of conscious rabbits exposed to room air (○—○) or 96% O_2 (●—●). ** = $p < 0.01$. Reprinted with permission from Dobuler, Catravas, and Gillis (1982).

TABLE 3. The Effect of Pulmonary Injury on Lung Metabolic Functions

Type of injury	Model	Effect on uptake			References
		5-HT	NE	PGs	
O_2 toxicity					
Normobaric hyperoxia	Isolated, perfused rat lung	↓	↓	↓	Block and Cannon, 1978; Block and Fisher, 1977; Klein et al., 1978
	Conscious rabbit	↓			Dobuler et al., 1982
Drugs or toxicants					
Alpha-naphthylthiourea	Isolated, perfused rat lung	↓			Block and Schoen, 1981
Paraquat	Isolated, perfused rat lung	↓			Block and Schoen, 1982; Roth et al., 1979
Monocrotaline	Isolated, perfused rat lung	↓	↓↔		Gillis et al., 1978; Huxtable et al., 1978
Bleomycin	Conscious rabbit	↓	↓		Catravas et al., 1981
Chlorphentermine	Isolated, perfused rat lung	↓	↓		Angevine and Mehendale, 1980; Angevine et al., 1981
Altered pulmonary vascular bed					
Cardiopulmonary bypass	Anesthetized dogs			↓	Pitt et al., 1982
	Humans	↓↑	↑		Gillis et al., 1974, 1979; Hammond et al., 1977
Microembolization	Anesthetized dogs	↓	↔		Flink et al., 1982
Pulmonary hypertension	Humans	↑	↑↓		Gewitz et al., 1982; Gillis et al., 1974; Sole et al., 1979
Miscellaneous					
Septic shock	Anesthetized baboons	↓			Kerstein et al., 1982
Elastase-induced emphysema	Isolated, perfused rat lung	↔			Ryerson et al., 1981
Pulmonary edema	Isolated, perfused guinea pig lung	↔			Steinberg et al., 1975
Streptozotocin-induced diabetes mellitus	Isolated, perfused rat lung	↔			Watkins and Rannels, 1982

injury cannot be reliably detected in the clinical setting, especially before it has resulted in extensive lung damage. Earlier detection of endothelial cell injury would be expected to allow earlier treatment, which should prevent the more advanced stages of lung injury and reduce morbidity and mortality. For these reasons, the development of clinically applicable methods to assess *in vivo* lung microvascular injury based on endothelial cell metabolic functions holds considerable promise for the future.

APPLICATION OF ENDOTHELIAL-CELL METABOLIC FUNCTIONS TO THE CLINICAL DETECTION OF LUNG MICROVASCULAR INJURY IN HUMANS

Gillis and colleagues (1979) developed a method of measuring the uptake of amines and PGs by the lung that can be used in intact animals, including humans. The method is based on the principles of multiple indicator dilution theory, and permits repeated measurement of uptake in patients with in-dwelling pulmonary arterial and radial artery catheters. The method involves the bolus injection of a small volume of saline solution containing an intravascular reference tracer—e.g., cardiogreen—and one or more radiolabeled test tracers, such as 5-HT, NE, or PGE_1 (Fig. 4). Simultaneously, arterial blood is sampled at 1-s or 2-s intervals from the radial artery catheter. The concentration of each tracer is measured, normalized to the total amount of the tracer injected, and plotted as a function of time. From this plot, the fraction of test tracer removed at any timepoint can be calculated (Fig. 4).

Using this technique, Gillis et al. (1979) evaluated the pulmonary clearance of $[^{14}C]$5-HT in a series of patients without pulmonary disease before and after cardiopulmonary bypass. Sequential measurements of 5-HT clearance were made before cardiopulmonary bypass and at various intervals after the completion of bypass. The results indicate that this *in vivo* technqiue can reproducibly measure amine uptake in patients with stable lung function (Fig. 5).

In a related study using a similar method, Hammond et al. (1977) evaluated pulmonary clearance of $[^{3}H]PGE_1$ in a series of patients undergoing cardiac surgery. As with the 5-HT clearance, the clearance of PGE_1 was reproducible and stable in patients without evidence of lung disease. However, a significant reduction in PGE_1 clearance was observed in one postoperative patient in whom clinical and radiographic evidence developed that was consistent with noncardiogenic pulmonary edema, i.e., adult respiratory distress syndrome (ARDS). Recently, Dargent et al. (1983) confirmed these observations and those of Gillis et al. (1979) and extended them to a series of patients with ARDS. In 8 patients with ARDS, pulmonary clearance of 5-HT was significantly reduced, and the magnitude of the reduction correlated with the severity of the ARDS.

The results of the studies reported to date indicate that alterations in metabolic functions of the lung reflect changes in integrity of the lung microvasculature in humans just as they do in experimental animals. Moreover, the studies suggest that the application of multiple indicator dilution methods may allow repeated bedside measurement of lung metabolic function to be performed in patients. Clearly, further investiga-

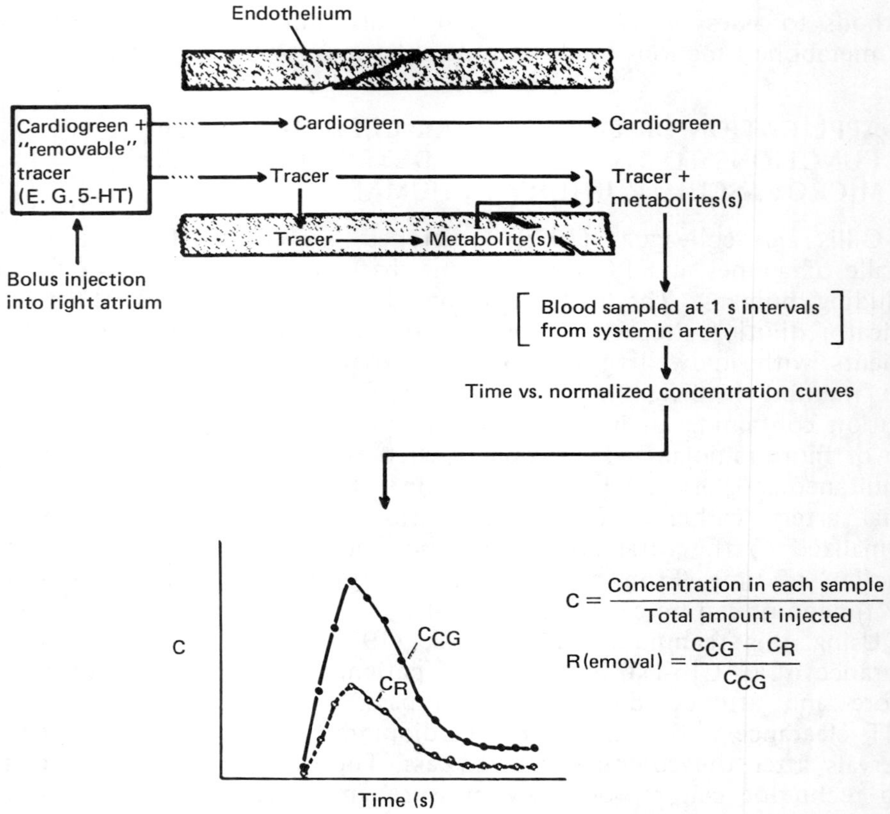

FIGURE 4. Diagram of the principles underlying the application of multiple-indicator dilution techniques to the measurement of lung removal processes. C_{CG} = normalized concentration of cardiogreen in arterial blood; C_R = normalized concentration of removable tracer in arterial blood. Reprinted with permission from Gillis and Catravas (1982).

tion is necessary to establish the validity and safety of the method. However, if it proves successful, it would provide the first test for *in vivo* detection of endothelial cell injury. In addition to being a valuable diagnostic tool, this method would be of great value in studying biochemical mechanisms of early lung injury and in evaluating potential new therapeutic regimens for ARDS and other acute lung injuries.

LIMITATIONS TO THE CLINICAL MEASUREMENT OF LUNG METABOLIC FUNCTIONS

From the preceding discussion, it seems clear that the mammalian pulmonary vascular bed has a highly effective capacity for extracting or degrading, or both, a number of biologically active substances by specific mechanisms residing on the vascular endothelial cells of the lung. In addition, quantifying these metabolic functions by multiple indicator dilution methods promises a useful means for assessing the presence and severity of pulmonary endothelial cell injury in humans. In fulfilling this promise, we must recognize and resolve several problems related to using this approach *in vivo* and extrapolating the methods from animal models to humans. These problems fall into three general categories, as shown in Table 4.

The most significant problems affecting *in vivo* measurements of lung metabolic function stem from limitations in methods presently employed. Removal of 5-HT, NE, and PGs by the lung is directly related to the integrity of the pulmonary endothelium, but removal is also influenced by perfusion flow rate, i.e., cardiac output, perfusion heterogeneity, pul-

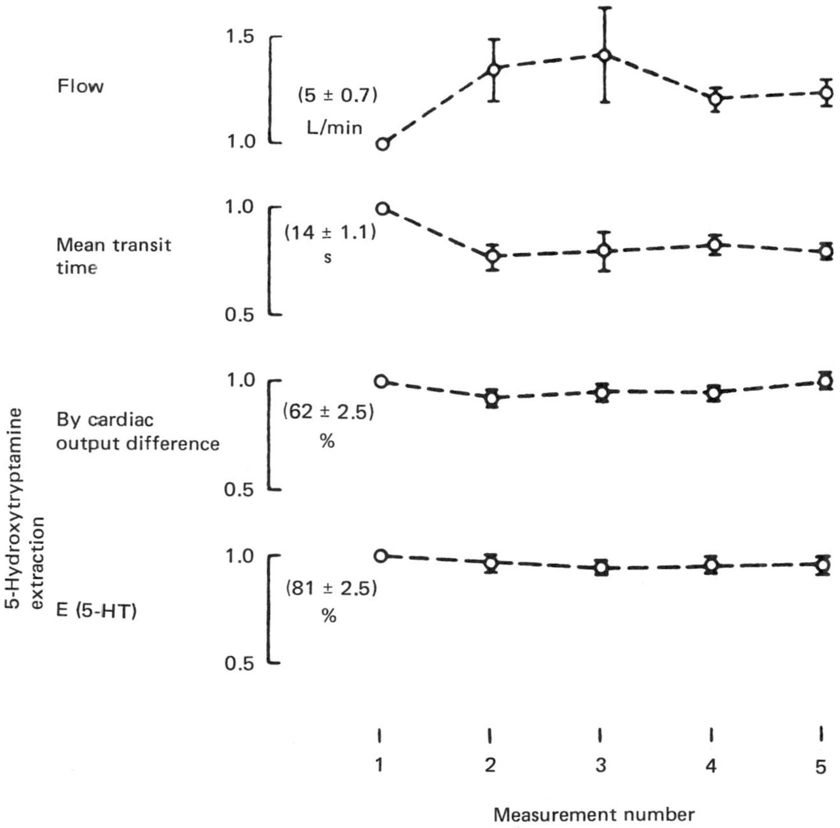

FIGURE 5. Flow, mean transit time, and 5-hydroxytryptamine (5-HT) extraction measured by cardiac output difference or multiple indicator dilution methods [E(5-HT)] during five successive measurements in each of five patients. All measurements were normalized by expression as fractions of the first (i.e., pre-bypass) measurements, which was set equal to 1. Values in parentheses refer to mean and SE of each parameter during the first of the five measurements. Reprinted with permission from Gillis et al. (1979).

TABLE 4. Limitations to the Measurement of Lung Metabolic Function in Human Patients

Methodologic variables
Perfusion heterogeneity
Pulmonary vascular surface area
Flow rate
Competition from endogenous substrates
Complexity of the clinical situation
Lack of an ideal substrate

monary vascular surface area and contact time, and substrate concentration (Block, 1983; Flink et al., 1982; Linehan and Dawson, 1979; Rickaby et al., 1981, 1982; Ryerson et al., 1981; Stalcup et al., 1982). For example, at any perfusion flow rate, the distribution of the perfusate—i.e., the perfusion heterogeneity—within the lung will determine the distribution of capillary transit times for each substrate, the time of contact between substrate and lung endothelium, and the pulmonary vascular surface area available for removing substrate. Decreases in the capillary transit time, the contact time, or the vascular surface area will decrease the pulmonary uptake of substrates like 5-HT, NE, and PGs. Similarly, at any given perfusion flow rate, if the concentration of substrate is less than that needed to saturate the transport process, the rate of removal of substrate by the lung will be decreased.

Because decreases in flow rate, perfusion homogeneity, and surface area are common in injured lungs, it is difficult to be certain that changes in a given metabolic function, e.g., 5-HT removal, are due to changes in endothelial cell function alone. When multiple indicators are used and uptakes are expressed in terms of an intravascular indicator substance, the results provide a measurement based on the perfused lung mass. As such, this compensates to some extent for the effects of perfusion heterogeneity and vascular surface area. However, even this approach does not fully correct for either large heterogeneity of flow or changes in surface area. Interpretation of the uptake results derived from multiple indicator dilution studies is complicated even more if this method is being used to evaluate carrier-mediated (i.e., saturable) transport processes for which there may be competition for the carriers from an endogenous pool of substrate (e.g., 5-HT, NE, and PGs). This problem exists because one critical assumption underlying the use of uptake equations is that the substrate being studied is not being added to the system at the same time that it is being removed. Violation of this assumption would have the effect of falsely decreasing the measured uptake of the test substrate (Stalcup et al., 1982). Since entrapment of platelets and release of 5-HT from these platelets into the lung microvasculature complicates many acute lung injuries, since the lung may release NE from nerve endings or other cellular sites, and since the lung generates and releases PGs, the use of uptake methods applied to these substrates must be viewed with caution in many clinical situations.

The inability to compensate completely for the effects of perfusion heterogeneity, for changes in vascular surface area, and for competition from endogenous substrate pools represents a major disadvantage of the *in vivo* measurement of lung metabolic function and underscores the need for more precise methods to monitor or control these variables. Recently, some of these problems were addressed in a series of experiments by Rickaby and co-workers (1981, 1982). Using an isolated, perfused lobe of a dog lung and multiple indicator dilution methods, these investigators developed a model, based on Michaelis–Menten kinetics, that represents the saturable uptake of 5-HT. By use of this model, they could determine the apparent affinity (K_m) and maximum velocity (V_{max}) of 5-HT uptake by multiple linear regression analysis of the time-concentration data derived from injection of two concentrations of 5-HT, one below the K_m and one designed to achieve a saturating level *in vivo*. Initial results suggest that this approach may be useful *in vivo* for separating changes in the kinetics of 5-HT uptake by endothelial cells from the effects of changes in surface area and perfusion heterogeneity. Although promising, this method requires further investigation to establish its validity and safety.

The second category of significant problems affecting *in vivo* measurements of lung metabolic function in humans relates to the complexity of the clinical situation surrounding the patient to be studied. In contrast to most animal models, in which a single, defined lung injury is induced in a controlled setting, studies in patients are complicated by uncontrolled differences in ventilatory patterns, nutritional status, degrees of shunting, medication regimens, inspired O_2 concentrations, and pulmonary hemodynamics. In addition, acute lung injury is often superimposed on antecedent lung disease that may or may not affect the measurement of the metabolic functions of the lung endothelium. Finally, because of the unstable nature of the patients, there are often minute-to-minute fluctuations in systemic and lung physiologic variables and therapeutic interventions. Such frequent and rapid changes would not be reflected by intermittent measurements of lung metabolic function. Because the impact of these many clinical variables on measurements of lung metabolic function is not known, application and interpretation of these measurements in human patients are limited and should be evaluated cautiously. To assess and predict the effect of these unique variables on the measurement of metabolic functions in humans, we need additional experiments in animal models that better represent clinical situations in humans.

The third major problem to be resolved is which substrate (or substrates) is most appropriate for human use. The ideal substrate is one that is nontoxic and may be administered repeatedly, selectively interacts with pulmonary endothelial cells, has no significant hemodynamic effects that would complicate data interpretation, and provides a sensitive and specific index of endothelial cell injury. At present, the ideal substrate has not been identified. Each of the three substrates studied to date and discussed in this review—5-HT, NE, and PGs—fails to fulfill at least one of the criteria for the ideal substrate. For example, although 5-HT and NE selectively interact with pulmonary endothelial cells and likely provide specific indices of injury to these cells, they are not free of hemodynamic effects and may not be safely administered in repeated doses or at near-saturating levels to patients with cardiopulmonary disease. In addition, endogenous 5-HT and NE may compete with exogenous radiolabeled 5-HT and NE, resulting in the aforementioned erroneous conclusions. Similarly, PGs are limited by potential hemodynamic effects and the lack of evidence to document that these agents selectively interact with pulmonary vascular endothelial cells *in vivo*. Clearly, the identification of a better substrate would greatly advance our efforts to develop and apply an accurate *in vivo* measure of the metabolic functions of the lung endothelium.

ROLE OF 5-HYDROXYTRYPTAMINE AND NOREPINEPHRINE UPTAKE AND METABOLISM IN THE CLINICAL MANIFESTATIONS AND PATHOGENESIS OF LUNG DISEASE

Aside from being potential markers for endothelial cell injury, alterations in pulmonary uptake and metabolism of 5-HT and NE may also be involved in the pathogenesis of various lung diseases and may contribute to the clinical manifestations of lung disease. To date, the best-studied example of this possibility is pulmonary O_2 toxicity. As discussed earlier, the pulmonary endothelial cell is one of the critical target cells and is the initial site of injury in normobaric O_2 toxicity. Destruction of the pulmonary capillary endothelium leads to interstitial and alveolar edema and platelet thrombi in the lung microvascular bed (Crapo et al., 1980). Besides this direct lung damage, pulmonary and systemic hypertension and altered peripheral and organ vascular resistances have been reported as frequent but later manifestations of O_2 toxicity (Wood et al., 1967, 1972). Several investigators using various animal models have reported that high O_2 tensions depress uptake and metabolism of 5-HT and NE by the pulmonary endothelium (Block and Cannon, 1978; Block and Fisher, 1977; Dobuler et al., 1982); this would be expected to result in increased circulating 5-HT and NE. In addition, Batcha et al. (1980) reported increased endogeneous levels of NE in unanesthetized dogs continuously exposed to 100% O_2 at 1 ATA for 3 d. Since 5-HT and NE are potent vasoconstrictors and 5-HT increases vascular permeability and promotes platelet aggregation, endothelial injury leading to impaired lung clearance of these amines may in part account for pulmonary edema and pulmonary and systemic hypertension, for changes in the peripheral vasculature, and for platelet aggregation in the pulmonary microvascular bed that commonly occur in the lungs of O_2-toxic animals. This sequence of events is schematically illustrated in Fig. 6.

Although O_2 toxicity is a good example of how altered metabolic processing of biogenic amines may be involved in the pathogenesis or manifestations of lung disease, a scenario similar to that depicted in Fig. 6 might be expected in any lung injury or disease state in which endothelial cell damage is prominent. As mentioned, Dargent et al. (1983) observed a decrease in the pulmonary uptake of 5-HT by the lungs of patients with the ARDS. Similarly, Kerstein et al. (1982) reported decreased pulmonary extraction of 5-HT in baboons during septic shock, and White et al. (1975) demonstrated a reduction in 5-HT uptake in the pump-perfused lobe of a dog lung with marked derangements in pulmonary physiology similar to those observed in ARDS. However, before we can fully accept this intriguing hypothetical sequence of events, we need to have reliable studies of circulating 5-HT and NE and of the effects of 5-HT and NE antagonists on systemic and pulmonary hemodynamics, microvascular

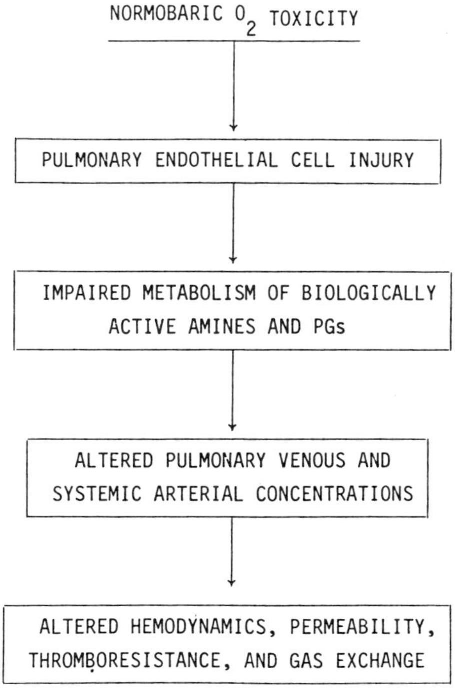

FIGURE 6. Hypothetical sequence of pathophysiologic events by which altered metabolic processing of biogenic amines may be involved in the pathogenesis or manifestations of pulmonary O_2 toxicity.

permeability, and platelet aggregation in the presence of specific pulmonary endothelial injury in humans.

CONCLUSION

Evidence from animal studies is mounting that the metabolic functions of the lung are important in health and disease, but there have been few studies in humans. Close correlations in the pattern and time course of metabolic alterations have been observed in the lungs of humans and experimental animals. Although these correlations are encouraging, they do not justify extrapolation of the data to humans. Until more reliable evidence is available from studies in humans, conclusions concerning the clinical relevance or application of the metabolic functions of the lung must remain somewhat speculative.

REFERENCES

Adamson, I. Y. R., and Bowden, D. H. 1974. The pathogenesis of bleomycin-induced pulmonary fibrosis in mice. *Am. J. Pathol.* 77:185–190.

Anderson, M. W., and Eling, T. E. 1976. Prostaglandin removal and metabolism by isolated perfused rat lung. *Prostaglandins* 11:645–677.

Angevine, L. S., and Mehendale, H. M. 1980. Effect of chlorphentermine on the pulmonary disposition of 5-hydroxytryptamine in the isolated perfused rabbit lung. *Am. Rev. Respir. Dis.* 122:891–898.

Angevine, L. S., Nabeshima, T., Ho, I. K., and Mehendale, H. M. 1981. Effect of chorphentermine on the pulmonary disposition of norepinephrine in the isolated perfused rabbit lung. *Exp. Lung Res.* 2:71–84.

Batcha, K., Stalcup, S. A., Mellins, R. B., and Turino, G. M. 1980. Hyperoxia alters concentrations of circulating vasoactive mediators and hemodynamics. *Circulation* 61:III-26 (abstr.).

Bito, L. Z., Baroody, R. A., and Reitz, M. E. 1977. Dependence of pulmonary prostaglandin metabolism on carrier-mediated transport processes. *Am. J. Physiol.* 232:E382–E387.

Block, E. R. 1983. Endotoxin protects against hyperoxic alterations in lung endothelial cell metabolism. *J. Appl. Physiol.: Respirat. Environ. Exercise Physiol.* 54:24–30.

Block, E. R., and Cannon, J. K. 1978. Effect of oxygen exposure on lung clearance of amines. *Lung* 155:287–295.

Block, E. R., and Fisher, A. B. 1977. Depression of serotonin clearance by rat lungs during oxygen exposure. *J. Appl. Physiol.* 42:33–38.

Block, E. R., and Schoen, F. J. 1981. Effect of alpha naphthylthiourea on uptake of 5-hydroxytryptamine from the pulmonary circulation. *Am. Rev. Respir. Dis.* 123:69–73.

Block, E. R., and Schoen, F. J. 1982. Depression of serotonin uptake by rat lungs exposed to paraquat. *J. Pharmacol. Exp. Ther.* 221:254–260.

Catravas, J. D., Lazo, J. S., and Gillis, C. N. 1981. Biochemical markers of bleomycin toxicity: Clearance of [^{14}C]-5-hydroxytryptamine and [^{3}H] norepinephrine by rabbit lung in vivo. *J. Pharmacol. Exp. Ther.* 217:524–529.

Chaudhari, A., Sivarajah, K., Warnock, R., Eling, T. E., and Anderson, M. W. 1979. Inhibition of pulmonary prostaglandin metabolism by exposure of animals to oxygen or nitrogen dioxide. *Biochem. J.* 184:51–57.

Clark, J. M., and Lambertsen, C. J. 1971. Pulmonary toxicity: A review. *Pharmacol. Rev.* 23:37–133.

Crapo, J. D., Barry, B. E., Foscue, H. A., and Shelburne, J. 1980. Structural and biochemical changes in rat lungs occurring during exposures to lethal and adaptive doses of oxygen. *Am. Rev. Respir. Dis.* 122:123–143.

Dargent, F., Bachman, M., Morel, D., Neidhart, P., Suter, P., and Junod, A. F. 1983. Assessment of pulmonary endothelial cell function and surface by the simultaneous measurement of serotonin and propranolol extraction rates in patients after extracorporeal circulation or with respiratory distress syndrome. *Am. Rev. Respir. Dis.* 127(Part 2):307 (abstr.).

Dobuler, K. J., Catravas, J. D., and Gillis, C. N. 1982. Early detection of oxygen-induced lung injury in conscious rabbits. *Am. Rev. Respir. Dis.* 126:534–539.

Fishman, A. P., and Pietra, G. G. 1974. Handling of vasoactive materials by the lung. *N. Engl. J. Med.* 291:884–890, 953–959.

Flink, J. R., Pitt, B. R., Hammond, G. L., and Gillis, C. N. 1982. Selective effect of microembolization on pulmonary removal of biogenic amines. *J. Appl. Physiol.: Respirat. Environ. Exercise Physiol.* 52:421–427.

Gaddum, J. H., Hebb, C. O., Silver, A., and Swan, A. A. B. 1953. 5-Hydroxytryptamine: Pharmacological action and destruction in perfused lungs. *J. Exp. Physiol.* 38:255–262.

Gewitz, M. H., Pitt, B. R., Laks, H., Hammond, G. L., Talner, N. S., and Gillis, C. N. 1982. Reversible changes in norepinephrine extraction by the lungs in children with pulmonary hypertension. *Ped. Pharmacol.* 2:57–63.

Gillis, C. N., and Catravas, J. D. 1982. Altered removal of vasoactive substances in the injured lung: Detection of lung microvascular injury. *Ann. N.Y. Acad. Sci.* 384:458–474.

Gillis, C. N., Cronau, L. H., Greene, N. M., and Hammond, G. L. 1974. Removal of 5-hydroxytryptamine and norepinephrine from the pulmonary vascular space of man: Influence of cardiopulmonary bypass and pulmonary arterial pressure of these processes. *Surgery* 76:608–616.

Gillis, C. N., Huxtable, R. J., and Roth, R. A. 1978. Effects of monocrotaline pretreatment of rats on removal of 5-hydroxytryptamine and noradrenaline by perfused lung. *Br. J. Pharmacol.* 63:435–443.

Gillis, C. N., Cronau, L. H., Mandel, S., and Hammond, G. L. 1979. Indicator dilution measurement of 5-hydroxytryptamine clearance by human lung. *J. Appl. Physiol.: Respirat. Environ. Exercise Physiol.* 46:1178–1183.

Ginn, R. W., and Vane, J. R. 1968. Disappearance of catecholamines from the circulation. *Nature (Lond.)* 219:740–742.

Gould, V. E., and Miller, J. 1975. Sclerosing alveolitis induced by cyclophosphamide. *Am. J. Pathol.* 81:513–520.

Gross, N. J. 1977. Pulmonary effects of radiation therapy. *Ann. Intern. Med.* 86:81–92.

Hammond, G. L., Cronau, L. H., Whittaker, D., and Gillis, C. N. 1977. Fate of prostaglandins E_1 and A_1 in the human pulmonary circulation. *Surgery* 81:716–722.

Hook, R., and Gillis, C. N. 1975. The removal and metabolism of prostaglandin E_1 by rabbit lung. *Prostaglandins* 9:193–201.

Huxtable, R., Ciaramitaro, D., and Eisenstein, D. 1978. The effect of a pyrrolizidine alkaloid, monocrotaline, and a gyrrole, dehydroretronecine, on the biochemical functions of the pulmonary endothelium. *Mol. Pharmacol.* 14:1189–1203.

Junod, A. F. 1975. Mechanism of uptake of biogenic amines in the pulmonary circulation. In *Lung Metabolism*, eds. A. F. Junod and R. de Haller, pp. 387–398. New York: Academic Press.

Kerstein, M. D., Kohler, J., Gould, S., and Moseley, P. 1982. Pulmonary extraction of biogenicamines during septic shock. *Am. Surg.* 48:552–554.

Kistler, G. S., Caldwell, P. R. B., and Weibel, E. R. 1967. Development of fine structural damage to alveolar and capillary lining cells in oxygen poisoned rat lungs. *J. Cell Biol.* 32:605–628.

Klein, L. S., Fisher, A. B., Soltoff, S., and Coburn, R. F. 1978. Effect of O_2 exposure on pulmonary metabolism of prostaglandin E_2. *Am. Rev. Respir. Dis.* 118:622–625.

Linehan, J. H., and Dawson, C. A. 1979. A kinetic model of prostaglandin metabolism in the lung. *J. Appl. Physiol.: Respirat. Environ. Exercise Physiol.* 47:404–411.

Meyrick, B., Miller, J., and Reid, L. 1972. Pulmonary edema induced by ANTU, or by high or low oxygen concentrations in rat—An electron microscopic study. *Br. J. Exp. Pathol.* 53:347–358.

National Institutes of Health. 1977. *Report on NHLBI Division of Lung Diseases Workshop on Mechanisms of Acute Respiratory Failure.* DHEW Publication No. (NIH)-77-981, pp. 1-12.

Pitt, B. R., Hammond, G. L., and Gillis, C. N. 1982. Depressed pulmonary removal of [^{3}H]prostaglandin E_1 after prolonged cardiopulmonary bypass. *J. Appl. Physiol.: Respirat. Environ. Exercise Physiol.* 52:887-892.

Rickaby, D. A., Linehan, J. H., Bronikowski, T. A., and Dawson, C. A. 1981. Kinetics of serotonin uptake in the dog lung. *J. Appl. Physiol.: Respirat. Environ. Exercise Physiol.* 51:405-414.

Rickaby, D. A., Dawson, C. A., and Linehan, J. H. 1982. Influence of blood and plasma flow rate on kinetics of serotonin uptake by lungs. *J. Appl. Physiol.: Respirat. Environ. Exercise Physiol.* 53:677-683.

Roth, R. A., Wallace, K. B., Alper, R. H., and Bailie, M. D. 1979. Effect of paraquat treatment of rats on disposition of 5-hydroxytryptamine and angiotensin I by perfused lung. *Biochem. Pharmacol.* 28:2349-2355.

Ryan, J. W., and Ryan, U. S. 1977. Pulmonary endothelial cells. *Fed. Proc.* 36:2683-2691.

Ryerson, G., Block, E. R., Harris, J. O., and Cannon, J. K. 1981. 5-hydroxytryptamine uptake by lungs of hamsters with pulmonary emphysema. *Exp. Lung Res.* 2:241-248.

Said, S. 1982. Metabolic functions of the pulmonary circulation. *Circ. Res.* 50:325-333.

Sole, M. J., Drobac, M., Schwartz, L., Hussain, M. N., and Vaughan-Neil, E. F. 1979. The extraction of circulating catecholamines by the lungs in normal man and in patients with pulmonary hypertension. *Circulation* 60:160-163.

Stalcup, S. A., Turino, G. M., and Mellins, R. B. 1982. Critical issues in the use of vasoactive substances to assess lung microvascular injury. *Ann. N.Y. Acad. Sci.* 384:435-457.

Steinberg, H., Bassett, D. J. P., and Fisher, A. B. 1975. Depression of pulmonary 5-hydroxytryptamine uptake by metabolic inhibitors. *Am. J. Physiol.* 228:1298-1303.

Watkins, C. A., and Rannels, D. E. 1982. Effect of diabetes on metabolism of 5-hydroxytryptamine by rat lungs perfused in situ. *Am. Rev. Respir. Dis.* 126:175-177.

White, M. K., Hechtman, H. B., and Shepro, D. 1975. Canine lung uptake of plasma and platelet serotonin. *Microvasc. Res.* 9:230-241.

Wood, C. D., Seager, L. D., and Perkins, G. 1967. Blood pressure changes and pulmonary edema in the rat associated with hyperbaric oxygen. *Aerospace Med.* 38:479-481.

Wood, C. D., Perkins, G. F., Smith, A. G., and Reux, J. M. 1972. Response of the cardiovascular system in oxygen toxicity. *Aerospace Med.* 43:162-167.

TOXICITY OF OXYGEN RADICALS IN CULTURED PULMONARY ENDOTHELIAL CELLS

Anne P. Autor

University of British Columbia Pulmonary Research Laboratory, St. Paul's Hospital, Vancouver, BC

Ann C. Bonham, Robert L. Thies

The Toxicology Center, Department of Pharmacology, The University of Iowa, Iowa City, Iowa

Superoxide dismutase and catalase, which catalytically remove O_2^- and H_2O_2, respectively, each separately protected cultured pulmonary artery endothelial cells from loss of membrane integrity after exposure to oxygen radicals generated either cellularly (polymorphonuclear leukocytes) or chemically (dihydroxyfumarate). Nicotinamide, a precursor of nicotinamide-adenine dinucleotide (NAD) and an inhibitor of ADP-ribose synthetase, also protected cultured endothelial cells from loss of membrane integrity in a concentration-dependent manner after exposure to DHF.

INTRODUCTION

Molecular oxygen, through the action of its metabolites, is capable of causing both acute and chronic tissue injury. The reactive metabolites of oxygen implicated in cytotoxicity include O_2^-, H_2O_2, and $HO^•$, as well as the hydroperoxides and hydroperoxy radicals of cellular macromolecules (Fridovich, 1982). Initiation of structural and functional damage to pulmonary tissue resulting from chronic lung inflammation and hyperoxic ventilation are attributed to oxygen metabolites. Lung inflammation apparently has its pathogenesis in the action of oxygen radical-producing leukocytes, which are frequently elevated and adhered in the lung microvasculature consequent to chronic infection or injury (Staub et al., 1982; Thommasen et al., 1984). Pulmonary pathology from hyperoxia appears to originate from enhanced autoxidation of certain cell components which are normally in the reduced state. Other factors, such as the metabolism of certain drugs and xenobiotics which injure lung tissue of animals and humans, have also been attributed to oxygen metabolites and radicals.

The earliest stages of histologically observable lung injury occur in the pulmonary capillary bed (Kistler et al., 1967). The endothelial cells that

This work was supported by U.S. Public Health Service grant GM12675.

line the capillaries are the initial site of attack (Crapo et al., 1978). Lung edema, commonly observed upon prolonged oxygen ventilation or with chronic inflammation, is the consequence (Staub, 1980). In order to study the initial molecular events that adversely affect the lining cells of the microvasculature, isolated endothelial cells were grown to confluence in primary monolayer culture and exposed to a flux of chemically generated oxygen radicals. Dihydroxyfumarate (DHF) in an oxygenated solution produces $O_2^{\overline{\cdot}}$, H_2O_2, and $HO^{\bullet}$ (Autor et al., 1980). Kinetics of the loss of endothelial cell membrane integrity by the action of oxygen radicals was measured, as well. Oxygen radical attack on cultured endothelial cells by activated leukocytes was also examined. The protective effects of superoxide dismutase (SOD) were assessed. Because alterations in the content of cellular pyridine nucleotides have been implicated as a causative factor in cell death consequent to oxidative stress, nicotinamide, a precursor for pyridine nucleotide synthesis (Yamamoto et al., 1981), was tested for its ability to prevent oxygen radical-induced endothelial cell injury.

METHODS

Dihydroxyfumarate (DHF) and nicotinamide were purchased from Sigma (St. Louis, Mo.). Catalase was purchased from Schwartz/Mann (Orangeburg, N.Y.).

Tissue Culture

Pulmonary artery endothelial cells were prepared for tissue culture by a modification of the method by Booyse et al. (1975). Briefly, bovine pulmonary arteries were resected under sterile conditions and rinsed with phosphate-buffered saline. CLS collagenase (75 mg%) (Worthington Biochemical, Freehold, N.J.) was infused into the lumen and retained for 20 min. The collagenase solution was removed from the lumen, and the endothelial cells were collected by infusing 3 X 10 ml modified M-199 medium containing 10% fetal calf serum (GIBCO, Grand Island, N.Y.) into the lumen and the cells loosened by gentle agitation. The endothelial cells were rinsed and then cultured in modified M-199 containing 20% fetal calf serum on plastic multiwell culture plates (Falcon Plastics Division, B-D Laboratories, Inc., Los Angeles, Calif.). After 4–6 h, most of the cells had adhered to the bottom surface of the culture plates. The plates were rinsed free of unadhered cells and debris. The endothelial cells were then allowed to grow to confluence (4–5 d).

Preparation of Dihydroxyfumarate

Dihydroxyfumarate stock solution was freshly prepared before use by dissolving 20 mg DHF in 1 ml nitrogenated 0.1 *N* NaOH. This solution was diluted to a final volume of 5 ml with nitrogenated water and the pH of the solution adjusted to approximately 6.8 with 3 *N* NAOH concentration of the final stock solution was 4 mg/ml.

Treatment of Endothelial Cells

After the endothelial cells reached confluence, they were washed with Hanks buffered salts solution containing 15 m*M* HEPES, pH 7.4. The cells were preincubated for 30 min at 34°C in a 95% O_2/5% CO_2 atmosphere prior to addition of DHF (400 μg/ml) and/or other additions. The treated cells were then incubated in a 95% O_2/5% CO_2 atmosphere at 34°C for 4 h. Endothelial cell membrane integrity was assessed either by dye exclusion, using erythrosin B (Phillips, 1973) or by the release of cytosolic lactate dehydrogenase (LDH) into the medium (Henry et al., 1960). DNA was measured by using the fluorescent dye 4,6-diamidino-2-phenylindole (DAPI) (Sigma Chemical Company, St. Louis, Mo.), (Brunk et al., 1979). Fluorescence was measured with an Aminco Spectrophotofluorometer. LDH activity is reported as units of LDH/μg endothelial cell DNA.

RESULTS

DHF, at a concentration of 400 μg/ml, produced O_2^- at the rate of approximately 45 nmol/min as assessed by the rate of superoxide dismutase-inhibitable reduction of nitroblue tetrazolium. When exposed to this rate of flux of oxygen radicals, only 15% of the cells in the monolayer retained intact cell membranes (Table 1). Both superoxide dismutase and catalase completely protected cell membranes against oxygen radical-induced injury. Heat-inactivated enzymes were ineffective as

TABLE 1. Time Course of Response of Pulmonary Artery Endothelial Cells in Culture to Oxygen Radical Flux: Intervention by Superoxide Dismutase and Catalase

Incubation time (h)	Percentage of endothelial cells excluding dye[a] (%)			
	No additions	DHF[b]	DHF + SOD[c]	DHF + catalase[d]
0	97.7 ± 1.0	93.9 ± 7.0	99.9 ± 0.1	99.9 ± 0.1
1	97.5 ± 2.4	96.9 ± 3.1	99.9 ± 0.1	95.5 ± 0.1
2	97.2 ± 2.7	85.0 ± 3.3	99.1 ± 1.1	99.1 ± 2.0
3	97.1 ± 5.7	42.2 ± 17.2	94.9 ± 3.3	91.9 ± 4.0
4	97.5 ± 2.6	15.2 ± 14.1	94.0 ± 5.6	92.3 ± 3.0

[a]Bovine pulmonary artery endothelial cells grown in monolayer culture as described were incubated for the indicated times and with the indicated additions. Cell integrity was evaluated by determining the percentage of cells which excluded erythrosin B dye; $n = 15$ for each group of experiments.

[b]Dihydroxyfumarate (400 μg/ml) was added at the beginning of each incubation period.

[c]Dihydroxyfumarate plus superoxide dismutase (300 units/ml) was added at the beginning of the incubation period.

[d]Dihydroxyfumarate plus catalase (30 units/ml) was added at the beginning of the incubation period.

protective agents. The loss of cell membrane integrity with autoxidizing dihydroxyfumarate, as assessed by dye exclusion, revealed a refractory period of approximately 2 h before damage was observed. Thereafter a rapid loss of membrane integrity was observed. These results contrast with the response of pulmonary macrophages to DHF. Although pulmonary macrophages are more resistant to oxygen radical injury induced by DHF, loss of membrane integrity was observed immediately and proceeded linearly with time (Autor et al., 1980).

Similar experiments were conducted in which monolayers of endothelial cells were incubated with conconavalin A-stimulated human polymorphonuclear leukocytes (PMN), which produced O_2 at the rate of 25 nmol/min per 10^6 cells. Assessed by dye exclusion, approximately 30% of the cells remained intact after 3 h. Addition of 300 units superoxide dismutase increased this number to 60%. Eighty-five percent of the cells remained intact with unstimulated PMN. Cells in the monolayer were unaffected by conconavalin A alone. Nicotinamide (10 m*M*) when added with DHF to the incubation medium gave virtually complete protection against the DHF-induced loss of endothelial cell membrane integrity (Table 2). In these experiments, cell intactness was assessed by measuring lactate dehydrogenase in the suspending medium released from damaged endothelial cells. Nicotinamide alone had no damaging effect on cell intactness. The concentration of nicotinamide to which the cultured endothelial cells were exposed during incubation in DHF was varied from 0.1 to 10 m*M* to assess the concentration dependence of the protection against oxygen radical stress. Using the measurement of lactate dehydrogenase released during cytolysis as an indicator of cell integrity, nicotinamide protection was observed to be directly dependent upon concentration in the range studied (Fig. 1).

Preliminary experiments were conducted to test the relationship between cell studies and the *in vivo* situation. Nicotinamide was administered subcutaneously (100 mg/kg) to rats immediately before and every

TABLE 2. Membrane Damage in Pulmonary Artery Endothelial Cells in Culture Exposed to Oxygen Radical Flux with Dihydroxyfumarate (DHF): Effect of Nicotinamide

Addition	Lactate dehydrogenase (LDH) release into medium (units/μg cell DNA × 10^4)[a]
No additions	15 ± 2
DHF	146 ± 14
DHF plus nicotinamide (10 m*M*)	21 ± 9
Nicotinamide (10 m*M*)	6 ± 6

[a]After 4 h of incubation in DHF (400 μg/ml), as described in methods.

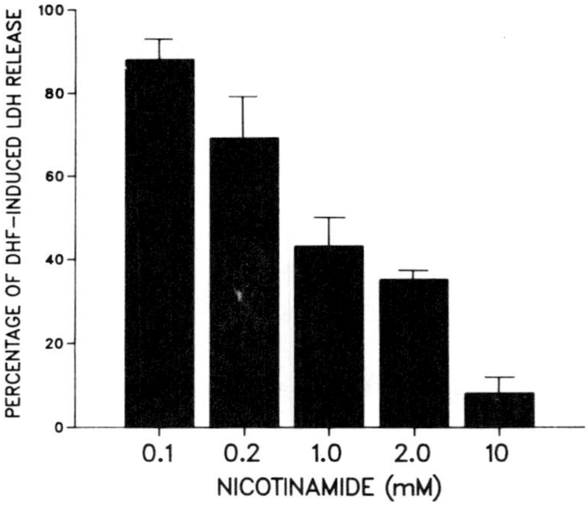

FIGURE 1. Effect of nicotinamide on dihydroxyfumarate (DHF) induced endothelial cell damage: concentration dependence. Endothelial cell monolayers were incubated in DHF (400 μg/ml) in the presence of various concentrations of nicotinamide. Cytotoxicity was assessed by measuring the release of lactate dehydrogenase (LDH) into the medium after 4 h of incubation at 34°C. The data are expressed as the percentage compared with LDH release of cells incubated in DHF alone, normalized to 100%. Incubation in DHF alone resulted in the release into the medium of $14 \pm 2 \times 10^{-4}$ units of LDH/μg cell DNA.

12 h during a 72-h exposure to an atmosphere of 95+% normobaric oxygen using previously developed exposure techniques (Stevens and Autor, 1977). Histologic examination of sectioned and stained lungs from these animals after hyperoxic exposure showed the recognizable lung pathology associated with hyperoxia; that is, interstitial edema, hypercellularity of the alveolar septa, and cellular infiltration into the alveoli. The lung parenchyma animals treated with nicotinamide, however, remained nearly normal in appearance even after 72 h of hyperoxia, although pleural effusions were seen on gross examination.

Because of the characteristic 2-h refractory period before gross injury occurred to endothelial cells exposed to oxygen radicals, a preliminary study of the time course of protection with nicotinamide and superoxide dismutase was undertaken (Table 3) by measuring LDH release upon loss of membrane integrity. Nicotinamide added as late as 2 h after the initial exposure to oxygen radicals provided complete protection to endothelial cells. Measurable cell damage occurred when nicotinamide was added at 180 min after DHF. Nicotinamide was ineffective when added later than 180 min after DHF. Superoxide dismutase, unlike nicotinamide, was not protective when added 60 min after DHF.

DISCUSSION

Use of the cultured cell system, consisting of isolated bovine pulmonary artery endothelial cells grown to confluence in multiwell culture

TABLE 3. Membrane Damage in Pulmonary Artery Endothelial Cells in Culture Exposed to Oxygen Radical Flux with Dihydroxyfumarate (DHF) for 4 h: Effect of Nicotinamide or Superoxide Dismutase Added at Various Times during Incubation

Addition	Time of addition after initiation of incubation (h)	Lactate dehydrogenase (LDH) release into medium[a] (%) (units/μg DNA × 10^4)[b]
No additions	—	24 ± 7
DHF	0	134 ± 12
DHF plus nicotinamide (10 m*M*)	0	43 ± 7
	1	40 ± 8
	2	45 ± 11
	3	123 ± 19
DHF plus superoxide dismutase	0	22 ± 5
(300 U)	60	130 ± 38

[a]After 4 h of incubation in DHF (400 μg/ml), as described in methods.
[b]$n \geqslant 4$.

dishes as described in this report, has allowed the collection of useful comparative data documenting oxidative cell injury. The flexibility and reproducibility of the model have permitted comparisons of the cytolytic effects of chemically (DHF) and cellularly (PMN) generated oxygen radicals. Protective effects of antioxidant enzymes as well as low-molecular-weight compounds were easily assessed and compared. Clearly, both SOD, and catalase, gave effective protection against oxygen radicals in the DHF system. The delay between oxygen radical exposure and measurable loss of cell integrity suggests either depletion of intrinsic protective factors or the operation of repair mechanisms. Unpublished studies from this laboratory (A. C. Bonham and A. P. Autor, unpublished data) showed that damage begins at 30 min after DHF incubation, although the damage as assessed by membrane integrity is not evident for an additional $1\frac{1}{2}$ h.

Oxygen radicals have been implicated as the toxic entity causing the characteristic alveolar septal damage resulting from hyperoxic ventilation. The effectiveness of SOD in protecting oxidatively stressed endothelial cells has a counterpart *in vivo*. Previous studies from this laboratory demonstrated that when superoxide dismutase was administered to rats by continuous intraperitoneal infusion, the loss of pulmonary function, structural damage, and, indeed, mortality typical of prolonged hyperoxic exposure was greatly diminished (McLennan and Autor, 1982).

The mechanism of PMN-induced cytolysis has not been clearly established as yet. There is evidence in the literature that implicates $O_2^{\overline{\cdot}}$ and $HO^{\cdot}$ (Shasby et al., 1982; Baehner et al., 1975), H_2O_2 (Weiss et al., 1981), leukocyte elastase (Harlan et al., 1981), and/or active chlorinating agents (Foote et al., 1983) in PMN-induced cell injury. The data included in this report emphasize the importance of oxygen radicals in PMN-induced cell injury. The protection afforded to endothelial cells by superoxide dismutase was not complete, nevertheless, it was substantial.

Although oxygen radical toxicity is ultimately an oxidative process (i.e., cell constituents are attacked by an oxidizing agent, probably $HO^{\bullet}$ or a related species), the first metabolic steps in the production of oxygen radicals are reductive and frequently require reduced pyridine nucleotides. In addition, one of the intracellular antioxidant enzymes systems, the glutathione peroxidase system, categorized as a protective system, consumes NADPH during oxidative stress. Since nicotinamide (or nicotinate) is the precursor for the pyridine nucleotides NAD(H) and NADP(H), it is of interest to determine whether alterations in the redox balance of pyridine nucleotides might be a factor in the toxic mechanism of oxygen radicals. Work from other laboratories suggested that the total cellular content of NAD(H) and NADP(H) may be depleted during oxidative stress (Boehme et al., 1982). It might be expected, therefore, that nicotinamide, as a precursor to putatively depleted pyridine nucleotides in oxidatively stressed endothelial cells, might provide protection. Such protection was indeed seen in the test system herein described. The question arises, however, as to the means by which pyridine nucleotides are depleted. Recently reported studies have implicated poly(ADP-ribose) synthetase, an enzyme which uses NAD as a substrate for the ribosylation of nuclear proteins and histones, as a significant means of depletion (Creissen and Sholl, 1982; Hayaishi and Ueda, 1982). The enzyme is activated by fragmented DNA, and its action is associated with DNA repair. Oxygen radicals are known to fragment DNA both *in vitro* (Brawn and Fridovich, 1981) and in isolated islets exposed to alloxan (Yamamoto et al., 1981). Oxidative damage induced by alloxan cleaves the DNA of the islets, resulting in depleted NAD and ultimate loss of function (Uchigata et al., 1982). In other studies, it was discovered that not only nicotinamide but also picolinamide, an analog of the former compound, prevents NAD depletion. Picolinamide is not a precursor for NAD but, like nicotinamide, is an inhibitor of poly(ADP-ribose) synthetase (Yamamoto and Okamoto, 1980; Uchigata et al., 1982). Superoxide dismutase also prevents cell damage in the islet system by inhibiting the fragmentation of DNA. Because superoxide dismutase, nicotinamide, and picolinamide (Thies and Autor, unpublished experiments) all prevent oxygen radical-induced damage to endothelial cells, it is tempting to speculate that poly(ADP-ribose) synthetase may be involved in the cytotoxic mechanism of oxygen radicals in these cells. The final answer awaits further experimental work, however.

REFERENCES

Autor, A. P., McLennan, G., and Fox, A. W. 1980. Oxygen free radicals generated by dihydroxyfumarate and ionizing radiation: Cytotoxic effect on isolated pulmonary macrophages. In *Molecular Basis of Environmental Toxicity*, ed. R. S. Bhatnagar, pp. 51-68. Ann Arbor, Mich.: Ann Arbor Science.

Baehner, R. L., Murrmann, S. K., Davis, J., and Johnson, R. B. 1975. The role of superoxide anion in phagocytosis-associated oxidative metabolic reactions. *J. Clin. Invest.* 56:571-576.

Boehme, D. E., Vincent, K., and Brown, O. R. 1976. Oxygen and toxicity inhibition of amino acid biosynthesis. *Nature (Lond.)* 262:418-420.

Booyse, F. M., Sedlak, B. J., and Rafelson, M. E. Jr. 1975. Culture of arterial endothelial cells: Characterization and growth of bovine aortic cells. *Thrombos. Diath. Haemorrh.* 34:835-839.

Brawn, K., and Fridovich, I. 1981. DNA strand scission by enzymatically generated oxygen radical. *Arch. Biochem. Biophys.* 206:414-419.

Brunk, C. F., Jones, K. C., and James, T. W. 1979. Assay for nanogram quantities of DNA in cellular homogenates. *Anal. Biochem.* 92:497-500.

Crapo, J. D., Peters-Golden, M., Marsh-Salin, J., and Shelburne, J. S. 1978. Pathologic changes in the lungs of oxygen-adapted rats: A morphometric analysis. *Lab. Invest.* 39:640-650.

Creissen, D., and Sholl, S. 1982. Regulation of DNA ligase activity by poly(ADP-ribose). *Nature (Lond.)* 296:271-272.

Foote, C. S., Gayne, T. E., and Lehrer, R. I. 1983. Assessment of chlorination by human neutrophils. *Nature (Lond.)* 301:715-716.

Fridovich, I. 1982. Superoxide dismutase in biology and medicine. In *Pathology of Oxygen*, ed. A. P. Autor, pp. 1-19. New York: Academic.

Harlan, J. M., Killen, P. D., Harker, L. A., and Striker, G. E. 1981. Neutrophil-mediated endotheial injury *in vitro*. Mechanism of cell detachment. *J. Clin. Invest.* 68:1394-1403.

Hayaishi, O., and Ueda, K. 1982. Poly- and mono(ADP-ribosyl)ation reactions: Their significance in molecular biology. In *ADP-Ribosylation Reactions*, eds. O. Hayaishi and K. Ueda, pp. 3-16. New York: Academic.

Henry, R. J., Chiamori, N., Goolub, O. J., and Berkman, S. 1960. Revised spectrophotometric methods for determination of glutamic-oxalacetic transaminase, glutamic-pyruvic transaminase and lactic acid dehydrogenase. *Am. J. Clin. Pathol.* 34:381-398.

Kistler, G. S., Caldwell, P. R. B., and Weibel, E. R. 1967. Development of fine structural damage to alveolar and capillary lining cells in oxygen-poisoned rat lungs. *J. Biol. Chem.* 33:605-628.

McLennan, G., and Autor, A. P. 1982. Effect of intraperitoneally administered superoxide dismutase on pulmonary damage resulting from hyperoxia. In *Pathology of Oxygen*, ed. A. P. Autor, pp. 85-97. New York: Academic.

Phillips, H. J. 1973. Dye exclusion tests for cell viability. In *Tissue Culture: Methods and Applications*, eds. P. F. Kruse, Jr. and M. K. Patterson, Jr., pp. 406-408. New York: Academic.

Shasby, D. M., Van Benthuysen, K. M., Tate, R. M., Shasby, S. S., McMurty, I., and Repine, J. E. 1982. Granulocytes mediate acute edematous lung injury in rabbits and isolated rabbit lungs perfused with phorbol myristate acetate: Role of oxygen radicals. *Am. Rev. Respir. Dis.* 125:443-447.

Staub, N. C. 1980. The pathogenesis of pulmonary edema. *Prog. Cardiovasc. Dis.* 23:53-80.

Staub, N. C., Schultz, E. L., and Albertine, K. H. 1982. Leucocytes and pulmonary microvascular injury. *Ann. N.Y. Acad. Sci.* (U.S.A.) 384:332-343.

Stevens, J. B., and Autor, A. P. 1977. Oxygen-induced synthesis of superoxide dismutase and catalase in pulmonary macrophages of neonatal rats. *Lab. Invest.* 37:470-478.

Thommasen, H. V., Russell, J. A., Boyko, W. J., and Hogg, J. C. 1980. *Lancet*, in press.

Uchigata, Y., Yamamoto, H., Kawamura, A., and Okamoto, H. 1982. Protection by superoxide dismutase, catalase, and poly (ADP-ribose) synthetase inhibitors against alloxan- and sterptozotocin-induced islet DNA strand breaks and against the inhibition of proinsulin synthesis. *J. Biol. Chem.* 255:6084-6088.

Weiss, S. J., Young, J., LoBuglio, A. F., Slivka, A., and Nimeh, N. 1981. Role of hydrogen peroxide in neutrophil-mediated destruction of cultured endothelial cells. *J. Clin. Invest.* 68:714-721.

Yamamoto, H., and Okamoto, H. 1980. Protection by picolinamide, a novel inhibitor of poly (ADP-ribose) synthetase, against both streptozotocin-induced depression of proinsulin synthesis and reduction of NAD content in pancreatic islets. *Biochem. Res. Commun.* 95:474–481.

Yamamoto, H., Uchigata, Y., and Okamoto, H. 1981. Sterptozotocin and alloxan induced DNA strand breaks and poly (ADP-ribose) synthetase in pancreatic islets. *Nature (Lond.)* 294:284–286.

IN VITRO SYSTEMS FOR EXPOSURE OF LUNG CELLS TO NO_2 and O_3

Ronald E. Rasmussen

Department of Community and Environmental Medicine,
University of California, Irvine, California

In vitro *studies of the effects of* NO_2 *and* O_3 *require development of methods for separation and culture of those lung cells that experience* in vivo *exposure, and also the design and construction of systems for controlled exposure of the cells to known concentrations of the gases. Separation of lung cell types has been accomplished by enzymatic dispersal of lung tissue and centrifugation of the mixed cells on media of various densities in order to separate the cells on the basis of buoyant density or sedimentation rate. The application of centrifugal elutriation has enabled separation of type II alveolar cells and Clara cells with a high degree of purity. Aveolar macrophages and endothelial cells have also been obtained in good yield.*

Exposure of cultured cells to test atmospheres requires precise control of pollutant levels, close contact of cells and gas without an intervening layer of medium, capability for prolonged exposure, and maintenance of sterile conditions, so that recovered cells can be cultured further or studied for other biological activity. Systems which meet these criteria include roller bottle cultures, petri dish cultures on rocker platforms, cell cultures on cellulose filters fed by perfusion of medium from the side opposite the cells, and cells grown in dishes with gas-permeable film bottoms. Systems that rely on solution of the gases in the overlaying medium do not resemble exposure conditions in vivo, *and may not be suitable for studying effects of the poorly soluble oxidant gases. The cell exposure systems have not been used extensively for studies of the effects of pollutants on freshly isolated specific lung cell types. Such studies should be encouraged.*

INTRODUCTION

The lung with its 40-odd cell types poses a formidable problem to the investigator who wishes to study toxic effects of oxidant gases on the structure and function of the individual cells. Various lesions involving certain cell groups have been identified morphologically after low-level exposures to NO_2 and O_3, and these lesions may be the precursors of chronic lung disease (Blair et al., 1960; Freeman et al., 1966). However,

Support for the studies described by Samuelsen et al. (1978) and Rasmussen and Crocker (1981, 1982) was provided in part by Environmental Protection Agency contract 68-02-1204, Air Force Office of Scientific Research contract 77-3343, and a grant from the National Institute for Environmental Health Sciences, number ES-01835.

We thank M. E. Witte, D. S. Swedberg, E. R. Fox, and M. Hawley for excellent technical assistance.

the relationships between the primary physiochemical interactions between the gases and cell components and the eventual development of disease are not well understood. The problem consists of two parts. First, methods must be developed for isolation, culture, and analysis of events in the various cell types that may be affected by oxidant gas exposure. Several laboratories have been active in this area, and it is now possible to obtain cultures of the major cell types from the lungs of the usual experimental animals. Less success has been attained with human lung, possibly because of difficulty in obtaining experimental material. For some experiments, it may be sufficient to isolate cells from animals following *in vivo* exposure to pollutants, and to then perform various measurements. However, for studies of the direct effects of the gases on specific cells, a second difficulty must be overcome, namely the development of methods for exposure of the cells to pollutant gases while maintaining viability, and if possible avoiding microbial contamination.

For the most part these two problems—isolation of specific lung cells and development of *in vitro* exposure methods—have been approached separately. This has been primarily the result of the complexity and expense of obtaining sufficient numbers of cells of a particular type for the *in vitro* exposures. Therefore, the many *in vitro* studies that have been done to date have used mostly established cell lines. A number of recent studies have demonstrated methods for obtaining good yields of specific lung cells, and experiments involving exposure of these cell types to various airborne pollutants can be expected in the near future. Representative examples of these cell isolation techniques are described in this report.

It may be argued that the response of the lung to oxidant gases is the sum of the responses of the individual cells and their intercellular communications, and once the cells have been separated, their behavior is no longer typical. While this is very likely to be true for many cellular functions, there are some which can best be studied in isolated cells. For example, the role of oxidant gases in the etiology of lung cancer is of particular interest. There is some evidence that O_3 and NO_2 may be genotoxic, and the susceptibility of lung cells in this regard is of interest. Effects on specific enzyme and secretory activities can be studied in cell cultures as well.

One of the early findings in studies of the effects of oxidant gases on cultured cells was that, unless the cell were directly contacted by the gas, there was little effect of the exposure except when very high concentrations of NO_2 and O_2 were used (Pace et al., 1961). Thus, systems that employ organ culture (Schiff, 1977) or organotypic culture (Douglas et al., 1976) or depend upon the solution of these relatively insoluble gases in the culture medium (Hagar et al., 1981) do not provide the conditions of exposure that obtain *in vivo*, where toxic effects are seen at low levels of the gases. Therefore, this review will focus on those systems in which monocellular layers have been exposed to polluted atmospheres, and in which an attempt has been made to maximize direct contact between cells and gas.

PREPARATION OF LUNG CELL CULTURES

As with many other tissues, the general method for preparation of dispersed lung cells involves treatment of the tissue with proteolytic enzymes and collection of the cells by filtration and centrifugation. A number of enzymes and mixtures have been used, including trypsin (Kikkawa and Yoneda, 1974), thermolysin (Frazier et al., 1975), collagenase (Douglas and Kaighn, 1974), and various proteases (Devereux and Fouts, 1981). Optimal cell yields are obtained with young animals. Direct culture of the unfractionated cells from normal experimental animals yields a mixture of cell types, which undergo several divisions but usually cease proliferation after a short time. Rodent lung cell cultures may undergo "spontaneous" transformation and give rise to cell lines that have abnormal karyotypes and unlimited growth *in vitro* (Mason et al., 1975). Such cells may retain differentiated characteristics (Kikkawa et al., 1975) and thus be useful for some studies. For example, cell lines derived from rat type II alveolar cells appear to retain the cytochrome P-450 enzymes involved in metabolism of polynuclear aromatic hydrocarbons (PAH) (Teel, 1979), and also the ability to synthesize and secrete surfactant (Mason and Williams, 1977).

Pulmonary Alveolar Macrophages

Because of the relative ease of isolation, pulmonary alveolar macrophages (PAM) have received most attention as an indicator cell for lung toxicity. The method of isolation has not changed significantly from that described by Myrvik et al. (1971), which consists of repeated lavage via the trachea with a warm balanced salt solution. A further purification step may be added by centrifugation of the cell suspension on a Ficoll–Hypaque step gradient, density 1.077, in the manner used for separation of blood lymphocytes (Boyum, 1964). Functional activity of the PAM can be measured *in vitro* by quantitation of phagocytic activity (Goldstein et al., 1976), motility (McAllen et al., 1981), and bactericidal activity (Goldstein et al., 1976). Nearly all of these activities have been measured after *in vivo* exposure; however, Voisin et al. (1977b) showed that PAM were susceptible to NO_2 inhibition of phagocytosis in a simple *in vitro* exposure system. Weissbecker et al. (1969) exposed PAM to oxidant gases in a hanging drop preparation and showed that toxic effects could be produced by O_3 at less than 1 ppm.

Type II Alveolar Pneumocytes

Because these cells perform certain critical functions in the lung, and because they may be susceptible to oncogenic transformation, they are of special interest. A principal target for toxic effects of oxidant gases is the type I alveolar cell. Death of this cell elicits a proliferative response in the type II cell which, itself, is protected from exposure because of its relatively compact geometry and location (Bertalanffy, 1964). The progeny of the type II cells then replace the intoxicated type I cells (Adamson and Bowden, 1974). In addition to being the precursor for the expendable type I cell, the type II cell is the principal source of the surfactant material that covers the alveolar surface (Mason and Williams, 1977).

Kikkawa and co-workers (1974, 1975) have described a method for isolation of type II cells from the rabbit lung. After perfusion of the vascular system to remove blood, and lavage via the trachea to collect macrophages, the lung was minced and incubated with trypsin, and free cells were collected by filtration and centrifugation. Colloidal barium sulfate was included in the enzyme mixture with the rationale that the PAM would engulf the particles and become more dense than other cells. The cell mixture was centrifuged on a Ficoll step gradient of density 1.047 at 1700 X g, which resulted in the type II cells banding at the interface, and the more dense PAM and bronchiolar epithelial cells in the pellet. Viability was greater than 90% and purity ranged up to 99% in some preparations.

Methods for obtaining type II alveolar cells from rats have been described by a number of investigators. In general, the procedures are similar to that used for rabbit lung using trypsin (Kikkawa et al., 1975) or elastase (Dobbs et al., 1980) for lung digestion and albumin or Ficoll discontinuous gradients for cell isolation. An important technique which has been used to further purify cell types from lung and other tissues is centrifugal elutriation. This method allows fractionation of cell suspensions on the basis of relative size and their rate of sedimentation under low g forces. This technique has been used for final purification of rat type II cells (Greenleaf et al., 1979) and of bronchiolar epithelial cells (Clara cells) from the rabbit (Devereux and Fouts, 1980).

Endothelial Cells

Movement of solubilized material and perhaps particulates from the airspaces to the circulatory system appears to require passage through the endothelial cells. Thus, these cells may be the target of inhaled toxins nearly to the extent of the more exposed cells lining the airways. Isolation of endothelial cells has been accomplished by collagenase treatment of blood vessels or perfusion of the pulmonary vasculature (Ryan and Ryan, 1980). Cells obtained by these methods can be subcultured and they retain characteristics of endothelial cells *in vivo*, including prostaglandin synthesis and angiotensin-converting enzyme synthesis. There is a large literature on *in vitro* studies of endothelial cell metabolism, but no reports were found concerned with *in vitro* exposure of endothelial cells to oxidant gases.

Tracheal and Bronchial Epithelial Cells

Airway epithelial cells are in nearly direct contact with the inhaled air, except for a variable covering of mucus. Thus, they may be most vulnerable to the effects of oxidant gases. Cultures of epithelial cells from rat or hamster trachea may be readily obtained as outgrowths of tracheal pieces (Marchok et al., 1977). Such cultures have a limited *in vitro* life and, unless treated with carcinogens or tumor promoters, they cannot be cloned (Steele et al., 1978). Isolation of the nonciliated Clara cells from the rabbit has been accomplished using centrifugal elutriation (Devereux and Fouts, 1980). The Clara cells show a higher level of xenobiotic-metabolizing activity than type II cells or PAM (Boyd et al., 1980), and also a higher content of cytochrome P-450 (Boyd, 1977). The behavior of these cells in culture has not been reported, a major problem being the maintenance of sterility during the complicated manipulations involved in the isolation procedure.

METHODS FOR EXPOSURE OF CELLS TO OXIDANT GASES

Requirements for *In Vitro* Exposure Systems

The aim of most systems is to approach as nearly as possible the exposure conditions that obtain *in vivo*. To this end specific criteria can be stated:

1. Precise control of pollutant generation and measurement is a first requirement. In most biologically oriented laboratories this will necessitate a cooperative effort with, perhaps, the school of engineering, or chemistry. Since the difficulty of generation and monitoring of oxidant gas atmospheres increases as the concentration decreases, attention must be given to the purity of the air supply as well as the capabilities of the monitoring equipment.
2. The test atmospheres should contact the cells as closely as possible. Oxidant gases will react with cell culture media components, and toxic effects may be masked.
3. The system should be designed to allow significant exposure times. Cells exposed to ambient atmospheres will be rapidly inactivated due to drying, and methods should be developed to maintain a humid atmosphere or to keep the cells moistened in some other way so that extended exposures may be done.
4. Once the exposure is completed, the cells should be easily recovered from the system for such tests as desired.
5. If possible, sterile conditions should be maintained during exposure so that cells can be continued in culture. Tests for mutagenesis or neoplastic transformation require days or weeks of culture following the exposure in order to detect possible effects. This precaution should

be observed, even for short-term studies such as for effects on enzyme activities. Gross bacterial or fungal contamination could easily affect cellular metabolism and lead to erroneous conclusions.

Specific studies may have additional requirements, but these seem to be of importance for all systems. The exposure methods described next have met these criteria with varying success.

Exposure of Flask Cultures of Cells to Oxidant Gases

Since many cell strains grow best when attached to a glass or plastic substrate, the most facile method for exposure of the cells to gases is to flush the sealed culture flask with the mixture of interest. Using this approach, Pace et al. (1961, 1969) measured the cytotoxic effects of NO_2 and O_3. For the NO_2 studies, mouse L cells or liver cells were cultured in specially designed flasks ("T-60" Pyrex) that were fitted with ground joints and tubes to permit continuous gas flow and exchange of culture medium without loss of sterility. When the cells were covered with serum-containing medium little effect was produced by the NO_2 at concentrations below 2400 ppm. At higher concentrations, up to 8600 ppm, cell growth was somewhat retarded, but not entirely inhibited. When the cells were washed free of growth medium and the flasks inverted so that the gas could contact the cells more directly, then as little as 5 ppm NO_2 produced significant inhibition of cell proliferation. The lack of effect when the cells were covered by growth medium was attributed to the physical protection afforded by the liquid layer, as well as to reaction of the NO_2 with serum or other components of the medium. A major conclusion from these studies was that thin layers of liquid are sufficient to protect cells from NO_2 and therefore studies of effects of this gas must employ systems that allow more direct contact between the gases and cells.

Using the same apparatus as for NO_2, Pace et al. (1969) exposed human fibroblasts to O_3, either periodically or continuously. Again it was found that the cells were protected by being covered with the culture medium or a balanced salt solution. At 4 ppm O_3, the lowest concentration tested, inhibition of cell proliferation was seen when the exposure was continuous. At 10 ppm O_3 and above, morphologic changes were noted consisting of vacuolization and rounding of the cells, suggesting an effect on cellular membranes. Mitosis appeared to be inhibited, but quantitative measurements were not done. Compared to NO_2, the toxic action of O_3 appeared to be somewhat greater, but since the experiments did not follow the same protocols, exact comparisons could not be made.

Exposure of Cell Cultures in Roller Bottles

Bolton et al. (1982) modified a commercially available system for roller-bottle cell culture to permit continuous O_3 exposure and monitoring of gas concentration. Conventional roller bottles (11 X 29 cm, New Brunswick Scientific, New Brunswick, N.J.) were fitted with modified rotating caps with tubing connections, to allow gas entrance and withdrawal and culture medium renewal. The control system included solenoid valves for gas control and an ultraviolet (UV) absorbance type O_3 monitor (Dasibi Environmental Corp., Glendale, Calif.). Reaction of O_3 passing through the bottles was measured at various flow rates and with various culture media. With dry, empty bottles, there was no loss of O_3 but if aqueous media were present, significant reaction (i.e., loss of O_3 during passage through the bottle) of the O_3 was seen. The reaction rates were dependent on flow rate and O_3 concentration, and appeared to approach saturation with prolonged exposures. It is of interest that addition of serum to the media did not affect the reaction rate.

Exposure of bovine kidney cells to 0.68 ppm O_3 for 24 h did not produce any striking effects on cell metabolism, although a slight decrease in RNA synthesis was noted. Since the medium present during exposure contained 2% fetal calf serum, there may have been some protective action. When a suspension of vesicular stomatitis virus (VSV) was exposed to O_3 in the system, a time- and concentration-dependent inactivation of plaque-forming units (pfu) was seen, with approximately a 4-log decrease in pfu after 12 h exposure to 0.64 ppm O_3. The mechanisms involved in the VSV inactivation are unknown.

This type of exposure system offers the capability of exposing large numbers of cells to various concentrations of oxidant gases while maintaining sterile conditions. However, the influence of various media components, especially serum, remains to be determined. The system may be especially useful for studies of chemical products resulting from oxidant-cell interactions.

Exposure of Cells in Dishes on Rocker Platforms

Most cultured mammalian cell lines will firmly attach to glass or tissue-culture-grade plastic petri dishes. Exposure to test atmospheres can be readily accomplished by placing the dishes in test chambers and tilting them back and forth to periodically expose the cell layer. The design of such a system is straightforward, although some components are expensive. An example of such a system is shown in Fig. 1 (Guerrero et al., 1979a). Major components are an O_3 generator (C), a Dasibi model 1003 AH O_3 monitor (D), and the rocker platform enclosed in an incubator (G,H,I). Cells exposed in such systems are sensitive to killing by concentrations of O_3 around 1 ppm, and consequently most experiments are conducted at this concentration or below. Hazards to operating personnel at this concentration are minimal.

Using this system, Guerrero et al. (1979a,b) have shown effects on alkaline phosphatase and some indication of genotoxic effects of O_3 in human fibroblasts. A dose response relationship was found for the

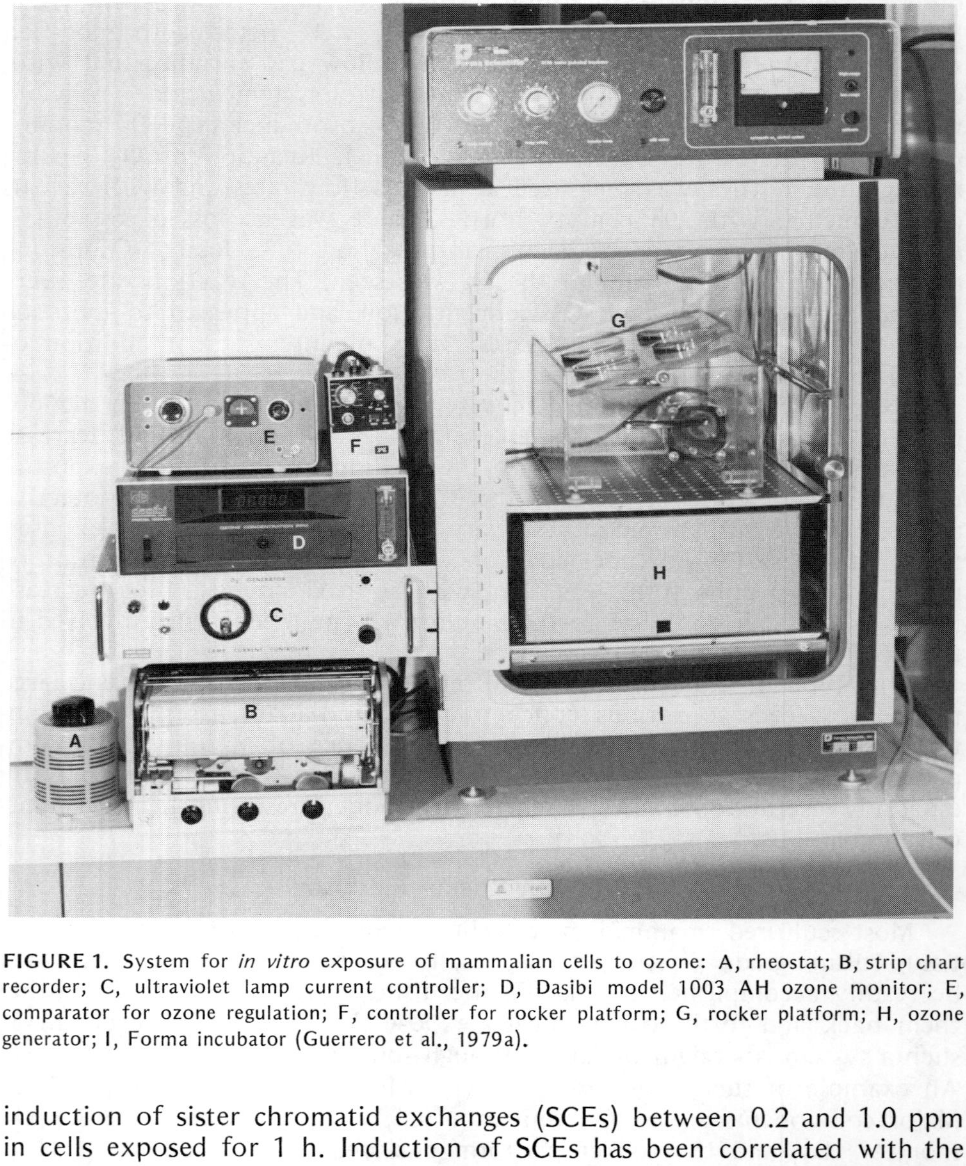

FIGURE 1. System for *in vitro* exposure of mammalian cells to ozone: A, rheostat; B, strip chart recorder; C, ultraviolet lamp current controller; D, Dasibi model 1003 AH ozone monitor; E, comparator for ozone regulation; F, controller for rocker platform; G, rocker platform; H, ozone generator; I, Forma incubator (Guerrero et al., 1979a).

induction of sister chromatid exchanges (SCEs) between 0.2 and 1.0 ppm in cells exposed for 1 h. Induction of SCEs has been correlated with the mutagenic or carcinogenic potential of a large number of chemical and physical agents (Latt et al., 1979).

Although exposure of fibroblasts to 0.5 ppm O_3 was clearly effective in producing SCEs, exposure of human volunteers for 2 h to this concentration produced no elevation of SCEs in peripheral lymphocytes (Guerrero et al., 1979b).

Exposure of Cells Using Gas-Permeable Membranes

Since mammalian cells will attach and proliferate on many surfaces, a possible mode of exposure might be to grow cells on very thin plastic or other film and allow gas to permeate through the film while maintaining the cells immersed in medium. Alink et al. (1979) described a specially designed culture dish in which the bottom consisted of a 25-μm-thick Telfon membrane (Fig. 2). After seeding the dish with V79 Chinese hamster lung fibroblasts, the dishes were placed in a chamber through which an O_3-containing atmosphere was directed. The O_3 dose to the cells was estimated by measurement of the concentration of O_3 in the medium using the KI method (Katz, 1968). External concentrations of 170 and 290 ppm O_3 produced medium concentrations of 0.19 ± 0.08 and 0.43 ± 0.05 μg/4 ml, respectively, during an exposure period of 2.5 h. The major effects seen were a decrease of 50–60% in plating efficiency and a 60% increase in superoxide dismutase. Lactate dehydrogenase and beta-glucuronidase were not affected.

The actual dose to the cells in this sytem is difficult to specify, since the cells on the membrane would encounter the O_3 nearly immediately after its permeation through the membrane. Also, the experimental conditions involved allowing the cells to form a confluent layer before exposure, and thus the O_3 concentration in the medium would be the result of that O_3 that escaped reaction with the cells. A further problem may lie in the variability among lots of the Teflon membranes, which may affect the rate of O_3 permeation (D. B. Menzel, 1983, personal communication). Contamination of the membrane with foreign materials—e.g., skin oil from handling—could also produce undesired effects. A possible hazard to personnel may be present because of the relatively high external concentrations needed to produce significant levels of O_3 in the cell culture

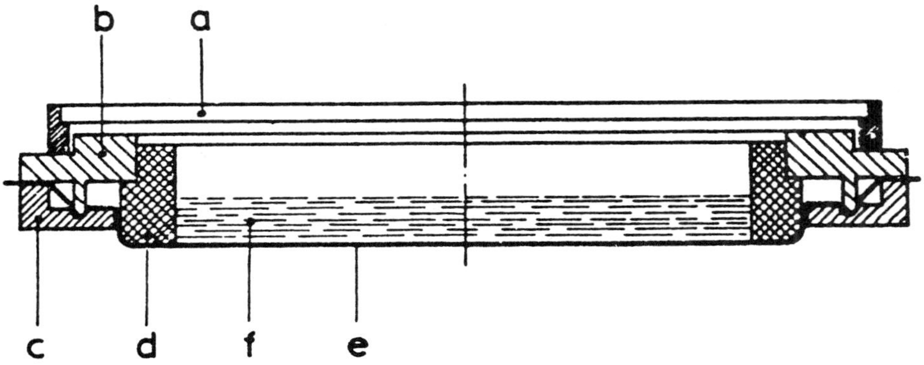

FIGURE 2. Culture dish with gas permeable film bottom: a, glass dish cover; b, top plate; c, base plate; d, Teflon ring; e, gas-permeable Teflon film; f, culture medium (Alink et al., 1979).

medium. In spite of these problems, the system offers ease of handling and maintenance of sterile conditions during exposure and could be useful in some applications.

In Vitro Models of the Respiratory Epithelium

The systems described above in which the cell layers are periodically wetted with saline solutions or culture medium have been very useful in obtaining some kinds of information regarding the effects of oxidant gases. However, they do not really resemble the respiratory epithelium *in vivo*. Bronchiolar and alveolar epithelial cells are maintained by nutrient and fluids from underlying capillaries, while the luminal surface of the cells is exposed nearly directly to the airstream, except for a mucus or surfactant layer of varying thickness. The ideal *in vitro* system would be an organotypic culture in which *in vivo* relationships were maintained, and that was accessible to experimental manipulation.

Most mammalian cells that can grow on glass or plastic surfaces will also attach and grow on cellulose acetate membrane filters. Voisin et al. (1977a,b) took advantage of this fact in their studies of the effect of NO_2 on PAM. They planted PAM from guinea pigs on cellulose acetate filters (Gelman, 0.2 μm pore) and then floated the filters atop nutrient medium in a petri dish. Diffusion of medium through the filter was sufficient to maintain cell viability during exposure, while not submerging the cells. Exposure of this preparation of NO_2 produced a reduction in phagocytic and bactericidal action of the PAM, with concentrations as low as 0.1 ppm producing a measurable effect after a 30-min exposure (Voisin et al., 1977b). The method offers simplicity of preparation, but the fluid level may be difficult to control so that the PAM do not become submerged during exposure. Cell types other than PAM could be used in this system, but no reports of such studies were found.

A system employing lung fibroblasts planted on cellulose acetate filters (Millipore, 0.45 μm pore) was described by Samuelsen et al. (1978). The filters were assembled into a specially designed holder that permitted the perfusion of medium through the filter from the side opposite the cells (Fig. 3). Excess medium was removed from atop the filter either by inverting the holder during exposure to the test gases (Samuelsen et al., 1978) or by using a peristaltic pump (Rasmussen and Crocker, 1981, 1982). In practice, the device is assembled with the filter in place and the lower chamber is filled with cell culture medium. A suspension of cells is pipetted into the upper well and the cells are allowed time to settle and attach to the filter. While still covered with culture medium, the assembly is placed in a gas-exposure chamber, and tubing connections are made for medium perfusion and removal. The chamber is then sealed, gas flow is started, and when the concentration of O_3 or NO_2 has reached the desired level, the medium in the upper well is pumped off and cell exposure is begun. Ozone concentrations are measured with a Dasibi model 1003 AH

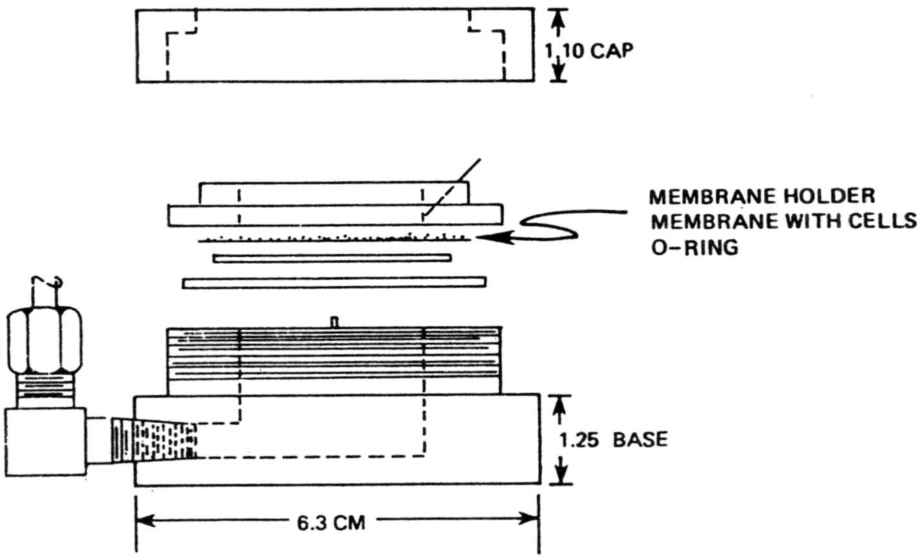

FIGURE 3. Diagram of holder for exposing cells on membrane filters to oxidant gases. Components were machined from 6.3-cm-diameter stainless steel or Lexan round stock. Dimensions are in cm (Rasmussen and Crocker, 1981).

monitor, and NO_2 concentrations with a Beckman 952A NO/NO_x instrument. After exposure, the process is reversed and the cells can be recovered for analysis as desired.

Studies of the cytotoxic effects of O_3 and NO_2 in the system have shown that most cell types are extremely sensitive to low concentrations of these gases. In experiments with V79 fibroblasts, exposure to 0.15 ppm NO_2 or 0.05 ppm O_3 for 6 h resulted in loss of about 90% of the cells from the filters, while clear air had little effect over the same period (Samuelsen et al., 1978; Rasmussen and Crocker, 1981). The effect of the gases was to cause lysis of the cells or detachment from the filters. The cells remaining on the filters were viable, as judged by Trypan blue dye exclusion or colony-forming ability.

Compared to cells exposed in roller bottles or in petri dishes on rocker platforms, the cells exposed on filters are much more sensitive to the toxic effects of oxidant gases. The reasons for this are not immediately obvious, since the intimacy of exposure to the atmosphere would seem to be about the same in all three cases. The major difference is in the nature of the substrate to which the cells are attached. The mechanism of attachment of cells to glass or plastic is through ionic bonds, while attachment to filters involves penetration of the membrane pores by cellular processes, and is therefore a physical attachment. Exposure of cells to O_3 or NO_2 results in a "rounding up" of the cells and a marked decrease in the area of cell surface in contact with the substrate.

In the case of cells attached to cellulose acetate membranes, oxidant-induced damage may result in breakage of the cellular attachments. On the other hand, some ionic bonds may be maintained when the substrate is glass or plastic, even though the cell has become deformed. Thus, the apparent greater sensitivity of cells exposed on membrane filters may be something of an artifact of the culture conditions.

DISCUSSION

The cell exposure systems that involve close contact between cells and test atmospheres can detect toxic effects of O_3 and NO_2 at concentrations near those encountered in the envrionment and known to produce toxic effects *in vivo*. Cell cultures in dishes on rocker platforms, in roller bottles, or on cellulose acetate membranes will be useful for various studies of oxidant gas effects. Systems requiring solution of the gases in overlying medium may not be as relevant. Although solution products of NO_2—i.e., NO_2^- or nitrous acid—may have some role in its toxicity, rather higher concentrations of these products must be reached before toxic effects are seen *in vitro* (Pace et al., 1961; Rounds and Bils, 1965; Vassallo et al., 1973). Similarly, cells covered with aqueous media are protected from the effects of O_3 in the gas phase.

The possible genotoxic effects of O_3 and NO_2 have been examined in various systems, and, although there is some evidence for mutagenic activity, the data should be accepted cautiously. Nitrite ion and nitrous acid are demonstrated mutagens for prokaryotes and cultured mammalian cells (Kodama et al., 1976; Zimmerman, 1977). Genotoxic effects in the form of chromosomal aberrations and SCEs have been reported in V79 cells after exposure to 50 m*M* $NaNO_2$ or NO_2 at 10 ppm or greater (Tsuda et al., 1981). The importance of these observations for *in vivo* exposures is not clear, given the high concentrations involved.

In vitro genotoxic effects of O_3 have been shown in the form of chromosomal deletions (Fetner, 1962) and SCEs (Guerrero et al., 1979b), but the relevance to *in vivo* exposure is in doubt since there are conflicting results from both human and experimental animal studies. Ozone has been supposed to act in a manner similar to ionizing radiation and should thus be mutagenic. However, no reports of direct mutagenic activity—i.e., alteration in DNA sequence leading to a change in phenotype—were found. Experiments in this laboratory (Rasmussen et al., unpublished) indicated that single or repeated 1-h exposures to O_3 at 5 ppm were not mutagenic in V79 cells at the HGPRT locus (Chu, 1971). The exposure system used was similar to that described by Guerrero et al. (1979a). Although the direct genotoxic effects of O_3 and NO_2 may be rather low, they may have comutagenic activity that could be tested for in the various exposure systems.

Among the most useful studies that could be done with the various *in vitro* cell exposure systems are those that focus on the cytotoxic effects of the oxidant gases rather than on the genotoxic effects, which may be less important in regard to lung injury. Such areas as effects on cell functions, recovery from injury, and effects of antioxidants could profitably be studied. Comparisons of *in vivo* and *in vitro* effects can now be made using specific lung cell types isolated using recently developed techniques.

REFERENCES

Adamson, I. Y. R., and Bowden, D. H. 1974. The type II cell as progenitor of epithelial regeneration. *Lab. Invest.* 30:35–42.

Alink, G. M., van der Hoeven, J. C. M., Debets, F. M. H., van de Ven, W. S. M., and Koeman, J. H. 1979. A new exposure model for *in vitro* testing of effects of gaseous pollutants on mammalian cells by means of gas diffusion through plastic films. *Chemosphere* 2:63–73.

Bertalanffy, F. D. 1964. Respiratory tissue: Structure, histophysiology, cytodynamics. *Int. Rev. Cytol.* 17:213–297.

Blair, W. H., Henry, M. C., and Ferris, G. B., Jr. 1960. Chronic toxicity of nitrogen dioxide. II. Effect of histopathology of lung tissue. *Arch. Environ. Health* 18:186–192.

Bolton, D. C., Tarkington, B. K., Zee, Y. C., and Osebold, J. W. 1982. An *in vitro* system for studying the effects of ozone on mammalian cell cultures and viruses. *Environ. Res.* 27:466–475.

Boyd, M. R. 1977. Evidence for the Clara cell as a site of cytochrome P-450 dependent mixed function oxidase activity in the lung. *Nature (Lond.)* 269:713–715.

Boyd, M. R., Statham, C. N., and Longo, N. S. 1980. The pulmonary Clara cell as a target for toxic chemicals requiring metabolic activation: Studies with carbon tetrachloride. *J. Pharmacol. Exp. Ther.* 212:109–114.

Boyum, A. 1964. Separation of white blood cells. *Nature (Lond.)* 204:793–794.

Chu, E. H. Y. 1971. Induction and analysis of gene mutations in mammalian cells in culture. In *Chemical Mutagens: Principles and Methods for Their Detection,* vol. 2, ed. A. Hollaender, pp. 411–444. New York: Plenum.

Devereux, T. R., and Fouts, J. R. 1980. Isolation and identification of Clara cells from rabbit lung. *In Vitro* 16:958–968.

Devereux, T. R., and Fouts, J. R. 1981. Xenobiotic metabolism by alveolar type II cells isolated from rabbit lung. *Biochem. Pharm.* 30:1231–1237.

Dobbs, L. G., Geppert, E. F., Williams, M. C., Greenleaf, R. D., and Mason, R. J. 1980. Metabolic properties and ultrastructure of alveolar type II cells isolated with elastase. *Biochim. Biophys. Acta* 618:510–523.

Douglas, W. H. J., and Kaighn, M. E. 1974. Clonal isolation of differentiated rat lung cells. *In Vitro* 10:230–237.

Douglas, W. H. J., Moorman, G., and Teel, R. 1976. The formation of histotypic structures from monodisperse fetal rat lung cells cultured on a three-dimensional substrate. *In Vitro* 12:373–381.

Fetner, R. H. 1962. Ozone-induced chromosome breakage in human cell cultures. *Nature (Lond.)* 194:793–794.

Freeman, G., Furiosi, N. J., and Haydon, G. B. 1966. Effects of continuous exposure of 0.8 ppm NO_2 on respiration of rats. *Arch. Environ. Health* 13:454–456.

Frazier, M. E., Hadley, J. G., Andrews, T. K., and Drucker, H. 1975. Use of thermolysin for the dissociation of lung tissue into cellular components. *Lab. Invest.* 33:231–238.

Goldstein, E., Jordan, G. W., MacKenzie, M. R., and Osebold, J. W. 1976. Methods for evaluating the toxicological effects of gaseous and particulate contaminants on pulmonary defense systems. *Annu. Rev. Pharmacol. Toxicol.* 16:447–463.

Greenleaf, R. D., Mason, R. J., and Williams, M. C. 1979. Isolation of alveolar type II cells by centrifugal elutriation. *In Vitro* 15:673-684.

Guerrero, R. R., Rounds, D. E., Booher, J., Olson, R. S., and Hackney, J. D. 1979a. Ozone sensitivity in aging WI-38 cells based on acid phosphatase content. *Arch. Environ. Health* 34:407-412.

Guerrero, R. R., Rounds, D. E., Olson, R. S., and Hackney, J. D. 1979b. Mutagenic effects of ozone on human cells exposed *in vivo* and *in vitro* based on sister chromatid exchange analysis. *Environ. Res.* 18:336-346.

Hagar, W. L., Sweet, W. E., and Sweet, F. 1981. An *in vitro* system for assessing lung cell response to ozone. *J. Air. Pollut. Control Assoc.* 31:993-995.

Katz, M. 1968. Analysis of inorganic gaseous pollutants. In *Air Pollution,* vol. 2, 2nd ed., ed. A. C. Stern, pp. 54-114. New York: Academic.

Kikkawa, Y., and Yoneda, K. 1974. The type II epithelial cell of the lung. I. Method of isolation. *Lab. Invest.* 30:76-84.

Kikkawa, Y., Yoneda, K., Smith, F., Packard, B., and Suzuki, K. 1975. The type II epithelial cells of the lung. II. Chemical composition and phospholipid synthesis. *Lab. Invest.* 32:295-302.

Kodama, F., Umeda, M., and Tsutsui, T. 1976. Mutagenic effects of sodium nitrite on cultured mouse cells. *Mutat. Res.* 40:119-124.

Latt, S. A., Schreck, R. R., Loveday, K. S., and Schuler, C. F. 1979. *In vitro* and *in vivo* analysis of sister chromatid exchange. *Pharmacol. Rev.* 30:501-535.

Marchok, A. C., Rhoton, J., Griesemer, R., and Nettesheim, P. 1977. Increased *in vitro* growth capacity of tracheal epithelium exposed *in vivo* to 7,12-dimethylbenz(a)anthracene. *Cancer Res.* 37:1811-1812.

Mason, R. J., and Williams, M. C. 1977. Type II alveolar cell. Defender of the alveolus. *Amer. Rev. Respir. Dis.* 115:81-91.

Mason, R., Williams, M. C., and Clements, J. A. 1975. Isolation and identification of type 2 alveolar epithelial cells. *Chest* 67:36S-37S (Suppl.).

McAllen, S. J., Chiu, S. P., Phalen, R. F., and Rasmussen, R. E. 1981. Effect of *in vivo* ozone exposure on in vitro pulmonary alveolar macrophage mobility. *J. Toxicol. Environ. Health* 7:373-381.

Myrvik, Q. N., Leake, E. S., and Farris, B. 1971. Studies on pulmonary alveolar macrophages from the normal rabbit. A technique to procure them in a high state of purity. *J. Immunol.* 86:128-132.

Pace, D. M., Thompson, J. R., Aftonomos, B. T., and Holck, H. G. O. 1961. The effects of NO_2 and salts of NO_2 upon established cell lines. *Can. J. Biochem. Physiol.* 39:1247-1255.

Pace, D. M., Landolt, P. A., and Aftonomos, B. T. 1969. Effects of ozone on cells in vitro. *Arch. Environ. Health* 18:165-170.

Rasmussen, R. E., and Crocker, T. T. 1981. A culture system for the direct exposure of mammalian cells to airborne pollutants. In *Short-term Bioassays in the Analysis of Complex Environmental Mixtures II,* eds. M. D. Waters, S. S. Sandhu, J. L. Huisingh, L. Claxton, and S. Nesnow, pp. 119-131. New York: Plenum.

Rasmussen, R. E., and Crocker, T. T. 1982. Lung cells grown on cellulose membrane filters as an in vitro model of the respiratory epithelium. In *Genotoxic Effects of Airborne Agents,* eds. R. R. Tice, D. L. Costa, and K. M. Schaich, pp. 105-120. New York: Plenum.

Rounds, D. E., and Bils, R. F. 1965. Effects of air pollutants on cells in culture. *Arch. Environ. Health* 10:251-264.

Ryan, U. S., and Ryan, J. W. 1980. Angiotensin converting enzyme: II. Pulmonary endothelial cells in culture. *Environ. Health Perspect.* 35:171-180.

Samuelsen, G. S., Rasmussen, R. E., Nair, B. K., and Crocker, T. T. 1978. Novel culture and exposure system for measurement of effects of airborne pollutants on mammalian cells. *Environ. Sci. Technol.* 12:426-430.

Schiff, L. J. 1977. Effect of nitrogen dioxide on influenza virus infection in hamster trachea organ culture. *Proc. Soc. Exp. Biol. Med.* 156:546-549.

Steele, V. E., Marchok, A. C., and Nettesheim, P. 1978. Establishment of epithelial cell lines following exposure of cultured tracheal epithelium to 12-*O*-tetradecanoyl-phorbol-13-acetate. *Cancer Res.* 38:3563–3565.

Teel, R. W. 1979. Induction of aryl hydrocarbon hydroxylase in primary cultures of type II alveolar lung cells and binding of metabolically activated benzo(a)pyrene to nuclear macromolecules. *Cancer Lett.* 7:349–355.

Tsuda, H., Kushi, A., Yoshida, D., and Goto, F. 1981. Chromosomal aberrations and sister chromatid exchanges induced by gaseous nitrogen dioxide in cultured Chinese hamster cells. *Muat. Res.* 89:303–309.

Vassallo, C. L., Domm, B. M., Poe, R. H., Duncombe, M. L., and Gee, J. B. L. 1973. NO_2 gas and NO_2^- effects on alveolar macrophage phagocytosis and metabolism. *Arch. Environ. Health* 26:270–275.

Voisin, C., Aerts, C., Jakubczak, E., and Tonnel, A. B. 1977a. La culture cellulaire en phase gazeuse. Un nouveau modele experimental d'étude *in vitro* des activites des macrophages alveolaires. *Bull. Eur. Physiopathol. Respir.* 13:69–82.

Voisin, C., Aerts, C., Jakubczak, E., Houdret, J. L., and Tonnel, A. B. 1977b. Effets du bioxyde d'azote sur les macrophages alveolaires en survie en phase gazeuse. Un nouveau modele experimental pour l'étude *in vitro* de lay cytotoxicité des gaz nocifs. *Bull. Eur. Physiopathol. Respir.* 13:137–144.

Weissbecker, L., Carpenter, R. D., Luchsinger, P. C., and Osdene, T. S. 1969. *In vitro* alveolar macrophage viability. Effect of gases. *Arch. Environ. Health* 18:756–759.

Zimmerman, F. K. 1977. Genetic effects of nitrous acid. *Mutat. Res.* 27:127–148.

Part Five

EXTRAPOLATION OF HOST DEFENSE RESPONSES AGAINST INFECTION

AN ASSESSMENT OF ANIMAL MODELS FOR TESTING THE EFFECT OF PHOTOCHEMICAL OXIDANTS ON PULMONARY SUSCEPTIBILITY TO BACTERIAL INFECTION

Elliot Goldstein

Division of Infectious and Immunologic Diseases, Department of Medicine, School of Medicine, University of California, Davis, California

In recent years, much information has accumulated proving that exposure to ambient levels of photochemical oxidants produce physiologic dysfunction and discomfort (DeLucia and Adams, 1977; Folinsbee et al., 1975; Linn et al., 1983). These studies have shown that exercising subjects exposed in environmental chambers to commonly occurring concentrations of ozone (0.15–0.30 ppm) have diminished respiratory function and symptoms of cough, dyspnea, and chest irritation. Exposure to higher and less frequently occurring ozone levels produce similar abnormalities in sedentary individuals (Hackney et al., 1975a, b). Although the evidence is less substantial, respiratory infections are another potential consequence of exposure to ambient concentrations of photochemical oxidants (Hammer et al., 1974; Durham, 1974; Melia et al., 1977; Speizer et al., 1980). Exposure to excessively high concentrations of ozone or nitrogen dioxide in industrial accidents causes pulmonary edema and pneumonia (Kleinfeld and Giel, 1956). Nursing and college students who are exposed to high ambient levels of photochemical pollution have increased incidences of transiently occurring pharyngitis and bronchitis (Hammer et al., 1974; Durham, 1974), and younger children exposed to excess nitrogen dioxide from gas or peat cooking stoves also have higher incidences of respiratory illness (Melia et al., 1977; Speizer et al., 1980). Even though these results do not prove a bacterial cause for pollutant-related respiratory illnesses, they have aroused interest in the possibility that pollutant exposure is a factor in respiratory bacterial infections. Moreover, the adverse effects attributable to pollutants are of particular concern in individuals with an underlying predisposition to bacterial infection, such as occurs in chronic obstructive pulmonary disease (Detels et al., 1981).

Ideally, definitive studies relating pollutant exposure to bacterial infections in humans should be performed using epidemiologic or volunteer methods of investigation. Unfortunately, the complexities inherent in epidemiologic investigations preclude attainment of the needed data in the immediate future. Volunteer studies are also unlikely to provide these data,

This work was supported by grant ES-00628 from the National Institute of Environmental Health Sciences and contract RP-2173-1 from the Electric Power Research Institute.

as ethical considerations restrict the induction of infection. Because of these investigational limitations, reliance must be placed on laboratory models of infection to determine the likelihood that ambient exposures to photochemical oxidants increase the risk of respiratory of bacterial infection (Goldstein et al., 1976).

As a surrogate for humans, rodent models relating the physiologic interaction between bacteria and host antibacterial defense systems have a number of virtues. Green's paper delineates the similarities of rodent and human antibacterial defenses that provide the basis for extrapolation (Green, 1984). Both defenses consist of an aerodynamic filtration system, a fluid layer lining the respiratory membranes, transport mechanisms for removing foreign particles, bacteria, and viruses, phagocytic cells (alveolar macrophages and when required polymorphonuclear leukocytes), and immune secretions of lymphocytes and plasma cells. In rodents and in humans, these components act in concert to maintain the lung free of bacteria.

The aerosol model in which rodents inhale bacteria and measurements are made of rates of physical removal and intrinsic bacterial killing is an excellent means of testing individual components as well as the entire antibacterial system. Moreover, as stated by Green, the model fufills the guidelines established by Weil for experiments to predict the safety of a material for man (Weil, 1972). When this rodent model is used, the animals are exposed for variable periods to different concentrations of photochemical oxidants and decreases in antibacterial activity represent quantitative impairments in function. Because of the similarities of human and rodent defense systems, similar impairments presumably occur in humans, and although quantitative dose-response extrapolations cannot be made from these rodent data, qualitative conclusions are justified. That is, the identification of a pollutant-induced inhibition in the ability of alveolar macrophages to kill inhaled bacteria, or identification of an impairment in mucociliary clearance, indicates that such derangements are to be expected in human exposures. Moreover, since these changes are forerunners of infection, pollutant testing in this physiologic model is a valid means of evaluating the propensity of a pollutant to produce bacterial infections in humans.

The importance of the alveolar macrophage, which is a principal component of mammalian antibacterial systems, is tested in the following manner. These critical cells are exposed directly to inhaled photochemical oxidants, and the effect of such exposures on chemotaxis, ingestion, and intracellular killing capacity is assessed *in vitro*. Again, deficits in function induced by the pollutant for rodent cells are likely to occur in human cells and to represent deviations from the heathly state. The advantage of investigations measuring cell function itself is that abnormalities observed with rodent cells are readily sought for in similar tests of alveolar macrophages obtained from nonhuman primates and humans by lavaged techniques. The disadvantage of *in vitro* tests is the lack of assurance that the pollutant exposure mimics the *in vitro* situation. As yet, correlative *in vitro* experiments testing the effect of photochemical oxidant exposures on rodents, nonhuman primates, and human alveolar macrophages have not been performed, but the methodologies are available and therefore future investigations of this relationship should provide valuable new data. Such information comparing the susceptibility of macrophages from different species to photochemical oxidants is a potentially sensitive means for determining pollutant toxicity and would provide quantitative information permitting extrapolation to humans.

Lippman and Schlesinger's paper describes the physiology of particle deposition and mucociliary clearance (Lippman and Schlesinger, 1984). This paper points out important differences in deposition pattern due to differing anatomy in humans and four-legged animals that limit intraspecies comparison of this parameter. In contrast, bronchial clearance measurements using radiolabelled particles show similar rates for humans and donkeys. Exposure to cigarette smoke also results in comparable reductions in bacterial clearance for donkeys and humans, providing further evidence that the donkey is a useful surrogate in determining the effect of toxins on this component of respiratory antibacterial defense. The effect of ozone exposure has not been tested in donkeys, but in view of studies showing substantial delays in mucociliary clearance in rats (Frager et al., 1979; Phalen et al., 1980), these investigations offer another means of quantitatively assessing the influence of photochemical oxidant exposure on pulmonary antibacterial defenses. In this regard, it is worth emphasizing that ozone (Miller, 1977), unlike NO_2, impacts maximally along the upper airways (Goldstein et al., 1977) where the mucociliary system is located, and that evidence exists linking impairments in mucociliary clearance to respiratory bacterial infections. Specifically, recurring bacterial bronchitis is a common accompaniment of Kartaganer's syndrome in which mucociliary clearance is abnormal due to an inherited absence of ciliated epithelium (Eliasson et al., 1977). Influenza also disturbs mucociliary function, predisposing the upper airways to bacterial superinfection (Wanner, 1977). Tests in donkeys showing that exposure to ambient levels of photochemical oxidants delay mucociliary transport rates would indicate an abnormality of potential consequence in humans.

The aforementioned papers emphasize the ability of photochemical oxidants to induce physiologic abnormalities that enhance susceptibility to bacterial infection. Whether such abnormalities result in disease or remain transient, self-limiting, and without further significance is discussed in Gardner's detailed review paper relating interference with antibacterial defenses by photochemical oxidants to the development of disease (Gardner, 1984). As noted by Gardner, the disease state is a complex biological phenomenon in which the uneradicated microbe elicits an injurious reaction. Injury is the critical pathophysiologic step in differentiating disease from disturbance when evaluating toxicty. This differentiation is of practical importance, since deviations from normal are sometimes inconsequential (such as the occurrence of asymptomatic and transient reductions in airway resistance). Because the aerosol model of rodent infection provides an excellent means for precisely measuring pollutant-induced physiologic abnormalities of antibacterial activity, substantiation of the likelihood of injury resulting from such physiologic impairments is important if these results are to be extrapolated to assessments of pollutant-induced illness. Gardner and Graham (Gardner and Graham, 1972; Coffin and Gardner, 1972). Coffin (Coffin and Blommer, 1967), Ehrlich (1963, 1966), and their colleagues have developed a rodent model often referred to as the "infectivity model" linking interference with antibacterial activity to disease and mortality. Instead of infecting rodents with minimally virulent staphylococci, which proliferate without provoking injurious responses (Goldstein et al., 1974; Pierce et al., 1977), they challenge rodents with highly virulent group C streptococci or kelbsiella species prior to or following the pollutant exposure. When these pathogens are used, the physiologic alterations in host antibacterial defenses induced by the pollutant result in bacterial proliferation and excessive mortality. Investigations with this infectivity model have shown that 2-h exposures to ozone concentrations as low as 0.1 ppm (Coffin and Blommer, 1965) or NO_2 concentrations as low as 3.5 ppm (Ehrlich, 1966) result in increased mortality. The infectivity model has also been used to show that exposure to NO_2 increases the susceptibility of nonhuman primates to a klebsiella infection (Henry et al., 1969). In these experiments, a 1-h exposure to 50 ppm nitrogen dioxide resulted in pneumonia and death. The difference between the intensity of exposure required to produce death in these two animal species emphasizes the qualitative and quantitative kinds of results obtained with these models. Because some comparable data exists in humans who have been exposed to high doses of NO_2 in industrial accidents (Ramirez and Dowell, 1971), one can conclude that NO_2-induced toxicity in the rodent model correlates with toxicity observed in humans, but that the human is approximately 10 times more resistant to the pollutant. These data also show that pollutant-induced physiologic abnormalities in antibacterial activity correlate with a predisposition to injury and death, confirming the importance of the physiologic disturbance.

Several conclusions are apparent from the abundance of information describing the effect of exposure to photochemical oxidants on the antibacterial defense mechanisms of the lung. First, sufficient similarity exists between defense mechanisms in rodents and humans to permit use of the rodent as a surrogate. Second, detection of pollutant-induced abnormalities in individual components of the pulmonary antibacterial system is a sensitive means of assessing potential toxicity. Third, pollutant-induced abnormalities in the individual components of the antibacterial defense systems result in diminished overall effectiveness of the system, and this in turn permits bacterial proliferation and the initiation of disease. Fourth, extrapolation from results in rodents to humans is qualitative. Quantitative relationships probably also exist, but the information required for such extrapolations is not available.

On the basis of the above conclusions, and assuming that no new epidemiologic or human volunteer studies defining the dose-response relationship for oxidants and bacterial infections will be forthcoming in the next few years, future experiments should attempt to quantify the dose-response relationships between photochemical oxidant induced disturbances in rodent and human components of the antibacterial defense system. The recent development of lavage techniques for procuring human alveolar macrophages, coupled with *in vitro* methods of testing, should permit a quantitative delineation of this relationship. Similarly, the dose-response relationship for photochemical oxidant-induced abnormalities in mucociliary transport rate can be inferred for humans by measuring the effect of photochemical oxidant exposure on mucociliary transport rates in donkeys, using the methods described by Lippman and Schlesinger. Methods for testing the influence of photochemical oxidant exposures and overall effectiveness of human pulmonary antibacterial defense systems are not possible with available methodologies. New and innovative means for determining the likelihood of a specific pollutant exposure provoking bacterial proliferation and disease in humans is necessary if this important relationship is to be defined. Pending the development of such methodologies, crude estimates can be inferred by comparing the relationships of components of the rodent and human system in regard to susceptibility to pollutant damage and, if the relationships are similar, using these relationships to extrapolate the rodent data to the human.

An additional area in which animal models may be useful is in assessing the influence of pollutant exposure to infection in humans with underlying disease. Animal surrogates for humans with chronic obstructive pulmonary diseases can be tested to determine if the condition predisposes to enhanced injury. Similarly, animal surrogates for humans with congestive heart failure, asthma, and other potentially predisposing illnesses can be tested to determine if these conditions render the host hypervulnerable to pollutant-induced infection. In summary, animal models provide important information delineating the pathophysiologic lesions accounting for pollutant induced enhancement in human susceptibility to pulmonary bacterial infection. At present, these are qualitative relationships, and quantitative extrapolations can be inferred but not made with certainty. Future studies should be directed at providing information quantifying these relationships, thereby permitting precise dose-response extrapolation from rodent models to humans.

Finally, two factors in this environmental problem deserve special emphasis. First, because the respiratory illnesses that result from exposure to photochemical oxidants may be due to irritative, viral, or bacterial etiologies, it is essential that volunteer and epidemiologic studies be performed, as well as animal investigations, to ensure that exposure to ambient levels of photochemical pollutants truly enhances bacterial infections in humans. Second, science progresses rapidly and in all likelihood newer technologies will be developed to provide more precise assessments of the individual components and overall function of pulmonary antibacterial defenses. These methodologies, particularly those applicable in humans, should be encouraged as a means of establishing newer and more definitive endpoints relating pollutant exposure to susceptibility to bacterial infection.

REFERENCES

Coffin, D. L., and Blommer, E J. 1965. The influence of cold on mortality from streptococci following ozone exposure. *J. Air Pollut. Control Assoc.* 15:523–524.

Coffin, D. L., and Blommer, E. J. 1967. Acute toxicity of irradiated auto exhaust. Its indication by enhancement of mortality from streptococcal pneumonia. *Arch. Environ. Health* 15:36–38.

Coffin, D. L., and Gardner, D. E. 1972. Interaction of biological agents and chemical air pollutants. *Ann. Occup. Hyg.* 15:219–234.

DeLucia, A. J., and Adams, W. C. 1977. Effects of O^3 inhalation during exercise on pulmonary function and blood biochemistry. *J. Appl. Physiol.* 43:82–85.

Detels, R., Sayre, J. W., Coulson, A. H., Rokaw, S. N., Massey, F. J. Jr., Tashikin, D. P., and Mei-Miau, Wu. 1981. The UCLA population studies of chronic obstructive respiratory disease. IV. Respiratory effect of long-term exposure to photochemical oxidants, nitrogen dioxide and sulfates on current and never smokers. *Am. Rev. Respir. Dis.* 124:673–680.

Durham, W. H. 1974. Air pollution and student health. *Arch. Environ. Health* 28:241–254.

Ehrlich, R. 1963. Effect of air pollutants on respiratory infection. *Arch. Environ. Health* 6:638–642.

Ehrlich, R. 1966. Effect of nitrogen dioxide on resistance to respiratory infection. *Bacteriol. Rev.* 30:604–614.

Eliasson, R., Mossberg, B., Cammer, P., and Afzelius, B. A. 1977. The immobile-cilia syndrome. A congenital ciliary abnormality as an etiologic factor in chronic airway infections and male sterility. *N. Eng. J. Med.* 297:1–6.

Folinsbee, N. J., Silverman, F., and Shepard, R. J. 1975. Responses following ozone exposure. *J. Appl. Physiol.* 38:996–1001.

Frager, N. B., Phalen, R. F., and Kenoyer, J. L. 1979. Adaptation to ozone in reference to mucociliary clearance. *Arch. Environ. Health* 34:51–57.

Gardner, D. E. 1984. Oxidant-induced enhanced sensitivity to infection in animal models and their extrapolations to man. *J. Toxicol. Environ. Health* 13:423–439.

Gardner, D. E., and Graham, J. A. 1972. Increased pulmonary disease mediated through altered bacterial defenses in pulmonary macrophage and epithelial cells. *Proceedings of the Sixteenth Annual Hartford Biology Symposium, eds. C. L. Sanders, R.* P. Schneider, G. E. Dogel, and H. A. Ragan, pp. 1–21. Richland, Washington: Technical Information Center.

Goldstein, E., Lippert, W. E., and Warshauer, D.. 1974. Pulmonary alveolar macrophage: Defender against bacterial infections of the lung. *J. Clin. Invest.* 54:519–528.

Goldstein, E., Jordan, G. W., MacKenzie, M. R., and Osebold, J. W. 1976. Methods of evaluating the toxicological effects of gaseous and particulate contaminants on pulmonary microbial defense systems. *Ann. Rev. Pharmacol. Toxicol.* 16:447–463.

Goldstein, E., Peek, N. F., Parks, N. J., Hines, H. H., Steffey, E. P., and Tarkington, B. 1977. Fate and distribution of inhaled nitrogen dioxide in rhesus monkeys. *Am. Rev. Respir. Dis.* 115:403–412.

Green, G. M. 1984. Similarities of host defense mechanisms against pulmonary infectious diseases in animals and man. *J. Toxicol. Environ. Health* 13:471–478.

Hackney, J. D., Linn, W. S., Law, D. C., Karrza, S. K., Greenberg, H., Buckley, R. D., and Pederson, E. E., 1975a. Experimental studies on human health effects of air pollutants III. Two hour exposure to ozone alone and in combination with other pollutant gases. *Arch. Environ. Health* 30:385–390.

Hackney, J. D., Linn, W. S., Mohler, J. G., Pederson, E. E., Breisacher, P., and Russo, A. 1975b. Experimental studies on human health effects of air pollutants II. Four-hour exposure to ozone alone and in combination with other pollutant gases. *Arch. Environ. Health* 30:379–384.

Hammer, D. I., Hasselblad, V., Portnoy, B., and Wehrle, P. F. 1974. Los Angeles student nurse study. Daily symptom reporting and photochemical oxidants. *Arch. Environ. Health* 28:255–260.

Henry, M. C., Ehrlich, R., and Blair, W. H. 1969. Effect of nitrogen dioxide on resistance of squirrel monkeys to *Klebsiella pneumoniae* infection. *Arch. Environ. Health* 18:580–587.

Kleinfeld, M., and Giel, C. P. 1956. Clinical manifestations of O_3 poisoning. *Am. J. Med. Sci.* 231: 638–643.

Linn, W. S., Avol, E. L., Shamoo, D. A., Anzar, U. T., Venet, T. G., and Hackney, J. D. 1983. Comparative respiratory effects of ozone (O_3) and ambient oxygen exposure during heavy exercise. *Am. Rev. Respir. Dis.* 127(Part II):159 (abstract).

Lippman, M., and Schlesinger, R. B. 1984. Interspecies comparisons of particle deposition and mucociliary clearance in trachcobronchial airways. *J. Toxicol. Environ. Health* 13:441–469.

Melia, R. J. W., Florey, C. V., Altman, D. G., and Swan, A. V. 1977. Association between gas cooking and respiratory disease in children. *Br. Med. J.* 2:149–152.

Miller, F. J. 1977. A mathematical model of transport and removal of ozone in mammalian lungs. Thesis, North Carolina State University, Raleigh.

Phalen, R. F., Kenoyer, J. L., Crocker, T. T., and McClure, T. R. 1980. Effects of sulfate aerosols in combination with ozone in elimination of tracer particles inhaled by rats. *J. Toxicol. Environ. Health* 6:797–810.

Pierce, A. K., Reynolds, R. C., and Harris, G. D. 1977. Leukocyte response to inhaled bacteria. *Am. Rev. Respir. Dis.* 116:679–684.

Ramirez, R. J., and Dowell, A. R. 1971. Silo-filler's disease: Nitrogen dioxide-induced lung injury. Long-term follow-up and review of the literature. *Ann. Int. Med.* 74:569–576.

Speizer, F. E., Ferris, B. Jr., Bishop, Y. M. M., and Spengler, J. 1980. Respiratory disease rates and pulmonary function in children associated with NO_2 exposure. *Am. Rev. Respir. Dis.* 121:3–10.

Wanner, A. 1977. Clinical aspects of mucociliary transport. *Am. Rev. Respir. Dis.* 116:73–125.

Weil, C. S. 1972. Statistics versus safety factors and scientific judgment in the evaluaton of safety for man. *Toxicol. Appl. Pharm.* 21:454–463.

OXIDANT-INDUCED ENHANCED SENSITIVITY TO INFECTION IN ANIMAL MODELS AND THEIR EXTRAPOLATIONS TO MAN

Donald E. Gardner

Northrop Services, Inc., Environmental Sciences,
Research Triangle Park, North Carolina

Animal studies have provided the toxicologist with useful and scientifically sound information indicating that exposure to oxidant gases can alter the functioning of the host's normal pulmonary defense system, resulting in an increased risk to infectious disease. Since the basic mechanisms of action of the human and the animal pulmonary defenses are similar, it is reasonable to relate these animal's biological responses to human exposures. This paper examines the possibility of quantitatively extrapolating such animal dose-response data to humans.

INTRODUCTION

The use of animal models for predicting and evaluating the response of humans to exposure to environmental chemicals has long been and will continue to be the basic tool used by the toxicologist. Although the uncertainties of using animals to assess the risk in humans are numerous and must also be carefully evaluated, these studies are irreplaceable in predictive toxicology.

The host's pulmonary defense mechanisms against infectious disease have been shown to be sensitive to a variety of gases and particles, especially ozone (O_3) and nitrogen dioxide (NO_2). The basic mechanism of action of these pulmonary defense systems are remarkably similar for humans and animals. The extensive data base on oxidant-induced enhanced sensitivity to infection in animals should provide a sound foundation for eventually extrapolating these effects to humans. Thus, to examine this possibility further, it would appear desirable to (1) identify those defense mechanisms that best lend themselves to analysis as well as interpretation of actual health risk; (2) determine to what extent oxidants significantly modify these functions; (3) explore theoretically the extent to which one can relate these animal studies to human effects; and (4) explore practical future research to provide species sensitivity data, which, when related to dosimetric data, would enable quantitative extrapolation.

To these ends, subsequent papers in this journal volume will address the major aspects of pulmonary host defenses in detail. The purpose of this discussion is to provide an overview and present the general perspective for the more specific discussions that will follow.

Various biological and physical factors contribute to interspecies differences, making any direct extrapolation difficult until their quantitative nature is determined. For example, the deposition of inhaled substances within the respiratory tract can be significantly different in various mammalian species depending on the respiratory tract morphology, the route of breathing, the depth and rate of airflow, the physiochemical properties of the mucus and surfactant layers, the physical processes that govern the transport of the inhaled gas or particle, and physiochemical properties of the test substance, such as solubility, particle size, shape, and diffusivity. Even when dosimetry data are more fully developed, the problems of species response to a given dose and possible species differences in manifestations of toxicity still exist. For example, within the lung, the deposited chemical can elicit a wide variety of pulmonary injury. These effects are often described in terms of standard physiological, biochemical, or morphological endpoints and result from the action of the inhaled substance on the pulmonary tissue. Although these endpoints are especially suitable for identifying toxic response in the lungs, they are often not easily extrapolated to humans.

The respiratory tract has a distinct position in toxicology. The lung is the only organ in which there is direct contact between the contents of environmental air and the vital functioning components of this organ. Thus, the total respiratory system serves as the site of initiation of deleterious responses to inhaled toxicants (Table 1). In addition to its vital role in the maintenance of oxygen supply and the excretion of carbon dioxide, the lung must also maintain an efficient and functioning complex system of several body defenses to ensure that unwanted, toxic material entering the body through this pathway will not cause numerous ramifications on other systems.

An area of inhalation toxicology that has undergone substantial expansion in recent years involves the understanding of the normal functioning of the lung defense mechanisms and how various noxious environmental substances can impair their ability to function properly. Elaborate defense mechanisms have evolved in humans and animals that enable the mammalian lung to protect itself from the constant assault of

TABLE 1. Pulmonary Injury Due to Environmental Agents

Mechanisms of the effect	Biological effect	Disease state
Proteolysis	Elastin and collagen destruction	Emphysema
Fibrosis	Increased connective tissue	Scarring, stiffness
Pulmonary edema, inflammation, altered alveolar stability	Increased fluid and serum components, altered surfactant, changes in cell number and type	Decreased gas exchange, increased susceptibility to infection
Altered host defenses	Cytotoxic effect on macrophages altered mucociliary transport	Altered susceptibility to microorganisms
Degenerative changes	Calcification and autolysis	Necrosis

potentially hazardous chemicals and infectious microorganisms. Despite great differences in body size and respiratory tract morphology, most mammals appear to have pulmonary clearance phenomena with remarkably similar mechanisms (Kenoyer et al., 1981). Thus, it is reasonable to reevaluate the existing data base on mucociliary clearance and alveolar macrophage functioning reported in animal toxicology studies to improve our understanding of the possible human risk associated with these observations. The available evidence from these investigations clearly indicates that oxidants such as O_3 and NO_2 can penetrate beyond the nasopharyngeal region and alter the normal functioning of these major pulmonary defenses to such a degree that the exposed subject is compromised and becomes vulnerable to secondary insults (Gardner, 1982a,b; Gardner et al., 1977a, 1979). Fortunately, considerable data are available for examination and comparison. Although most of this data base is based on animal studies, the strong possibility exists that a similar response could occur in humans.

The critical questions that still exist are, how far can these animal studies be extrapolated for predicting human health effects, and what is the precise dose of the chemical needed to elicit a similar response in the human adult or child? Perhaps the answers to these questions will be forthcoming from the research efforts generated by this group of papers.

APPROACH

To meet the unique goals and objectives of this report, we should not be overly concerned with the actual concentrations of O_3 and NO_2 required to elicit these responses, but instead we should examine the basic similarities and differences in how the individual host's defense system reacts to the oxidant assault. One can measure both the direct effect of the environmental agent on lung tissue or on specific pulmonary cells, as well as measure the indirect effects resulting from this initial response. Such an examination of the relevant data will improve our understanding of how such an alteration may be involved in the expression and pathogenesis of human disease.

Host-environment interaction will be examined first by evaluating multiple effects of oxidants on two vital clearance systems present in the mammalian lung. Two major phases of pulmonary clearance have been identified by kinetic studies: the first, or rapid, phase is equated with tracheobronchial clearance; the second, which is much slower and involves the alveolar macrophage, is the transfer of particles out of the deeper portions of the lung. Such biphasic clearance has also been demonstrated in humans (Lippmann et al., 1971). Pulmonary defense ultimately depends upon the closely integrated functioning of the mucociliary clearance mechanism and the phagocytic activity of the alveolar macrophage.

The mucociliary system is one of the lung's primary defense mechanisms. It protects the conducting airways by trapping and quickly sweeping away various cellular elements, debris, and inhaled material that have been deposited or are being cleared from the alveolar region. The effectiveness of this protective mechanism can be determined by measuring such biological parameters as the rate of transport of deposited material, the frequency of ciliary beating, the structural integrity of the ciliated cells, and the size, number, and distribution of the mucus-secreting cells. Once this defense has been altered, a build-up of viable and nonviable inhaled substances would be expected to occur, which may jeopardize the host's health depending on the nature of the substance.

The alveolar macrophage is the first line of defense within the gaseous exchange region of the lung. In addition to acting as scavengers in removing cellular debris, these cells also detoxify and remove inhaled particles, maintain pulmonary sterility, and interact with lymphoid cells for immunological protection. To adequately fulfill their purpose, these defense cells must maintain active mobility, a high degree of phagocytic activity, an integrated membrane structure, and a well-developed and functioning enzyme system.

If normal functioning of either of these two defense mechanisms is suppressed by the inhaled substance, then the resulting effect would be expected to have serious consequences on the health of the impaired individual. The following two sections will review the specific effects of O_3 and NO_2 on these two systems so that this information can be used to draw certain correlations between small and human respiratory defense impairment.

MUCOCILIARY DEFENSE

In humans and animals, the effectiveness of the mucociliary system depends on the integrity of the cilia and the respiratory epithelia, the chemicophysical properties of the mucus produced, and the rate of mucus transport. Exposure to a variety of noxious substances including formaldehyde, cigarette smoke, airborne particles, trace metals, and sulfuric acid can cause effects such as dysfunction, slowing of the transport rate, epithelial cell damage, or excessive mucus production. If this system is unbalanced or impaired, clearance of any viable or toxic deposited particle would be delayed, increasing the probability of disease.

Strong evidence exists that oxidant gases such as O_3 and NO_2 disrupt the normal activity of this system. A number of studies using mice, rats, and cats have reported morphological damage to the ciliated cells of the tracheobronchial tree (Boatman et al., 1974; Castleman et al., 1977; Freeman et al., 1974; Stephens et al., 1978). After such an exposure, the cilia are either completely absent or had become noticeably shorter. These effects appear to be dose-related and were reversible after the exposure ended.

A method for measuring ciliary activity of the tracheal ciliary epithelium is by directly observing and measuring the ciliary beating frequency in normal and exposed animals (Dalhamn, 1960; Adler et al., 1973; Tanaka, 1967). In these studies, a number of experimental animals have been used, and the data indicate that the beat rates are different, suggesting the existence of species difference in the animals (Table 2). However, the standard deviations of the ciliary beat rate are relatively small, which signifies that the reliability of the data is high although the methods employed by each author are different.

Only a few investigators have studied the influence of oxidant pollutants on the beating activity of the tracheal cilia. *In vitro* exposure to O_3, NO_2, NO, HCHO, or SO_2 reduced or completely stopped the cilia beating activity (Schiff, 1977; Oomichi and Kita, 1974). Unfortunately, the exposure levels employed in these studies were extremely high. In the latter study, the authors compared the ability of these five chemicals to affect cilia activity and found that SO_2 was the most toxic, followed by HCHO, NO_2, and NO, with the lowest being O_3. These authors suggested that SO_2 and HCHO are dissolved easily in the mucus and thus reacted quickly with the cilia, but the less soluble gases passed over the mucus surface of the upper airway, and thus less is dissolved in the mucus. Therefore, the cilia beneath the mucus surface are influenced little by these gases.

Grose et al. (1980) measured *in vitro* ciliary activity in hamsters following *in vivo* exposure to O_3. Exposure to 196 μg O_3/m^3 for 3 h did not alter the ciliary beating frequency. However, an acute sequential exposure to O_3 (196 $\mu g/m^3$ and H_2SO_4 (1090 $\mu g/m^3$) significantly reduced ciliary activity immediately after the exposure. The effect could be attributed entirely to the acid aerosol. However, the complexity of this issue is apparent from a similar study (Grose et al., 1982), where one can compare the sequential exposure effect to the exposure to a O_3 and H_2SO_4 mixture, which produced an antagonistic response.

Although the morphological studies visually indicated that oxidant pollutants caused significant epithelial changes including loss of cilia, depressed exfoliation, blebbing of cells, hypertrophy, and focal hyperplasia, only *in vitro* measurement data exist for O_3 and NO_2 that provide evidence of such exposure actually slowing the beating ability of these cells.

TABLE 2. Example of Ciliary Activity (beats/min) in Various Animal Species

Animal	Ciliary beat rate/min
Rabbit (*in vivo*)	1099 ± 171
Cats (*in vivo*)	871 ± 151
Rabbits (*in vivo*)	1140 ± 95
Guinea pigs (*in vivo*)	667 ± 100
Hamsters (*in vitro*)	1004 ± 71
Humans (*in vivo*)	1000 —
Rats (*in vitro*)	996 ± 96

Tracheobronchial epithelium is generally stimulated to produce more mucus after exposure to respiratory irritants. This induction varies with dose, host factors, climatic conditions, and pollutant. Such an effect can be a significance in the genesis of obstructive airway disease. Mucus hypersecretion has been morphologically correlated with airway mucosal cell hypertrophy and hyperplasia, as well as submucosal gland hypertrophy, and has been histochemically correlated with increases in sulfomucins and sialidase-resistant sialomucins relative to sialidose-sensitive and neutral glycoprotein (Last and Cross, 1978). A biochemical model for assessing biological effects of atmospheric pollutants on respiratory tract mucus production has demonstrated the potentiating effects of ozone on airway mucus production. This method allowed for observation of quantitative and qualitative modifications of secreted mucus during controlled environmental exposures (Last et al., 1977; Last and Kaizu, 1980). Last et al. exposed rats to O_3 at 1568 $\mu g/m^3$ for up to 90 d. Data indicate that secretion rates of glycoprotein were significantly decreased initially (2–3 d). As O_3 exposure continued, the secretion rates became significantly greater. Monkeys failed to exhibit such alteration, indicating that there may be a species difference in either sensitivity or dosimetry of O_3 between rats and monkeys.

Most studies do not characterize ciliary motility and the quantity of the mucus separately. In such studies, the interaction of the two processes, mucociliary transport, is the variable measured. However, the data would indicate that ciliary beat frequency and amplitude, as well as viscosity and depth of the serous fluid, correlate with the total transport clearance rate. This mucociliary transport rate can be determined by visually monitoring the rate of movement of particles placed within excised tracheobronchial system or an incised and externalized *in situ* system. Better physiological measurements may be obtained by externally monitoring the movement of inhaled radioactive particles in the intact animals and humans.

Evidence indicates that O_3 and NO_2 exposures can cause a substantial delay in rat, rabbit, and sheep mucociliary clearance rates (Kensler and Battista, 1966; Giordano and Morrow, 1972; Abraham et al., 1980; Frager et al., 1979; Kenoyer et al., 1981; Phalen et al., 1980). Concentration-response data indicate that the clearance impairment was dose-related and that the effect could be significantly enhanced by elevated humidity and the interaction of O_3 with H_2SO_4 (Phalen et al., 1980). These effects were probably reversible, since the animals that recovered in clean air had a clearance rate similar to that of controls. Unfortunately, there are essentially no data for effects of prolonged exposure to O_3 and NO_2 on ciliary functional activity or for mucociliary transport rates measured in the intact animal. Although some of these effects are reversible, one should not minimize their significance. Such an effect in highly susceptible individuals, particularly the newborn and aged and those who have cardiopulmonary infirmities and chronic pulmonary disease, may constitute an unacceptable risk to their well-being. In addition, the absence of clearance data after chronic exposure does not necessarily mean an absence of effects. As mentioned earlier, long-term exposure to oxidants causes morphological changes in ciliated epithelium that probably would have an influence on clearance. Table 3 summaries the effects of oxidants on mucociliary defense.

ALVEOLAR MACROPHAGE

As the first line of defense for the alveolar surface, the pulmonary macrophage must (1) isolate ingested particles by phagocytosis; (2) act as a vehicle for physical transport from the lung; and (3) inactivate or detoxify inhaled and ingested microbes or chemicals. The sequence of events that must take place for this protective process to function is very complex and involves a number of interrelated biological functions (i.e., physiology, biochemistry, immunology). Also, chemicals can interfere with this function at many sites. Numerous studies have reported that both gaseous and particulate airborne chemicals can affect the ability of the macrophage to perform these functions adequately, and consequently, these effects could be expected to increase the host's risk of pulmonary disease (Table 4).

The response of the alveolar macrophage to oxidant gases provides a sensitive indicator of functional changes in the lung. Inhalation of O_3 and NO_2 results in an influx of polymorphonuclear leukocytes and a reciprocal decrease in the number of macrophages recovered from the lung by

TABLE 3. Effects of Oxidants on Mucociliary Defense

Parameter	Effects	Animal	Reference
Morphological damage *in vivo*	Cilia absent or shorter, abnormal shapes, hyperplasia (dose-related)	Mice Rats Cats	Boatman et al. (1974); Castleman et al. (1977); Freeman et al. (1974); Stephens et al. (1978)
Ciliary beating *In vitro*	Reduced or eliminated	Hamsters Guinea pigs	Schiff (1977); Oomichi and Kita (1974)
In vivo	No effect with O_3 alone but reduced with H_2SO_4 NO_2 decreased activity	Hamsters	Schiff (1977); Grose et al. (1980)
Mucus production *in vivo*	O_3 altered normal secretion rates; no effect on monkeys	Rats Monkeys	Last et al. (1977); Last and Cross (1978); Last and Kaizu (1980)
Mucociliary transport	Delayed clearance (dose-related)	Rats Sheep	Kenoyer et al. (1981); Phalen et al. (1980); Fragen et al. (1979); Giordano and Morrow (1972); Abraham et al. (1980)

TABLE 4. Environmental Chemicals that Alter Normal Pulmonary Cellular Defenses

Particles	Gases
Diesel exhaust	CO
PbO, Pb_2O_3	NO_2
$NiCl_2$, NiO	O_3
$CdCl_2$, CdO	
Cigarette smoke	
Chrysotile	
Fly ash	
V_2O_5	
M_3O_4	

pulmonary lavage (Coffin and Gardner, 1972; Gardner, 1982a,b). Such a rapid decline in viable macrophages suggests that the cells are either being destroyed or rendered more susceptible to lysis, which is supported by increased osmotic fragility (Dowell et al., 1970), increased lysis (Gardner and Graham, 1977), or reduction in the number of alveolar macrophages recovered from O_3-exposed animals. Other examples of the disruption of macrophages membrane integrity by oxidant gases have been described by Hadley et al. (1977), Gardner et al. (1971), and Sherwin et al. (1968). Table 5 illustrates the effects of oxidants on the alveolar macrophage.

The depression of macrophage phagocytic and bacteriocidal function as measured by the ability of these cells to handle bacteria and viruses has also been studied (Gardner, 1982b). These functional effects of decreased phagocytosis and bacteriocidal activity are supplemented by a quantitative reduction of macrophage acid hydrolases. Even at concentrations as low as

TABLE 5. Effects of Oxidant Gases on Alveolar Macrophages

Parameter	Effects	Animal	Reference
Viability	More susceptible to lysis; reduction in numbers, increased fragility, cytotoxic	Rabbit	Coffin et al. (1968); Sherwin et al. (1968); Dowell et al. (1970); Gardner et al. (1971)
Phagocytosis	Decreased ability to ingest bacteria, particles; reduced motility	Rabbit Rat	Coffin and Gardner (1972); Goldstein et al. (1978); McAllen et al. (1981); Schwartz and Christman (1979)
Defense against bacterial and viral infection	Diminished rate of bacterial killing, decreased lysosomal activity, reduced interferon production	Rabbit Rat	Kimura and Goldstein (1981); Goldstein et al. (1973); Hurst et al. (1970); Hurst and Coffin (1971); Amoruso et al. (1981); Ibrahim et al. (1976); Nakajima et al. (1980)

0.25 ppm, O_3 induced marked depression of acid phosphatase, beta-glucuronidase, and lysozyme (Hurst and Coffin, 1970; Hurst et al., 1971). The depression of macrophage ability to produce interferon, a substance that defends the host against viral infections, has also been attributed to the effects of O_3 and NO_2. Nakajima et al. (1980), Acton and Myrvik (1972), and Ibrahim et al. (1976) have reported that after exposure to these oxidants, macrophages and tracheal epithelial cells produce significantly less interferon than controls. Continuing investigations of the interaction of virus and macrophage in the alveolus may provide important information for future clinical studies.

CONSEQUENCES FROM ALTERED HOST DEFENSES

The information just provided clearly indicates that the respiratory tract has an array of complex, intricate, and interlocking defense mechanisms to detoxify and physically remove inhaled material by cellular and acellular processes. Exposure to numerous environmental agents, including O_3 and NO_2, can make these defenses less effective in many ways. From these results, one could predict that the possible consequences resulting from this altered host's defense may be sufficient enough to increase the host's susceptibility to infections. These predictions then would have a much stronger scientific base if there were supporting evidence that these "impaired" animals were also more susceptible to pulmonary infectious disease.

The establishment of a disease state is very complex. For example, the interaction of host and microorganism is an extremely common biological phenomenon. Some factors involved in disease establishment are: the host; the causative agent; several other characterized modifiers, such as stress, preexisting disease, dietary deficiency; and interaction with numerous noxious substances such as smoking or environmental pollutants, which can modify and complicate the etiology and pathogenesis of the disease. An infection is the establishment of a host-microorganism interaction, and it requires only the presence of the host and organism. When clinically manifested injury to the host results from this interaction, an overt infectious disease has occurred. At the other extreme, colonization occurs when the organism is deposited in the host and proliferates enough to maintain its original numbers, but does not incite any host reaction. Between colonization and overt infection lies covert or subclinical infection. The establishment of the disease state is not a result of a simple one-to-one cause and effect relationship, but instead is a reflection of several different influences acting on the host simultaneously. The establishment of an overt infectious disease requires a specific potential pathogen and additional factors that render a host vulnerable to microbial invasion. Figure 1 illustrates the complexity of the establishment of

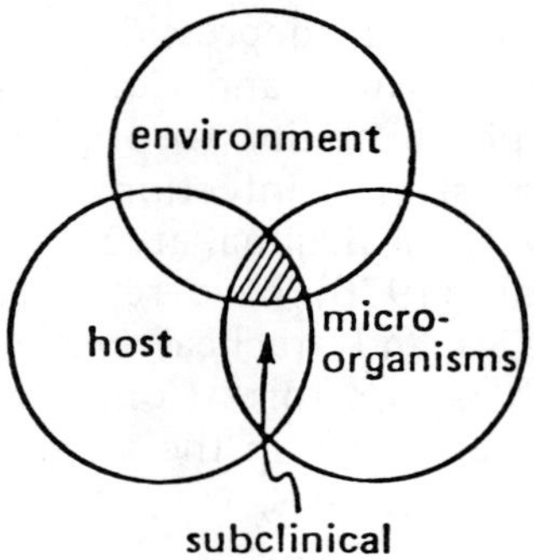

FIGURE 1. Biological interaction of host–microorganism–environment in the disease process.

disease. Thus, the question arises, do small decrements in the functioning of these various host defense mechanisms compromise the host so that it is unable to defend itself against an opportunistic pathogen?

In addition, a multitude of host functions are interrelated in defense. Often when these indicators are "dissected" for mechanism studies, for example, only a few representative lysosomal enzymes of alveolar macrophages are routinely examined in biochemical studies and potential compensatory mechanisms are not evaluated, making it extremely difficult to predict the actual health risk, even for the animal species studied.

To answer this question, models have been developed to determine the effectiveness of the host defenses by introducing an infectious microorganism into the respiratory tract of the test animals, thereby establishing a laboratory-induced infection. This microbial infection alone does not cause a high incidence of mortality. The organism must proliferate to cause disease; that is, the internal milieu of the host must be nutritionally suitable and the microbe must be able to withstand the defenses. Any biologically significant alteration in these defenses should assist the microbe in prolonging its life within the pulmonary milieu and permit its multiplication. Ultimately, if uncontrolled, death of the test animal would result (Miller et al., 1978; Gardner, 1982a,b). This infectious disease test would indicate weakened defense against the microbe, reflecting the summation of various deleterious changes within the lung including edema, reduced phagocytic or microcidal activity, inflammation, and immunosuppression. Such a model has been developed for bacteria and has been shown to be valid across species, to allow for reproducible data, to be a sensitive indicator of the effects of a wide range of pollutants, to be supported by mechanistic experimental information, and to be capable of epidemiological confirmation.

A number of investigators have shown that exposure to O_3 and NO_2 so alters the defense system of these test animals that they can no longer cope with an infectious agent (Gardner et al., 1977b, 1979; Ehrlich et al., 1978; Abraham et al., 1982). This is in contrast to the control animals that were not exposed to the gases but received the identical pulmonary infection. These effects were supported by numerous mechanistic studies that indicated the ineffective functioning of the bacteriocidal capabilities of the lungs exposed to the pollutant gas.

Hence, animal studies have provided the environmental toxicologist with some very useful and scientifically sound information clearly indicating that oxidant exposure increases the incidence of pulmonary infectious disease. This concept is strongly supported by *in vivo* and *in vitro* data describing the various mechanisms of action that account for this increased susceptibility to infection. Thus, little doubt should exist concerning the validity and reliability of this animal data. However, these animal findings need to be related to human effects. This can be accomplished either by human clinical studies, epidemiological studies, or through some form of extrapolation.

As usual, the total available toxicological data base represented by clinical and epidemiological studies is quite small compared to the data available from animal studies. The animal experimentation provides information using a wide range of concentrations, exposure regimens, biological parameters, and animal species. Epidemiological studies suggest human health effects but cannot provide direct "cause-and-effect proof." Community studies must cope with a host of complex covariates, are restricted to a limited range of exposures, and often provide information too late for preventive measures. With controlled human studies, the safety of the volunteer is of the greatest importance, chronic exposures cannot be performed, and only limited biological parameters can be measured using noninvasive techniques. Thus, in clinical studies, the evaluation of toxic effects must be restricted of acute and reversible effects. Trying to relate dose-response results obtained in animal studies to the human situation is currently very difficult, even qualitatively.

A HUMAN COUNTERPART TO THE INFECTIOUS DISEASE MODEL

Under usual epidemiological study conditions, it is immensely difficult to quantitatively identify chemicals that cause disease because of many confounding variables. To make maximum use of animal infectivity data, one should seek a human exposure situation that closely mimics the environmental conditions most influential in predisposing the animals to the infectious disease.

For this presentation, we can use the available epidemiological evidence suggesting that in homes where gas stoves are used for cooking, respiratory infections increase when compared to homes with electric stoves. Although several similar studies are on-going, published data can still be useful in this discussion.

The emissions from gas-cooking ranges are exhausted directly into the household atmosphere. Numerous studies have shown that the average indoor concentrations may be at least five times higher than outdoor NO_2 levels.

Epidemiological studies conducted in the United States and Great Britain indicate that the incidence of respiratory illness, bronchitis, and cough is higher in children living in homes with gas stoves. Melia et al. (1977) reported that children between ages 6 and 11 in British homes using gas stoves experienced an increase in prevalence of bronchitis and increases of "day or night cough" and "colds going to the chest." In a more recent study, Melia et al. (1979) found that the incidence of respiratory illness increased in children aged 5 to 10 living in urban homes with gas stoves. There also appeared to be an association between illness and gas stove homes for girls under the age of eight who lived in rural areas. However, the NO_2 levels associated with these effects cannot be determined precisely since homes other than those of the study subjects were investigated (Goldstein et al., 1979). Thus, the Melia studies can only provide qualitative support for similar investigations.

Speizer et al. (1980) studied 6 communities in the United States and found that children under age 2 living in homes with gas stoves had an increase in serious respiratory illness and a decrease in pulmonary function. This study and a related one (Spengler et al., 1979) reported NO_2 levels in homes in excess of 470 $\mu g/m^3$ (0.25 ppm) and as high as 940 $\mu g/m^3$ (0.5 ppm). The 95th percentile of the 24-h average NO_2 concentration ranged from 38 to 113 $\mu g/m^3$ (0.02 to 0.06 ppm) for gas-stove homes and 19 to 94 $\mu g/m^3$ (0.01 to 0.05 ppm) for electric-stove homes.

Major uncertainties associated with these various epidemiological studies were what specific chemical agent(s) (CO, NO, HCHO, particles, or NO_2) caused the response and what exposure level (concentration and averaging time) was associated with the reported effects. The animal studies previously described can be helpful in addressing these uncertainties.

If this animal model is to be used to reflect the toxicological response occurring in humans, then the endpoint for comparison of such studies should not be mortality, since today few individuals die of bacterial pneumonias. A better comparison in humans would be the increased prevalence of respiratory illness in the community. Such a comparison is proper since both mortality (animals) and morbidity (humans) result from a loss in pulmonary defenses.

In an early study, Gardner et al. (1979) examined the interaction between concentration C and time T of exposure to determine if the product $C \times T$ is directly proportional to the toxic effect. Although the $C \times T$ product remained constant, the response was significantly different when C and T were varied. The data indicate that concentration is more important than the length of exposure. Such information illustrates the need for obtaining and examining detailed monitoring data in epidemiological studies to understand the quantitative cause of the effect. Thus, a simple arithmetic average (i.e., the product of $C \times T$) could not predict the response.

It is important to compare the intermittent and continuous exposure regimens, because many individuals are exposed intermittently rather than continuously. Therefore, an experiment was designed in which mice were exposed either continuously or 7 h/d to two concentrations (2800 or 660 μg NO/m^3; 1.5 or 3.5 ppm) (Gardner, 1979; Gardner et al., 1979). At the higher concentration, streptococdal-induced mortality increased significantly for each of the experimental groups with increasing lengths of exposure, and the effect observed for the two exposure modes was not significantly different at any time period tested. Thus, complete recovery did not occur during the clean air portion of the intermittent exposure. At the lower concentration after approximately 100 h of intermittent exposure (administered over about 2 wk), the response was again similar regardless of the mode of exposure.

We also must know how the subject responds to single and multiple peak exposures of various lengths and what effect these peaks have when superimposed on a lower basal level of exposure. Graham et al. (1979) and Gardner et al. (1983) exposed the animals to a single peak concentration of 8100 $\mu g/m^3$ (4.5 ppm) for various lengths of time (1, 3.5, and 7 h) and then immediately challenged with the infectious agent. This exposure regimen resulted in an increase in percentage mortality with increasing lengths of the spike exposures. However, if the laboratory infection was induced 18 h after cessation of the NO_2 exposure, mortality was not enhanced, which indicated that NO_2 damage to host defense systems was repaired during the 18 h of clean air. When these same spikes were superimposed instead on a constant background level of NO_2 (2800 $\mu g/m^3$; 1.5 ppm), a different pattern of response was observed. The ratio of peak to basal concentration was maintained at a 3:1 ratio. This ratio was selected because it is similar to that frequently seen in urban communities. The test animals in this study failed to recover from the spike exposure, even though they were placed in an atmosphere containing much less NO_2.

Gardner et al. (1982, 1983) studied the effect of multiple spikes on a lower continuous background concentration of NO_2 (2800 $\mu g/m^3$, 1.5 ppm). Spikes were superimposed twice a day, in the morning and afternoon, and the bacterial challenge was administered either immediately before or after the morning spike. These studies indicated the complexity involved in trying to define the toxicological response associated with various but realistic exposure patterns. These studies confirmed that NO_2 causes significant increases in mortality (i.e., loss of host defenses); however, the pattern and magnitude of the response was affected significantly by the specific exposure patterns.

Although the gas-stove studies alone provide useful information, the animal infectivity model data add support and elucidate significant uncertainties. Thus, the evidence from animal infectivity studies shows that NO_2 was probably the causative agent in the epidemiological studies and that concentration of NO_2 is more important than length of exposure in increasing susceptibility to infectious pulmonary illness. The short-term peaks observed in the gas-stove homes were probably the major causative factor in the effects observed.

SUMMARY

Ozone and NO_2 cause decrements in host defenses of several animal species. Because of the major importance of properly functioning defenses and the potential for oxidants to affect these systems in humans, quantitative extrapolation of these effects from animals to humans becomes essential.

When the data from a toxicological experiment are used to assess human risk, special consideration must be given to experimental design to provide the maximal useful information. In addition to being carefully designed and properly carried out, one should follow certain other guidelines that would make the data more relevant for extrapolation to humans:

1. Use the species whose biological and physiological handling of the material most closely resembles that of humans.
2. Use several animal species. If an effect is seen in a number of species, a similar effect will probably be seen in humans.
3. Use several dose levels, since toxicological and pharmacological action in humans and animals are dose-related.
4. Whenever possible, try to determine the dose delivered to the target organ under experimental conditions to predict human dose under similar exposure conditions.

REFERENCES

Abraham, W. M., Januszkiewicz, A. J., Mingle, M., Welker, M., Wanner, A., and Sackner, M. A. 1980. Sensitivity of bronchoprovocation and tracheal mucous velocity in detecting airway responses to O_3. *J. Appl. Physiol.* 48:789–793.

Abraham, W. M., Lauredo, I., Sielczak, M., Verger, L., King, M. M., and Ratzan, K. 1982. Enhancement of bacterial pneumonia in sheep by ozone exposure. Abstract. *Am. Rev. Respir. Dis.* (Supplement)126:148.

Acton, J. D., and Myrvik, Q. N. 1972. Nitrogen dioxide effects on alveolar. *Arch. Environ. Health* 24:48–52.

Adler, K. B., Wooten, O., and Fulfano, M. J. 1973. Mammalian respiratory mucociliary clearance. *Arch. Environ. Health* 27:364–369.

Amoruso, M. A., Witz, G., and Goldstein, B. 1981. Decreased superoxide anion radical production by rat alveolar macrophages following inhalation of ozone or nitrogen dioxide. *Life Sci.* 28:2215–2221.

Boatman, E. S., Sato, S., and Frank, R. 1974. Acute effects of ozone on cat lungs. *Am. Rev. Respir. Dis.* 110:157–169.

Castleman, W. L., Tyler, W. S., and Dungworth, D. L. 1977. Lesions in respiratory bronchioles and conducting airways in monkeys exposed to ambient levels of ozone. *Exp. Mol. Pathol.* 26:384–400.

Coffin, D. L., and Gardner, D. E. 1972. Interaction of biological agents and chemical air pollutants. *Ann. Occup. Hyg.* 15:219–235.

Coffin, D. L., Gardner, D. E., Holzman, R. S., and Wolock, F. J. 1968. Influence of ozone on pulmonary cells. *Arch. Environ. Health* 16:633–636.

Coffin, D. L., and Gardner, D. E., and Blommer, E. J. 1976. Time-dose response for nitrogen dioxide exposure in an infectivity model system. *Environ. Health Perspect.* 13:11–15.

Dalhamn, J. 1960. The determination *in vivo* of the rate of ciliary beating in the trachea. *Acta Physiol. Scand.* 49:242–250.

Dowell, A. R., Lohrbaner, L. A., Hurst, D., and Lee, S. D. 1970. Rabbit alveolar macrophages damage caused by *in vivo* O_3 inhalation. *Arch. Environ. Health* 21:121–127.

Ehrlich, R., Findlay, J. C., and Gardner, D. E. 1978. Susceptibility to bacterial pneumonia in animals exposed to sulfates. *Toxicol. Lett.* 1:325–330.

Frager, N. B., Phalen, R. F., and Kenoyer, J. L. 1979. Adaptation to ozone in reference in mucociliary clearance. *Arch. Environ. Health* 34:51–57.

Freeman, G., Juhos, L. T., Furios, N. J., Mussenden, R., Stephen, R. J., and Evans, M. J. 1974. Pathology of pulmonary disease from exposure to independent ambient gases. *Arch. Environ. Health* 29:203–210.

Gardner, D. E. 1979. Influences of exposure patterns of nitrogen dioxide on susceptibility to infectious respiratory disease. In *Biological Studies of Environmental Pollutants*, ed. S. D. Lee, pp. 267–288. Ann Arbor, Mich.: Ann Arbor Press.

Gardner, D. E. 1982a. Use of experimental airborne infections for monitoring altered host defenses. *Environ. Health Perspect.* 43:99–107.

Gardner, D. E. 1982b. Effects of gases and airborne particles on lung infections. In *Air Pollution—Physiological Effects*, eds. J. J. McGrath and E. B. Barnes, pp. 47–79. New York: Academic.

Gardner, D. E., and Graham, J. A. 1977. Increased pulmonary disease mediated through altered bacterial defenses. In *Pulmonary Macrophages and Epithelial Cells*, eds. C. L. Sanders, R. P. Schneider, G. E. Dagle, and H. A. Ragen, pp. 1–21. ERDA Symposium, Series 43, Washington, D.C.

Gardner, D. E., Pfitzer, E. A., Christian, R. T., and Coffin, D. L. 1971. Loss of protective factor for alveolar macrophages when exposed to ozone. *Arch. Intern. Med.* 127:1078–1084.

Gardner, D. E., Coffin, D. L., Sidorenko, G. I., and Pinigin, M. A. 1977a. The role of time as a factor in toxicity of chemical compounds in intermittent and continuous exposure. *J. Toxicol. Environ. Health* 3:811–828.

Gardner, D. E., Miller, F. J., Illing, J. W., and Kirtz, J. M. 1977b. Increased infectivity with exposure to ozone and sulfuric acid. *Toxicol. Lett.* 1(2):59–64.

Gardner, D. E., Miller, F. J., Blommer, E. J., and Coffin, D. L. 1979. Influence of exposure mode on toxicity of NO_2. *Environ. Health Perspect.* 30:23–29.

Gardner, D. E., Miller, F. J., Illing, J. W., and Graham, J. A. 1982. Non-respiratory function of the lungs: Host defenses against infection. In *Air Pollution by Nitrogen Oxides*, eds. T. Schneider and L. Grant, pp. 401–417. New York: Elsevier.

Gardner, D. E., Graham, J. A., Illing, J. W., Blommer, E. J., and Miller, F. J. 1983. Impact of exposure patterns on the toxicological response to NO_2 and modifications by added stressors. *1980. Proc. U.S.-U.S.S.R. 3rd Joint Symp. Problems of Environmental Health.* Suzdal, U.S.S.R., Oct. 1979, pp. 17–41.

Giordano, A. M., and Morrow, P. E. 1972. Chronic low level nitrogen dioxide exposure and mucociliary clearance. *Arch. Environ. Health* 25:443–449.

Goldstein, E., Bartlema, H. C., van der Ploeg, van der Duijn, van der Stap, and Lippert W. 1978.

Effect of ozone on lysosomal enzymes of alveolar macrophages engaged in phagocytosis and killing of *Staphylococcus aureus*. *J. Infect. Dis.* 139:299–311.

Goldstein, B. D., Melia, R. J., Chinn, S., Glorey, C. V., Clark, D., and John, H. H. 1979. The relation between respiratory illness in primary school children and the use of gas for cooking. *Int. J. Epidemiol.* 8:339–345.

Graham, J. A., Gardner, D. E., and Menzel, D. B. 1979. Health effects issues of NO_X emissions. *Proceed. EPRI Central Technology Symposium*, ed. J. E. Cichanowicz, pp. 2-1-2-46. EPRI FP-1109-SR. Palo Alto, Calif.: Electric Power Research Institute.

Grose, E. C., Gardner, D. E., and Miller, F. J. 1980. Response of ciliated epithelium to ozone and sulfuric acid. *Environ. Res.* 22:377–385.

Grose, E. C., Richards, J. H., Illing, J. W., Miller, F. J., Davies, D. W., Graham, J. A., and Gardner, D. E. 1982. Pulmonary host defense response to inhalation of sulfuric acid and ozone. *J. Toxicol. Environ. Health* 10:351–362.

Hadley, J. G., Gardner, D. E., Coffin, D. L., and Menzel, D. B. 1977. Effects of ozone and nitrogen dioxide exposure on rabbits on the binding of autologous red cells to alveolar macrophages. In *International Conference on Photochemical Oxidants and Its Control*, vol. 1, ed. B. Dimitriades, pp. 505–512. EPA-600-377-001. U.S. Environmental Protection Agency, Washington, D.C.

Hurst, D. J., and Coffin, D. L. 1971. Ozone effect on lysosomal hydrolases of alveolar macrophages. *In Vitro Arch. Intern. Med.* 127:1059–1063.

Hurst, D. J., Gardner, D. E., and Coffin, D. L. 1970. Effects of O_3 on acid hydrolases of the pulmonary alveolar macrophage. *Res. J. Reticuloendothel. Sol.* 8:288–300.

Ibrahim, A. L., Zeo, Y. C., and Osebold, J. W. 1976. The effects of ozone on the respiratory epithelium and alveolar macrophages of mice: I. Interferon production. *Proc. Soc. Exp. Biol. Med.* 152:483–488.

Kenoyer, J. L., Phalen, R. F., and Davis, J. R. 1981. Particle clearance from the respiratory tract as a test of toxicity. *Exp. Lung Res.* 2:111–120.

Kensler, C. J., and Battista, S. D. 1966. Chemical and physical factors affecting mammalian ciliary activity. *Am. Rev. Respir. Dis. (Suppl.)* 93:93–102.

Kimura, A., and Goldstein, E. 1981. Effect of ozone on concentration of lysozyme in phagocytic alveolar macrophages. *J. Inf. Dis.* 143:247–251.

Last, J. A., and Cross, C. E. 1978. A new model for health effects of air pollutants: Evidence for synergistic effects of mixtures of ozone and sulfuric acid aerosols on rat lungs. *J. Lab. Clin. Med.* 91:328–339.

Last, J. A., and Kaizu, T. 1980. Mucus glycoprotein secretion by tracheal explants: Effects of pollutants. In *Experimental Models for Pulmonary Research*, eds. D. E. Gardner, E. P. Hu, and J. A. Graham, *Environ. Health Perspectives* 35:131–138.

Last, J. A., Jennings, M., Schwartz, L. W., and Cross, C. E. 1977. Glycoprotein secretion by tracheal explants cultured from rats exposed to ozone. *Am. Rev. Respir. Dis.* 116:695–703.

Lippmann, M., Albert, R. E., and Peterson, H. T. 1971. The regional deposition of inhaled aerosols in man. In *Inhaled Particles* III, ed. W. H. Walton, pp. 105–122. Old Woking, Surrey, England: Unwin Bros.

McAllen, S. J., Chiu, S. P., Phalen, R. F., and Rasmussen, R. E. 1981. Effects of ozone, nitrogen dioxide, and sulfur dioxide on the experimentally induced allergic respiratory disorders in guinea pigs. IV. Effects on respiratory sensitivity to inhaled acetylcholine. *Am. Rev. Respir. Dis.* 120:429–439.

Melia, R. J., Glorey, C. V., Chinn, S., Altman, D. S., and Swan, A. V. 1977. Association between gas cooking and respiratory disease in children. *Br. Med. J.* 2:149–152.

Melia, R. J., Glorey, C. V., and Chinn, S. 1979. The relation between respiratory illness in primary school children and the use of gas for cooking. *Int. J. Epidemiol.* 8:333–388.

Miller, F. J., Illing, J. W., and Gardner, D. E. 1978. Effects of urban ozone levels on laboratory-induced respiratory infections. *Toxicol. Lett.* 2:163–169.

Nakajima, T., Kusumoto, S., Oda, H., Nogami, H., and Shingu, H. 1980. Biological effects of exposure to photochemical reaction mixture. In *Proc. 5th Clean Air Congress*, pp. 785–789, Buenos Aires, Argentina.

Oomichi, S., and Kita, H. 1974. Effect of air pollutants on ciliary activity of respiratory tract. *Bull. Tokyo. Med. Dent. Univ.* 21:327–343.

Phalen, R. F., Kenoyer, J. L., Crocker, T. T., and McClure, T. R. 1980. Effects of sulfate aerosols in combination with ozone on elimination of tracer particles inhaled by rats. *J. Toxicol. Environ. Health* 6:797–810.

Schiff, L. J. 1977. Effects of nitrogen dioxide on influenza virus infection in hamster trachea organ culture. *Proc. Soc. Exp. Biol. Med.* 156:546–549.

Schwartz, L. W., and Christman, C. A. 1979. Alveolar macrophage migration. Influence of lugn lining material and acute lung insult. *Am. Rev. Respir. Dis.* 120:429–439.

Sherwin, R. P., Richters, V., Brooks, M., and Buckley, R. D. 1968. The phenomenon of macrophage congregation *in vitro* and its relationship to *in vivo* NO_2 exposure of guinea pigs. *Lab. Invest.* 18:269–277.

Speizer, F. E., Ferris, B. G., Bishop, Y. M., and Spengler, J. 1980. Respiratory disease rates and pulmonary function in children associated with NO_2 exposure. *Am. Rev. Respir. Dis.* 121:3–10.

Spengler, J. D., Ferris, B. G., and Dockery, D. W. 1979. Sulfur dioxide and nitrogen dioxide levels inside and outside homes and the implications on health effects research. *Environ. Sci. Techno.* 13:1276–1281.

Stephens, R. K., Dzubay, T. G., Russwurm, G., and Rickel, D. 1978. Sampling and analysis at atmospheric sulfates and related species. *Atmos. Environ.* 12:55–68.

Tanaka, T. 1967. Experimental studies on the ciliary activity in the trachea. *J. Otolaryngol. Jap.* 70:2091–2107.

INTERSPECIES COMPARISONS OF PARTICLE DEPOSITION AND MUCOCILIARY CLEARANCE IN TRACHEOBRONCHIAL AIRWAYS

Morton Lippmann, Richard B. Schlesinger

Institute of Environmental Medicine, New York University Medical Center, New York, New York

Inhaled insoluble particles that deposit along normal healthy tracheobronchial airways of humans and other mammals are transported on the proximally moving mucous lining to the larynx, where they are swallowed. The transit time from the most distal ciliated airways varies from 0.1 to 1 d, with each individual having a relatively constant, characteristic time. The exact time course of clearance depends on the distributions of both particle deposition and mucus velocities along the airways.

There are too few data on intrabronchial deposition and mucociliary transport rates for laboratory animals to permit a thorough intercomparison among species. However, enough is known about the relative lung sizes and anatomical differences among the various species to make some preliminary, but important, distinctions. As compared to commonly used experimental animals, humans have larger lungs and a more symmetric upper bronchial airway branching pattern. In addition, humans do considerable oral breathing, thus bypassing the effective air cleaning capability of the nasal airways. These differences contribute to a greater amount of upper bronchial airway particle deposition in humans, as well as to greater concentrations of deposition on localized surfaces near airway bifurcations.

Airborne irritants that deposit in small ciliated airways may produce marked changes in mucociliary transport. Such materials include cigarette smoke, submicrometer-sized sulfuric acid mist, nitrogen dioxide, and ozone. For cigarette smoke and sulfuric acid, which have been studied for transient effects following single brief exposures in both humans and animals, the responses are similar. Upon repetitive exposures in animals, both of these irritants produce persistant alterations in clearance rates and airway morphometry. Studies of the effects of ozone on mucociliary clearance have, up to now, been limited to tests of the responses of rats to single exposures. The similarities between the known effects of various irritants suggests a nonspecific response.

INTRODUCTION

One of the important host defense mechanisms against chronic disease and acute infection is mucociliary clearance. It acts to remove insoluble particles from the conductive airways by surface transport to the

The research described in this paper was supported by research grant ES-00881 from the National Institute of Environmental Health Sciences (NIEHS), by research contract RP-1157 with the Electric Power Research Institute, and by research contract 68-01-1726 from the U.S. Environmental Protection Agency. It is part of Center program ES-00260 of NIEHS, and Center program CA-13343 of the National Cancer Institute.

esophagus, where they are swallowed. The particles are transported on a superficial gel component of the mucous layer of the airways by the momentum transferred to it via the synchronous beating of cilia situated within a serous hypophase. Mucus transport velocities decrease distally, and the residence time within the lungs for a particle deposited on the airway mucus depends on both the site of initial particle deposition, and the transport velocities at this site and along the more proximal airways. The total time required for the completion of bronchial clearance varies considerably among healthy humans, ranging from about 2.5 to 24 h (Albert et al., 1973). Overall bronchial clearance times for most experimental animal species are not known. However, at least for donkeys (Albert et al., 1968) and rabbits (Chen and Schlesinger, 1983), they are similar to those in humans.

Although noninvasive measurements of mucus transport velocity can be performed in the trachea and main bronchi, this is not possible in more distal airways, due to the complex geometrical distribution of these airways and the limited spatial resolution of available imaging systems. The only practical indices of mucociliary clearance are those based on serial measurements of total thoracic retention. Since clearance of particles deposited in the nonciliated airways and airspaces beyond the terminal bronchioles is so slow, removal from that region during the first day after exposure is usually negligible. Thus, the decline in particle retention within the thorax during the first day represents the cumulative removal, via the trachea, of particles deposited at various sites along the bronchial tree.

Since the ultimate shape of the retention curve depends on the distribution both of particles and of mucus transport velocities along the airways, it is not possible to determine either distribution from the overall curve without knowing the characteristics of the other. For example, rapid clearance could be due to a concentration of particle deposition in proximal airways, or to relatively rapid surface transport from more distal airways. Thus, interpretation of bronchial clearance data requires an appreciation of the influence of the intrabronchial deposition pattern.

The pattern of deposition within the airways is dependent on airway size and branching structure, and on breathing pattern during aerosol inhalation. These factors vary significantly within a given species and, to a much greater extent, between species. The differences between humans and the commonly used experimental animals in airway size and branching pattern alone can account for substantial differences in upper bronchial tree particle deposition. The larger size of human airways results in larger Reynolds numbers for the same flow velocities. As a result, turbulent flow, which occurs in the larger airways in humans at moderate to high rates of ventilation and can contribute greatly to particle deposition, is rare or absent in animal airways, especially in sedentary animals. Another anatomic difference between humans and the common experimental animals may also contribute to relatively less upper bronchial airways particle deposition in the latter. The animals' tracheas are much longer in relation to their diameters than are human tracheas. Thus, the turbulence introduced by jet flow through the larynx is much less likely to persist into the bronchi and contribute to particle deposition in the lungs of the former; the long trachea facilitates the establishment of a more parabolic slow profile.

The branching pattern of the airways in humans is substantially different from that in all other animals (Schlesinger, 1980), with the possible exception of the higher primates. As shown in Fig. 1, humans have a relatively symmetrically dichotomous branching pattern, while the four-legged animals have a highly asymmetric branching pattern, known as monopodal. In human airways, each bifurcation is close to the centerline of the parent airway, where the particles have the highest velocities. This favors deposition "hot-spots" on the carinal ridges of bifurcations during inhalation (Fig. 2*a*). On the other hand, in monopodal branching, the bulk of the flow continues into the major daughter with little change in velocity or direction (Fig. 2*b*). The smaller flow fraction enters the minor daughter from slower moving streamlines nearer the wall of the parent airway. In such a system, there is much less tendency for concentrated areas of deposition.

In human upper bronchial airways, there are also deposition "hot spots" associated with exhalation flow (Fig. 3*a*). These occur on opposite sides of the parent airway directly downstream from the two daughter branches (Schlesinger et al., 1983). Such "hot spots" are less likely to occur in monopodal branching systems (Fig. 3*b*), since the major daughter and the parent airways are closely aligned, and the small jet coming out of the minor daughter is not likely to persist across the full diameter of the parent.

Humans are also subject to more upper bronchial deposition than are most animals because they do much more oral breathing; the commonly used experimental animals are, except for canines, obligate nasal breathers.

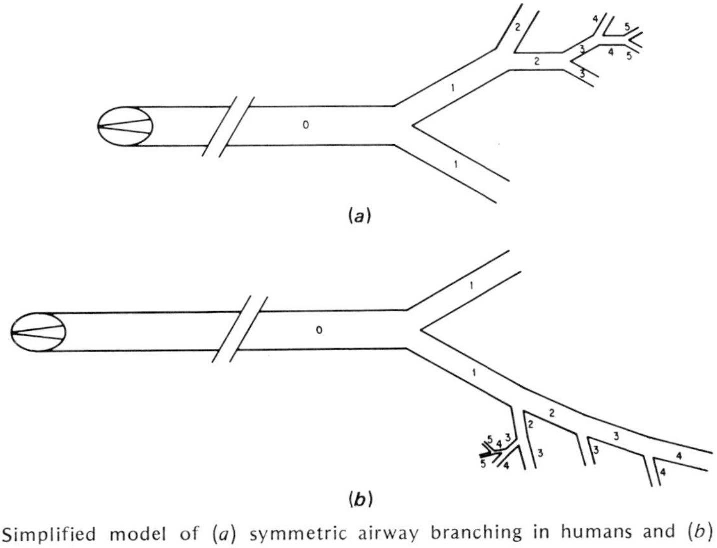

FIGURE 1. Simplified model of (*a*) symmetric airway branching in humans and (*b*) a symmetric branching pattern in experimental animals.

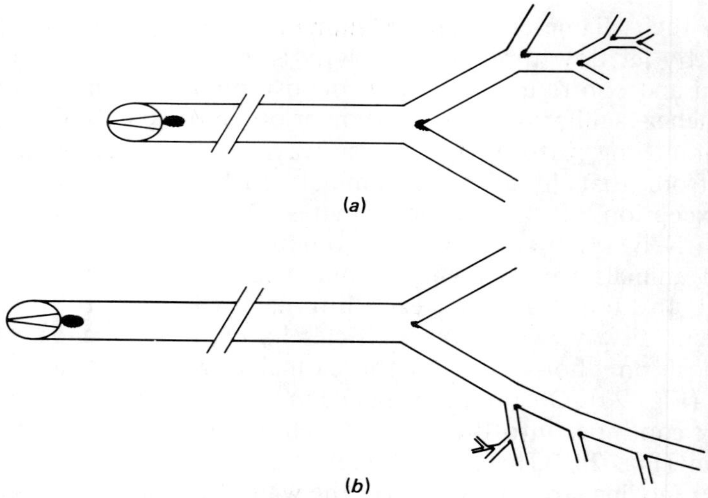

FIGURE 2. Location of deposition "hot spots" for inspiratory flow in (*a*) humans and (*b*) experimental animals.

Oral breathing permits greater particle penetration to the lungs than does nasal breathing. In addition, animal nasal passages are, in general, more efficient particle filters than are those of humans.

All of these deposition-influencing factors limit the applicability of animal models for particle-related diseases of the human upper bronchial airways. Interspecies differences in bronchial airway particle deposition

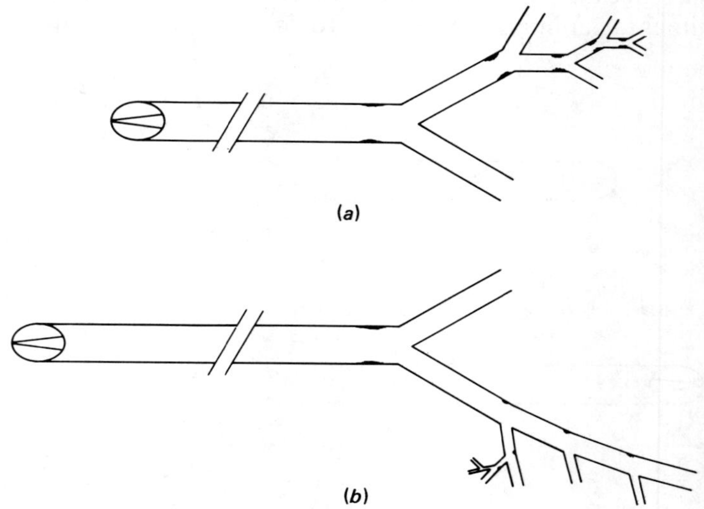

FIGURE 3. Locations of deposition "hot spots" for expiratory flow in (*a*) humans and (*b*) experimental animals.

also limit interspecies comparisons of bronchial mucociliary clearance, or will until we can differentiate between the contribution of particle distribution and mucus velocities to clearance patterns.

Interspecies differences in mucociliary clearance can also be attributed to differences in epithelial structure and cell distribution. The mucus glands present in humans and other large animals do not occur in some smaller animals, and the distribution of ciliated and secretory cells may also differ.

The discussion that follows reviews the structural and functional factors affecting mucociliary clearance in humans and experimental animals, as well as the limited amount of experimental data on deposition and clearance and the influence of airborne irritants on these functions.

PARTICLE DEPOSITION

Particles deposit at various sites within the bronchial tree by several mechanisms. Deposition efficiency at each airway level depends on the aerodynamic properties of the particles, the anatomy of the airways, and the geometric and temporal patterns of flow.

Deposition Mechanisms

Figure 4 illustrates the five mechanisms by which significant particle deposition can occur: impaction, sedimentation, diffusion, interception, and electrostatic precipitation. In most cases, however, only impaction, sedimentation, and diffusion will be important.

The inhaled air follows a tortuous path through the branching airways in the lung. Each time the air changes direction, the momentum of particles tends to keep them on their preestablished trajectories, which can cause them to impact on airway surfaces. The most likely deposition sites are at or near airway bifurcations.

Gravitational sedimentation is an important mechanism for deposition

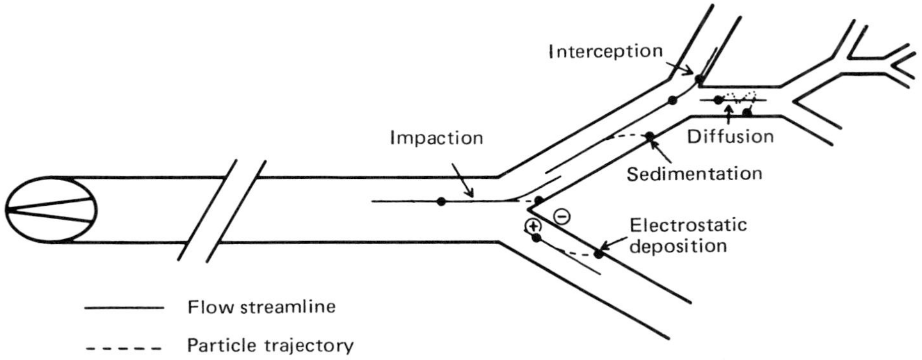

FIGURE 4. Mechanisms for particle deposition in the respiratory tract.

in the smaller bronchi, the bronchioles, and the alveolar spaces, where the airways are small and the air velocity is low. Sedimentation becomes less effective than diffusion when the terminal settling velocity of the particles falls below 0.001 cm/s, which for unit-density spheres is equivalent to a diameter of 0.5 μm.

Submicrometer-sized particles in air undergo a random motion caused by the impact of gas molecules. This Brownian motion increases with decreasing particle size and becomes an effective mechanism for particle deposition in the lung as the root-mean-square displacement approaches the size of the air space. Diffusional deposition is important in small airways and alveoli, and at airway bifurcations for particle sizes below 0.5 μm. For radon and thoron progeny, where the particle size is molecular, diffusional deposition efficiency can be high in the head and in large airways such as the trachea.

Interception is usually significant only for fibrous particles. It occurs when the trajectory of a particle brings it close enough to a surface so that an edge contacts the surface. Thus, the particle size must be a significant fraction of the airway diameter.

Particles with high electric mobility can have an enhanced respiratory tract deposition even though no external field is applied across the chest. Deposition results from the image charges induced on the surface of the airways by the charged particles. Test aerosols from evaporation of aqueous droplets can have substantial mobilities, and the results of some experimental deposition studies using such aerosols without charge neutralization are accordingly suspect. Freshly fractured mineral-dust particles may also be highly charged. Most aerosols have lower charge levels, however, and the deposition due to charge is usually small in comparison to deposition by the preceding mechanisms.

There are several ways of expressing particle size. Nonspherical particles are often characterized in terms of equivalent spheres—for example, on the basis of equal volumes, equal masses, or aerodynamic drag. The most commonly used unit is called aerodynamic diameter (D), which incorporates both particle density and drag. Aerodynamic diameter is the most appropriate unit in terms of particle deposition by impaction and sedimentation; diffusional displacement depends only on particle size and not on density or shape. Interception also depends on the linear dimensions of the particle, including its shape, since aerodynamic drag can affect orientation within the airway.

Influence of Airway Anatomy on Deposition

The influence of airway branching pattern on particle deposition was discussed briefly in the introduction. Unfortunately, there are virtually no data available on flow distribution in monopodal airways to extend the discussion or to provide any quantitative insights on deposition patterns.

Among humans and experimental animals, individual variations in airway anatomy affect particle deposition in several ways: (1) the diameter of the airway influences the displacement required by the particle before it contacts the airway surface; (2) the cross section of the airway determines the flow velocity for a given volumetric flow rate; and (3) the variations in diameter and branching patterns along the bronchial tree affect the mixing characteristics between the tidal and reserve air in the lungs. For particles with aerodynamic diameters below 2 μm, convective mixing can be the most important factor determining deposition efficiency.

There are also significant individual differences in respiratory-tract anatomy. For example, the average alveolar-zone air-space size in humans has a substantial coefficient of variation, when measured either post-mortem on lung sections or *in vivo* by aerosol persistence during breath holding. In the former, Matsuba and Thurlbeck (1971) reported a mean size (±SD) of 0.678 mm (±0.236). In the latter, Lapp et al. (1975) found values of 0.535 mm (±0.211). The tracheobronchial airways have a comparable degree of intersubject variability. Chan and Lippmann (1980) developed an index of average bronchial airway size based on *in vivo* deposition, and reported a coefficient of variation for nonsmoking healthy adults of 0.23.

Influence of Respiratory Flow Factors

Increasing air velocity enhances impaction deposition, but decreases sedimentation and diffusion by reducing residence time. Respiratory flow is cylical and reverses many times per minute. At its peak, it may be turbulent in the human trachea, but the Reynolds number decreases with increasing lung depth, so that in the smaller conducting airways it is always laminar and, in the avelolar region, viscous (Davies, 1973).

Since the flow is laminar in most of the anatomical dead space, the central core velocity approaches twice the average velocity. Thus, even in very shallow breathing, a substantial fraction of the inhaled volume penetrates beyond the anatomical dead space, and particles with appreciable sedimentation rates—e.g., diameter > 2 μm—or larger diffusional displacements—e.g., diameter < 0.1 μm—can deposit efficiently in peripheral airways.

The tidal volume is an important respiratory factor affecting deposition. The air inhaled at the start of each breath goes deeper into the lung and remains there longer than the air inhaled later in the breath. It follows that the deeper the air goes and the longer it stays, the greater its depletion of inhaled particles. Thus, for quiescent breathing, when the air velocity is low, mixing is minimal, and the tidal volume is only two to three times the dead space volume, a larger proportion of the inhaled particles can be exhaled. Conversely, for heavy exertion when greater volumes are inhaled at higher velocities, both impaction in the large airways and sedimentation and diffusion in the smaller airways and alveoli will increase.

Overall Tracheobronchial Deposition

The only published *in vivo* measurements of human tracheobronchial deposition are those of Lippmann and Albert (1969), Lippmann (1977), Foord et al. (1978), Stahlhofen et al. (1980), Chan and Lippmann (1980), and Emmett et al. (1982). Data for nonsmokers are shown in Fig. 5. Tracheobronchial deposition for a given particle size varies greatly from subject to subject, but each individual has a characteristic and reproducible relationship between particle size and deposition. Tracheobronchial deposition includes that occurring both by impaction in the larger airways and by sedimentation in the smaller airways. Impaction deposition predominates for larger particles ($D > 3$ μm) and high flow rates ($>\sim 20$ l/min), while sedimentation deposition becomes a larger fraction of a diminishing tracheobronchial component for smaller particles and lower flows.

For the tracheobronchial zone, the pattern of deposition within hollow airway casts made from normal human lungs shows that the airway bifurcations accumulate a disproportionate share of the tracheobronchial deposit (Schlesinger and Lippmann, 1972; Schlesinger et al., 1977). Furthermore, there appears to be a close correspondence between the density of the surface deposition on the bifurcation regions and the origin of primary bronchogenic cancer (Schlesinger and Lippmann, 1978). For example, among the lobar bronchi, the largest number of tumors are found in the right upper bronchus and the lowest in the right middle bronchus. Figure 6 shows cast deposition data for the lobar bronchi for various particle sizes and flow rates. While the absolute amounts of deposition vary considerably with particle size and flow rate, the relative amounts on the

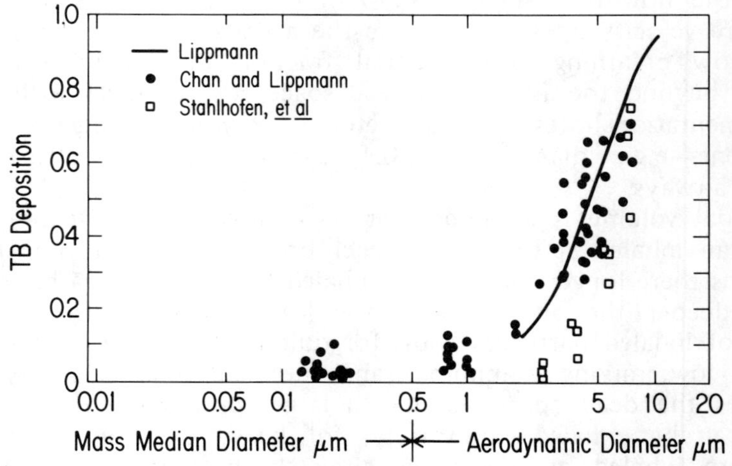

FIGURE 5. Tracheobronchial zone deposition efficiency, from data of Lippmann (1977), Chan and Lippmann (1980), and Stahlhofen et al. (1980).

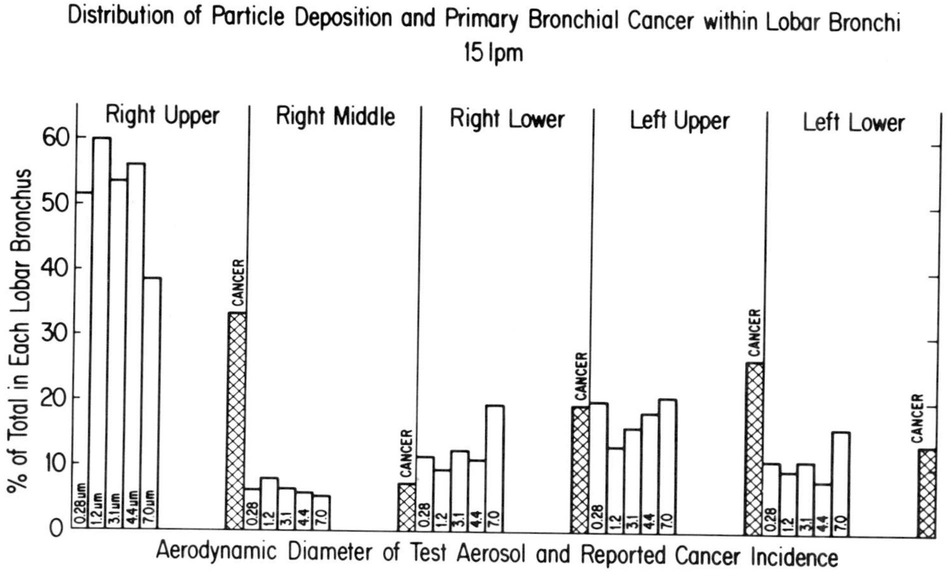

FIGURE 6. Percent of total deposition of lobar bronchi occurring in each lobar bronchus of a hollow human airway cast for a constant inspiratory flow of 15 l/min, and reported distribution of primary bronchial cancer among the lobar bronchi. [From Schlesinger and Lippmann (1978).]

various lobar bronchi remain relatively constant and appear to be closely associated with reported cancer incidence.

As an impactor, the human TB region is much more efficient than the oral airways, but somewhat less efficient than the nasal airways. Thus, particle deposition within the lungs for 1–10-μm particles is very much dependent on whether the individual breathes through the nose or mouth. For example, at an average inspiratory flow rate of 30 l/min, there would be approximately 80% penetration of 5-μm particles to the lungs with mouth breathing, and only approximately 25% with nasal breathing (Lippmann, 1977).

There are very few data on interspecies differences in nasal deposition efficiency. Those that involve comparisons of different species under identical conditions indicate substantial interspecies variation in penetration to the lungs. For example, deposition in the upper respiratory tract of dogs, guinea pigs, and rats averaged 16, 28, and 44% (of total respiratory tract deposition), respectively, for aerosols of 0.4–1 μm inhaled via the nasal passages (Raabe et al., 1977). Table 1 presents available data for tracheobronchial deposition of comparable-sized particles in both humans and experimental animals.

Intrabronchial Deposition Patterns

Since 1970, the Institute of Environmental Medicine has used hollow airways casts for quantitative determination of deposition within specific

TABLE 1. Tracheobronchial Deposition in Various Species

Species	Aerodynamic diameter (μm)	Percent of total lung deposition occurring in the tracheobronchial tree			Reference
		Median	Mean	Range	
Human	0.8	19	—	8–29	Chan and Lippmann (1980)
	3	30	—	10–57	Lippmann (1977)
	6	65	—	28–92	Lippmann (1977)
Dog	0.5–0.6	~4	—	—	Cuddihy et al. (1973)
	2.5–3	~11	—	—	Cuddihy et al. (1973)
	6.6	~44	—	—	Cuddihy et al. (1973)
Donkey	0.3	25	—	18–33	Schlesinger (1975)
	3	58	—	49–68	Albert et al. (1968)
	5	70	—	68–75	Schlesinger et al. (1978)
	5	64	—	55–70	Schlesinger et al. (1979)
Hamster	0.5	—	6	—	Raabe et al. (1977)
	3.1	—	25	—	Raabe et al. (1977)
Rabbit	0.5	33	—	20–64	Leikauf (1978)
	3	29	—	17–55	Holma (1967)
	4.5	72	—	66–79	Chen and Schlesinger (1983)
	6	60	—	40–93	Holma (1967)
	6	77	—	54–80	Leikauf (1978)
Rat	0.5	—	5	—	Raabe et al. (1977)
	3.1	—	50	—	Raabe et al. (1977)

airways of the tracheobronchial tree, providing the first detailed quantitative descriptions of the effects, in a realistic airway geometry of:

1. Jet flow through the larynx on deposition pattern within the trachea and on the tracheal bifurcation, and deposition efficiency in subsequent airway generations (Chan et al., 1980; Schlesinger and Lippmann, 1976; Schlesinger et al., 1977).
2. Effects of particle size, particle electrical charge, inspiratory flow rate, and flow pattern on deposition efficiencies within the trachea and bronchi during inspiratory flow (Chan et al., 1980; Schlesinger et al., 1977; Schlesinger et al., 1982a).
3. Effects of particle size and flow rate on deposition patterns and efficiencies within the trachea and larger bronchi during expiratory flow (Schlesinger et al., 1982b).

These kinds of data cannot be obtained from *in vivo* tests, where the airways are inaccessible for detailed measurements and where clearance processes would move the particles before they could be measured. Nor can they be obtained from tests in idealized tubular airway models, where the walls are smooth, and where the cross-sectional area changes at and near the airway bifurcations cannot realistically be simulated (Chan et al., 1980; Schlesinger, 1980). Our work on hollow cast systems began with the production of casts extending to 5-mm-diameter airways. Subsequent studies used a cast extending a 2-mm-diameter airways, and utilized casts of the larynx to produce more realistic entry conditions. Our most recent studies have utilized replicate casts, which could be cut up into specific bifurcation and length segments, allowing more precise measurement of distribution by segment as well as microscopic analysis of the surface density of deposition in the vicinity of the first bifurcation. In these studies, we also examined differences in deposition between constant and cyclic inspiratory flows, and deposition during constant expiratory flows.

Influence of cyclic flow on deposition efficiency and pattern in casts of the human tracheobronchial tree. Gurman (1983) performed a series of hollow cast inhalation studies with 3-μm and 8-μm mass median aerodynamic diameter (MMAD) ^{99m}Tc-tagged Fe_2O_3 microspheres produced by a spinning disc aerosol generator (Lippmann and Albert, 1967). Tests were performed for each particle size at constant inspiratory flows of 15, 30, and 60 l/min, and with cyclic inspiratory flows that averaged 15, 30, and 60 l/min. All tests were performed using a variable-orifice larynx (Gurman et al., 1980), with a replicate, hollow airway cast extending to 3-mm-diameter airways. Each replicate was made from the same solid master tracheobronchial (TB) cast, which was itself made from a normal human lung. Each cast was cut up after the inhalation into appropriate airway bifurcation and length segments, permitting accurate quantification of deposition efficiency in each segment by gamma-counting in a scintillation-well counter, and, in addition, deposition surface pattern at bifurcations by microscopic counting of particles. The results of the surface density analyses will be discussed in the next section.

Figure 7 shows a summary of deposition efficiency by airway generation for some of these tests. The efficiency at each generation increases with both flow rate and particle size, as would be expected for this system and these particle sizes, where inertial effects have a major influence. In addition, deposition efficiencies are substantially higher, in all of the bronchial airways, for cyclic flow than for constant flow. However, the reverse is true for deposition in the trachea for the 3-μm particles at 30 and 60 l/min.

Particle deposition during exhalation in casts of the human TB tree. Very little information is available regarding the deposition of inhaled particles in the human TB tree during exhalation. A simple dual hollow cast system was developed to initiate such studies (Schlesinger et al., 1983). The major findings were:

1. Exhalation deposition is less efficient than inhalation deposition, with the ratio of exhalation to inhalation efficiency varying from 0.62 to 0.82.

2. Deposition in the inhalation cast was enhanced for several centimeters distal from the larynx, and at the tracheal bifurcation.
3. Deposition on exhalation is enhanced for several centimeters proximal to the tracheal bifurcation, and in the main bronchi proximal to the main bronchial bifurcation. Thus, particle deposition by inertial impaction is apparently enhanced downstream of bifurcations during both inhalation and exhalation.

It is likely that differences in branching patterns between humans and experimental animals would result in different patterns of intrabronchial deposition. However, the only direct investigation of intrabronchial particle deposition patterns in hollow casts of bronchial trees which show clearly different geometries was performed by Schlesinger (1980). Figure 8 shows deposition efficiencies obtained in generations 0 through 5 of the tracheobronchial tree of the donkey during inspiratory flow at 10 l/min.

MUCOCILIARY CLEARANCE

The following discussion reviews the important anatomical, histological, and physiological features of the tracheobronchial tree that relate to the clearance of secretions and inhaled deposited substances.

Morphological Description of the Tracheobronchial Airways

The tracheobronchial tree extends from the larynx to the terminal bronchioles, the airways where proximal mucociliary transport begins. The epithelium at this distal level in humans consists of ciliated cells dispersed among Clara cells, brush cells, epithelial serous cells, K cells, globule leukocytes, and basal cells (Breeze and Wheeldon, 1975). The cilia are shorter and sparser here than in the larger airways (Reid, 1973).

Of all regions of the bronchial tree, the bronchioles may be at special risk. There are several reasons for this. A small increase in the volume of cellular secretions or fluid transudation in distal airways could lead to a rapid increase in the thickness of the fluid layer in more proximal airways. Also, the airway fluid lining in small airways may be much more permeable to inhaled particles and vapors than is the more continuous, thicker, and more viscous mucus of the larger airways. Additionally, retention times for particles deposited in small airways are greater.

In airways between approximately 0.75 and 3 mm, the ciliated epithelium becomes continuous, but is still only 20 to 30 μm deep. The ciliated cells are more numerous and occupy approximately half of the surface. Clara cells are sparser, and goblet cells, which secrete a relatively viscous mucus, are more numerous. According to Ebert and Terracio (1975) and Sturgess (1977), these airways are covered with a continuous mucus sheath, a view that is disputed by Van As (1977; Van As and Williams, 1983). The viscous mucus contains loosely intertwined glyco-

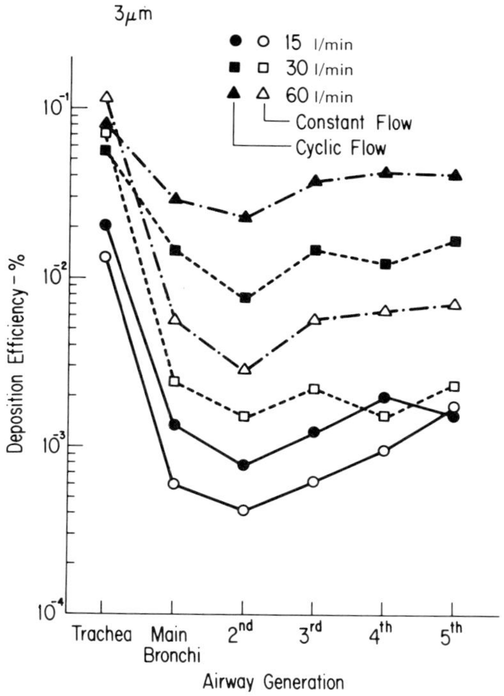

FIGURE 7. Deposition efficiencies during inspiratory flow in a human upper tracheobronchial hollow lung cast at various airway generations for 3-μm monodisperse particles at various flow rates. [From Lippmann et al. (1983).]

poprotein strands and is transported on the tips of the cilia, which beat in an underlying, less viscous fluid. This lower sol layer is thought to consist of serum transudate and secretions from Clara cells.

The ciliated epithelium from the subsegmental bronchi proximally is fully developed pseudostratified columnar, 50 to 70 μm thick (Gastineau et al., 1972). Mucus is produced largely by goblet cells and submucosal glands, the latter lying in juxtaposition to the cartilaginous plates and rings of these airways; the ducts from these glands perforate the epithelium, with a surface density of about 1/mm^2. It has been postulated that most of the mucus is secreted by these glands, as they have 40 times the volume of the goblet cells (Reid, 1970). The mucus in these airways is about 5 μm thick, overlying the tips of the cilia, and is transported at 1 to 5 mm/min in the trachea (Yeates et al., 1975; Foster et al., 1976, 1980).

The relative numbers and distributions of the various tracheobronchial epithelial cells and submucosal glands differ between species (Jeffery, 1982; Mariassys and Plopper, 1983). Such variation may result in differences in the nature of the mucous lining—e.g., relative ratio of acidic to neutral glycoproteins—of various species, as well as differences in the TB

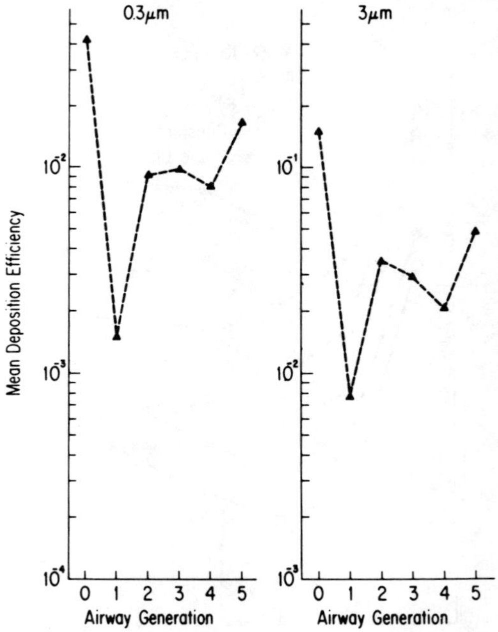

FIGURE 8. Mean fractional deposition efficiencies obtained in a cast of a donkey, for two particle sizes at a flow rate of 10 l/min.

response to toxic agents. Thus, for example, Mariassy and Plopper (1983) suggest that differences in the distribution of basal cells could result in response differences, since the cells are the primary stem cell for other epithelial cells.

Control of Mucociliary Transport

The rate of mucus transport depends on the viscosity, elasticity, and depth of the upper viscoelastic mucus layer and on the viscosity and depth of the periciliary layer (Ross and Corrsin, 1974). Additionally, the beat rate of the cilia and the effectiveness of each cycle are important. Each of these variables has a number of control mechanisms that can alter its function. Drugs that affect the autonomic nervous system have been demonstrated to alter mucus transport (Yeates et al., 1975; Foster et al., 1976; Camner et al., 1974, 1976; Pavia and Thomson, 1971; Forbes, 1976; Landa et al., 1975). Mucus transport may also be altered by agents that change epithelial permeability via changes in chloride active transport (Yeates et al., 1973). Other agents may have a direct toxic effect on the cilia or the secretory apparatus. The extent to which mucus transport is reversibly or irreversibly changed as a result of either disease or chronic exposure to air contaminants is only beginning to be elucidated.

Kinetics of Mucociliary Clearance

Since the mucociliary transport system is sensitive to inhaled irritants and to many other agents, normal mucociliary clearance can only be determined using techniques that do not themselves perturb the system. The best data have been generated using external radiation detectors to measure the retention and transport of inhaled radionuclide labelled aerosols, which are insoluble and delivered without extraneous carrier gases. Clearance rates vary widely among nonsmoking, healthy humans, even when the amounts cleared are comparable. On the other hand, they are quite reproducible in a given individual when the same particle size is inhaled, and they vary systematically with changes in particle size and concomitant changes in deposition pattern.

Normal tracheal mucociliary transport rates (TMTRs) in humans observed by Yeates et al. (1981) were 4.8 ± 3 mm/min, while Foster et al. (1980) observed mean tracheal velocities of 5.5 mm/min and a mean velocity of 2.4 mm/min in the major bronchi. TMTRs in a variety of experimental animal species are greater, with reported means in rats, cats, dogs, and donkeys being in the range of 10–15 mm/min (Barclay and Franklin, 1937; Dalhamn, 1956; Carson et al., 1966; Martin and Morrow, 1969).

The wide interindividual variation of mucus transport is not likely to be due to variations in ciliary beat, which in almost all measurements falls within the range of 1000 to 1200 beats/min. If the effective stroke of a cilia is 5 μm and it beats at 1000/min, mucus will be transported at 5 mm/min. This rate corresponds to the average upper airway values measured by Yeates et al. (1981) and Foster et al. (1980).

Inter- and intraindividual variations in tracheal transport rates of 75% and 25% could be related to similar inter- and intraindividual differences in mucus viscosity, which are 60% and 30%, respectively (Lopez-Vidriero and Reid, 1978). This, in turn, may be related to the variations of mucus type within cells. However, Ross and Corrsin (1974) have shown, theoretically, that relatively large changes in mucus viscosity result in small changes in transport rates, but that mucociliary transport is very sensitive to changes in the depth and viscosity of the periciliary layer. Thus, substances that optimize this layer may increase mucus transport.

It is clear that there is a wide variation in the rate at which seemingly healthy individuals clear deposited radioaerosols from the lungs. In cigarette smokers and persons with chronic obstructive pulmonary disease, there is a somewhat wider variation, as shown by Albert et al. (1973) and Gongora et al. (1981). In such subjects, the within-subject variation is also greatly increased. This suggests that loss of control of mucociliary transport could cause and/or result from chronic obstructive lung disease.

Pollutants Affecting Mucociliary Transport

Cigarette smoke. Cigarette smoke affects mucociliary transport in humans (Albert et al., 1971a, 1973, 1975; Camner and Philipson, 1972;

Camner et al., 1973; Sanchis et al., 1971; Pavia et al., 1971), but the reported effects have included both accelerated and slowed clearance. Some confusion has arisen because of the differences in dose and dose rate in studies focussed on the short-term effects of recent smoke exposures. Also, previous chronic smoking can affect airway structure and the capacity of the mucociliary system to respond to further short-term exposures. Since cigarette smoke is a complex mixture of particles, gases, and condensible vapors, its physical and chemical composition changes rapidly before and after it is inhaled. Furthermore, it has physiological and toxic effects on the airways, such as bronchoconstriction, increased mucus secretion, reduced ciliary motility, desquamation of surface epithelium, etc. Thus, in studying the effects of smoking on bronchial mucociliary clearance, it is important to differentiate the short-term effects produced by recent smoking from the effects resulting from chronic exposure.

Work performed in our laboratory has described the short-term effects of the inhalation of cigarette smoke on both humans and donkeys. In humans, exposure to the smoke from 2 to 7 cigarettes resulted in a twofold reduction in the time required to complete 90% of the bronchial clearance ($T_{90\%}$) in both 9 nonsmokers and 6 smokers (Albert et al., 1975). In 3 donkeys, exposure to the whole fresh smoke from 2 cigarettes had a similar effect, while short-term exposures to the whole smoke from 10 cigarettes produced a transient slowing of upper bronchial mucus transport lasting less than 15 min, and whole smoke from more than 15 cigarettes extended the duration of bronchial clearance by a factor of 2 or more (Albert et al., 1974). With still higher doses of cigarette smoke in both short- and long-term exposures, clearance can be temporarily stopped (Albert et al., 1971a,b).

Chronic high-level exposure to whole smoke from 30 cigarettes, 3 times/wk, for 6–8 mo, produced some of the same bronchial clearance abnormalities in donkeys that had previously been observed in humans who were long-term heavy smokers. These included: (1) intermittent retrograde mucus flow in the trachea, (2) intermittent clearance, in which periods of clearance stasis alternate with abrupt drops in retention, and (3) long delays in the onset of clearance followed by rapid clearance (Albert et al., 1971a,b, 1973). After the end of the exposure series, mucociliary clearance from the deep lung was faster than in preexposure control tests in two animals, but slower in the third.

In human studies, we have also observed that cigarette smokers in abstinence had significantly slower bronchial clearance as compared to their clearance when they smoked *ad libitum*. This suggests that they have become dependent on cigarettes for the maintenance of their "normal" clearance rate, or their clearance rate, when in abstinence, is closer to the rate they would have had if they had never smoked. Camner et al. (1973; Camner and Philipson, 1972) have reported similar findings.

Ozone. Frager et al. (1979) exposed rats for 4 h to 1.2 ppm of ozone following tagged tracer aerosol particle deposition. The O_3 exposure caused a substantial delay in mucociliary clearance of the tracer. This delay was, however, eliminated by preexposure to 0.8 ppm of ozone 3 d prior to deposition of particles. When a 13-d period intervened between the ozone preexposure and challenge, a substantial delay was again seen. Thus, preexposure to ozone appeared to afford essentially complete protection at 3 d, and no protection by 13 d. In subsequent tests with 4-h exposures to 0.8 ppm O_3, Phalen et al. (1980) observed a similar effect, which was amplified by high humidity.

Sulfur oxides: *Sulfur dioxide.* SO_2 is largely removed from the inspired air during its passage through the upper respiratory tract, and *in vivo* exposures to sulfur dioxide at concentrations that have been measured in ambient air are not likely to affect mucociliary clearance. At higher levels, more typical of some occupational exposures, effects have been observed. Wolff et al. (1975) exposed 9 nonsmokers to 5 ppm (13 mg/m^3) SO_2 for 3 h after exposure to a tagged albumin aerosol. The tracheobronchial mucociliary clearance of the tagged aerosol was essentially the same as in control tests, except for a transient acceleration at 1 h after the start of the SO_2 exposure. In further tests by Wolff et al. (1977), 5 ppm SO_2 during exercise speeded clearance significantly beyond that produced by the exercise alone.

High concentrations of $>$300 ppm SO_2 can slow bronchial clearance. In donkeys, 30-min exposures to SO_2 via nasal catheters produced delayed bronchial clearance, severe coughing, and mucus discharge via the nose (Spiegelman et al., 1968). Mean residence times of a tracer aerosol following exposure to 53 to 300 ppm were not significantly different from control levels. The one test performed at a lower concentration (27 ppm) produced an acceleration in bronchial clearance, which would be consistent with the clearance acceleration seen by Wolff et al. (1975, 1977) with 5 ppm exposures. These results—an acceleration at low concentrations and a slowing at higher levels of exposure—are similar to those produced by cigarette smoke.

Sulfur oxides: *Sulfuric acid.* Fairchild et al. (1975) showed that 4-h exposures to high concentrations of sulfuric acid (15 mg/m^3 of 3.2-μm CMD droplets) reduced the rate of mucociliary clearance of a tagged streptococcal aerosol from the lungs and noses of mice. At 1.5 mg/m^3 (0.6 μm CMD), there were no significant effects.

Schlesinger et al. (1978) demonstrated that 1-h exposures to submicrometer sulfuric acid mist at concentrations in the range of 200 to 1000 $\mu g/m^3$ produced transient slowing of bronchial mucociliary particle clearance in 3 of 4 donkeys. In addition, 2 of the 4 animals developed slowed baseline clearance rates after about 6 acid aerosol exposures, and slowed clearance persisted for several months after the last exposure. Corresponding exposures to ammonium sulfate had no measurable effect.

In subsequent tests, the 2 animals showing only transient responses, and 2 previously unexposed animals were exposed 1 h daily, 5 d/wk, to submicrometer sulfuric acid mist at 100 $\mu g/m^3$ (Schlesinger et al., 1979). Within the first few weeks of exposure, all four animals developed erratic clearance rates, i.e., rates that on specific test days were either significantly slower than, or significantly faster than, those in the preexposure period. The 2 previously unexposed animals developed persistently slowed bronchial clearance during the second 3 mo of exposure and during 4 mo of follow-up clearance measurements, while the 2 previously exposed animals adapted to the exposures in the sense that their clearance times fell consistently within the normal range during the last few months of exposure. However, after the end of the exposure series, their clearance rates were significantly faster than in their preexposure control tests (Lippmann et al., 1982). These effects are illustrated in Fig. 9.

The effects of H_2SO_4 on bronchial mucociliary clearance have also been studied in rabbits (Chen and Schlesinger, 1983). Single 1-h oral exposures to submicrometer droplets produced a clear-cut exposure-response relationship, as illustrated in Fig. 10, characterized by substantial and significant acceleration of clearance at low concentrations and significant retardation at high levels.

In additional tests performed using rabbits to assess the relative irritant potency of NH_4HSO_4, $(NH_4)_2SO_4$, and Na_2SO_2 on bronchial mucociliary clearance, only NH_4HSO_4 produced a significant change in clearance time, and only at an exposure concentration of ~2,000 $\mu g/m^3$ (Schlesinger, 1984). These results indicate that H_2SO_4 is the most potent of the three major ambient sulfur oxide aerosols—i.e., H_2SO_4, NH_4HSO_4, and $(NH_4)_2SO_4$—in terms of producing a change in the time of bronchial mucociliary clearance of tracer particles. In addition, the results suggest a retention between the hydrogen-ion concentration $[H^+]$ and extent of bronchial mucociliary clearance alteration.

The effective functioning of mucoiary transport depends upon optimal beating of the cilia and the presence of mucus having appropriate physicochemical properties. Both ciliary beating as well as mucus viscosity may be affected by the deposition of acid on the conducting airways. Normally, tracheobronchial mucus has a pH of 6.5–8.2 (Kwart et al., 1963; Guerrin et al., 1971; Gatto, 1981; Holma et al., 1977). *In vitro* studies have shown that at alkaline pH mucus is more fluid than at acid pH (Breuninger, 1964). Thus, an increase in viscosity that may occur due to deposited acid could "stiffen" the mucus blanket, promoting the clearance mechanism and increasing the efficiency of removal of the tracer particles (Holma et al., 1977). This may occur at low acid exposure concentrations, where ciliary activity would not be affected, and is consistent with the clearance acceleration observed at low concentrations of H_2SO_4.

At higher exposure concentrations, H_2SO_4 may affect ciliary beating. Schiff et al. (1979) and Grose et al. (1980) found that 2–3-h *in vivo*

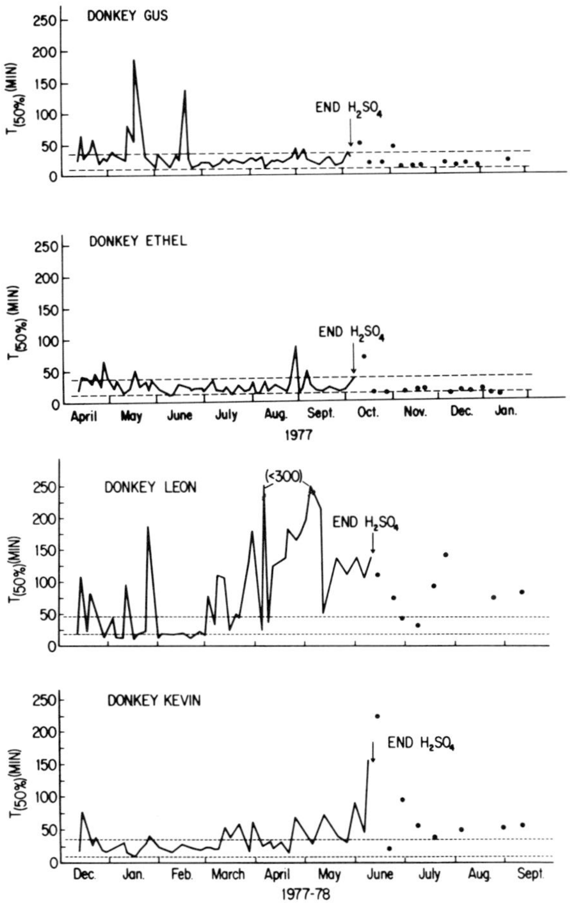

FIGURE 9. Effect of daily 1-h exposures to 0.5 μm H_2SO_4 at 100 $\mu g/m^3$ on $T_{\frac{1}{2}}$ in four donkeys. The dashed lines indicate the 95% tolerance intervals for $T_{\frac{1}{2}}$ based on 15 control tests on each animal, which preceded the H_2SO_4 exposures.

exposures of hamsters to 900–1000 $\mu g\ H_2SO_4/m^3$ resulted in a depression of ciliary beating frequency in tracheal explants prepared after exposure. *In vitro* studies clearly indicate that complete ciliostasis will occur if the pH is low enough, however, ciliostasis in only some regions, presumably with a change in clearance function, may occur at pH values above this critical value. Thus, $[H^+]$-induced depression of ciliary beating, with or without a change in mucus viscosity, could account for the observed retardation of clearance of the tracer aerosol at higher exposure concentrations of H_2SO_4 or NH_4HSO_4. The lack of any significant effect upon mucociliary clearance due to $(NH_4)_2SO_4$ or Na_2SO_4 at concentrations that would provide an amount of $[SO_4^{-2}]$ equivalent to the effective

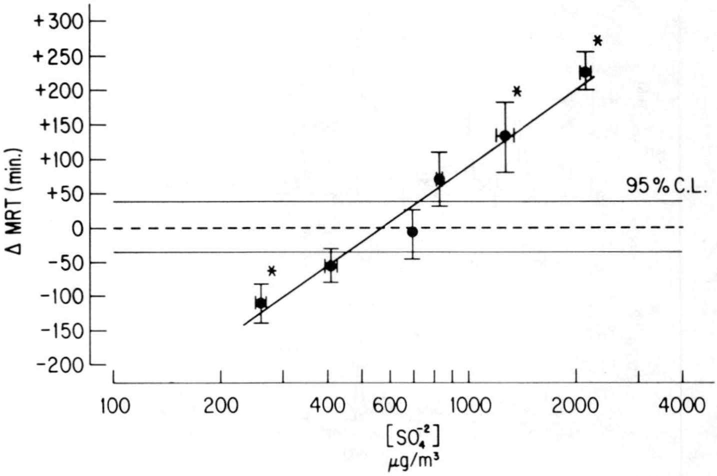

FIGURE 10. Effects of 1-h exposures to submicrometer-sized H_2SO_4 via mouthpiece at the indicated concentrations on the mean residence time (MRT) of γ-tagged Fe_2O_3 particles in the lungs of rabbits. The solid lines indicate the 95% tolerance intervals for MRT, based on control tests.

concentrations of the acid sulfates demonstrates that the SO_4^{-2} component is not responsible for the observed effects of NH_4HSO_4 or H_2SO_4, and that NH_4^+ does not mask or mitigate any potential response that may have been due to SO_4^{-2}.

The rabbit model was also used to examine the response to repetitive H_2SO_4 exposures (Schlesinger et al., 1983). Groups of 5 rabbits were exposed to submicrometer H_2SO_4 for 1 h/d, 5 d/wk, for 4 wk, during which time mucociliary clearance was monitored by external *in vivo* measurements of tagged tracer aerosol retention. One group was exposed orally to 250 $\mu g/m^3$, another to the same concentration via the nose, and a third to 500 $\mu g/m^3$, also via nasal breathing. Clearance was accelerated on specific individual days during the course of the acid exposures, especially at 500 $\mu g/m^3$. In all exposure series, clearance was significantly faster, compared to preexposure controls, during a 2-wk follow-up period after acid exposures had ceased. At the end of this period, the rabbits were sacrificed, and histological sections were obtained from the tracheobronchial tree. Significantly increased epithelial thickness of small conducting airways, compared to sham exposure controls, occurred in rabbits exposed orally at 250 $\mu g/m^3$ or nasally at 500 $\mu g/m^3$, and additionally the lumen of the smallest airways of the former group was narrower than control. The number of airways containing epithelial secretory cells was also significantly greater in these acid exposure groups compared to sham controls. The only change in the rabbits exposed nasally at 250 $\mu g/m^3$ was a significant increase in the number of small airways having epithelial secretory cells. The histological alterations provide a basis for observed changes in clearance, and are similar to those found in chronic bronchitis in man and experimental animals. Differences in site and degree of histological response and degree of physiological change between the two groups exposed to identical acid concentrations appear to have been due to differences in exposure mode, and resultant effects on breathing pattern, aerosol size distribution, and H_2SO_4 concentration penetrating beyond the upper respiratory tract.

Phalen et al. (1980) exposed rats to 3.5 mg/m^3 of submicrometer $(NH_4)_2SO_4$, $Fe_2(SO_4)_3$, and H_2SO_4 at both a low relative humidity (RH) of 30–40% and a high RHS (>80%), with and without simultaneous exposure to 0.8 ppm O_3. High RH alone slowed bronchial clearance. Further slowing was seen with O_3 at high RH, while the largest effect was seen with O_3 and H_2SO_4 at high RH. Comparison of the results of these tests with those of Schlesinger (1984) in rabbits suggest that either the rat is less sensitive than the rabbit, or that the test procedure of Phalen et al. is less sensitive to irritant-induced changes in mucociliary clearance rates.

Tests on humans are, of course, limited to short-term exposures that produce transient responses. Newhouse et al. (1978) exposed healthy young adult nonsmokers to 0.5 μm H_2SO_4 at 1 mg/m^3 for 2.5 h in an exposure chamber, with 5 4-min periods of exercise during the first hour. The clearance of 3-μm tagged albumin aerosol at 2 h was increased by an average of approximately 15% in comparison to that observed in sham exposures to water mist. However, much of the difference was attributable to major changes among six subjects, and there was little or no response in the other four.

In a study of the effects of transient H_2SO_4 inhalation exposures on humans, Leikauf et al. (1981) exposed 10 healthy nonsmokers to 0.5-μm H_2SO_4 aerosol for 1 h at concentrations of 100, 300, and 1000 μg/m^3 via nasal breathing. Tagged Fe_2O_3 aerosol with a mass median aerodynamic diameter of 7.6 μm was inhaled approximately 10 min before each H_2SO_4 exposure. Lung retention of the deposited radioactivity was monitored by external scintillation detectors, while a tracheal probe containing six rectangular collimated scintillation detectors was used to determine the tracheal mucus transport rates (TMTRs) or boil of activity. No significant changes in group mean TMTRs were observed following H_2SO_4 exposure at any level, but bronchial mucociliary clearance was markedly altered in 6 of the 10 subjects. The results for one subject are illustrated in Fig. 11.

Following the exposure at 100 μg/m^3, the mean bronchial clearance half-time ($TB_{\frac{1}{2}}$) was decreased from 80 to 50 min, indicating a significant increase in the rate of particle clearance. Following exposure to 1000 μg/m^3, the mean $TB_{\frac{1}{2}}$ was significantly greater than the control mean. At both 300 and 1000 μg/m^3, the $TB_{\frac{1}{2}}$ values were much more variable than those in the sham and 100-μg/m^3 exposures, and the responses at 300 μg/m^3 included equal numbers showing substantially faster and substantially slower $TB_{\frac{1}{2}}$ times than in the control tests. There was a correlation ($r^2 = 0.81$) between the control values of TMTR and $TB_{\frac{1}{2}}$ when paired by subject, but this relationship was disrupted following the H_2SO_4 exposures. Mucociliary transport in the airways distal to the trachea was, therefore, affected more by the H_2SO_4 exposures than was transport in the trachea. The 4 individuals whose clearance times were not significantly affected by the initial series of H_2SO_4 exposures had the fastest bronchial clearance among the 10, and they were each given an additional test with an H_2SO_4 exposure of 1000 μg/m³ preceding the tagged Fe_2O_3 aerosol. Three of them responded with threefold or greater increases in $TB_{\frac{1}{2}}$. Thus, 9 of 10 subjects had substantial changes in bronchial clearance times following H_2SO_4 exposures.

The 7.6-μm Fe_2O_3 particles were primarily deposited in the larger bronchial airways, where submicrometer H_2SO_4 has very little deposition. To determine the extent that submicrometer H_2SO_4 aerosol affects clearance from the more distal ciliated airways, Leikauf et al. (1984) measured the clearance of a monodisperse 4.2-μm Fe_2O_3 aerosol, which deposits a greater fraction of particles in distal conductive airways, in 8 other healthy nonsmoking subjects. Bronchial mucociliary clearance was slower following 1-h nasal H_2SO_4 inhalations at 100, 300, and 1000 μg/m³ than after sham exposures, while mucociliary transport rates within the trachea and indices of respiratory mechanics were unchanged. A comparison of the effects of 1-h exposures at 100 μg/m³ on the clearance of 7.6- and 4.2-μm particles

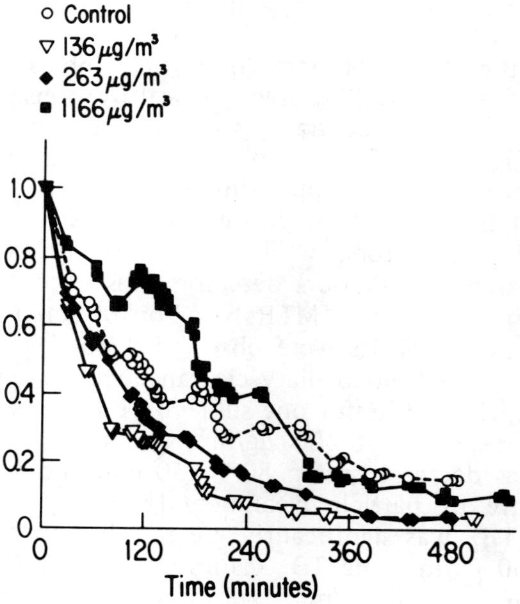

FIGURE 11. Effects of 1-h exposures to submicrometer-sized H_2SO_4 via nasal mask to a human nonsmoker on the tracheobronchial retention of γ-tagged Fe_2O_3 particles.

suggests greater physiological response in distal ciliated airways than in larger central airways.

The human studies on the effects of 1-h nasal inhalations of submicrometer H_2SO_4 aerosols on tracheobronchial mucociliary particle clearance were extended to a group of 10 asthmatic subjects, using 4-μm tagged Fe_2O_3 aerosol (Leikauf, 1981). The 6 subjects not on routine medication had similar mucociliary clearance response to those of healthy nonsmokers, with 1000 $\mu g/m^3$ producing a transient slowing of mucociliary clearance. The four asthmatics who were on daily medication exhibited stepwise mucociliary clearance that was too variable to permit detection of any H_2SO_4 effect on clearance. Although the magnitude of the mucociliary clearance response in the nonmedicated asthmatics was not greater than for the healthy nonsmoking volunteers, the asthmatics had lower baseline rates for bronchial clearance, which, combined with the observations of an effect on respiratory mechanics indices not seen in healthy people, suggests that asthmatic subjects may be more sensitive to H_2SO_4 aerosols than healthy persons.

As noted earlier, Newhouse et al. (1978) exposed exercising nonsmokers to H_2SO_4 at 1000 $\mu g/m^3$ and reported that the exposure produced an acceleration of bronchial clearance. The discrepancy may be partially accounted for by differences in the method of exposure and level of activity. In the Newhouse et al. study, the subjects were mouth breathing in an exposure chamber. Ammonia released by the chamber occupants could have resulted in partial neutralization of the H_2SO_4 to the more inert $(NH_4)_2SO_4$, as has been shown by Kleinman et al. (1981).

The absence of a consistent effect of H_2SO_4 on tracheal transport rates is supported by the results of other studies. Sackner et al. (1978) failed to find significant changes in tracheal mucus velocity following short-term exposures to 14 mg/m^3 (0.12 μm) in sheep, but did see reduced TMTRs for 1.5-μm H_2SO_4. We saw no effect on tracheal transport for 0.5-μm H_2SO_4 up to 1.4 mg/m^3 in donkeys (Schlesinger et al., 1978). On the other hand, Wolff et al. (1981) reported that tracheal transport rate was depressed in dogs exposed in a chamber for 1 h to 1.0 mg/m^3 (0.9 μm). Measurements of TMTR were made at 30 min, 1 d, 1 wk, and 5 wk after the end of the H_2SO_4 exposure by following the movement of a 10-μl droplet containing 50 μCi ^{99m}Tc tagged albumin inserted with a fiberoptic bronchoscope after administration of halothane. The TMTR at 30 min, 1 d and 1 wk were significantly depressed, but had returned to normal at 5 wk. At 0.5 mg/m^3, a significant depression was only seen at 1 wk. For 0.3-μm H_2SO_4 at 1.0 and at 5.0 mg/m^3, there were no significant effects on TMTR. Leikauf et al. (1981) recorded no significant change in the mean tracheal mucociliary transport rates in human studies with 0.5-μm H_2SO_4. Thus, effects on mucociliary transport in the trachea may depend more on particle size than on concentration, with the critical droplet size being >0.5 μm.

The effects of ambient H_2SO_4 exposures on mucociliary clearance may be influenced by the action of atmospheric co-contaminants. Last and Cross (1978) exposed rats for 3 and 24 d (23.5 h/d) to 0.5 μm H_2SO_4 at 1.1 mg/m^3 and to O_3 at 0.5 ppm and produced a significant increase (approximately 30%) in the secretion of mucus glycoproteins into the trachea, while exposures to each alone at these concentrations did not. When the H_2SO_4 concentration was lowered to 11 μg/m^3, the 3-d exposure to this plus 0.5 ppm O_3 still produced a 15% increase in secretion.

Significance of Changes in Tracheobronchial Clearance

The significance of the wide range of tracheobronchial clearance rates found in apparently healthy humans is not known. It has often been supposed that slow clearance in a healthy subject is detrimental, but such individuals may have greater reserve for coping with an increase in secretions resulting from the inhalation of irritant chemicals.

The mucociliary system presumably can cope with a considerable increase in secretions above the normal baseline level, but the relationship between mucus volume transport and surface transport rates is not known. Refluxing of mucus has been reported (Albert et al., 1973; Yeates et al., 1975). This could have been due to the mucus becoming too thick for the cilia to propel, or to loss of cilia, as has been observed in bronchitic rats (Iravani, 1971). In individuals with chronic bronchitis, mucus in the trachea and larger bronchi can be cleared by coughing, in contrast to healthy nonsmoking adults in whom cough appears to have little effect.

Interpretation of clearance alterations in terms of protential health problems is, at this time, still somewhat speculative. Although dysfunction of mucociliary transport may be involved in the pathogenesis of both acute and chronic respiratory disease, the clinical relevance of mucociliary changes is only recently being elucidated and its importance to pulmonary disease development is only beginning to be experimentally established, e.g., Schlesinger et al. (1983). Clinical studies of individuals with a disease condition characterized by impaired clearance—i.e., Kartagener's syndrome—may also be used to assess the effect of mucociliary dysfunction upon both acute and chronic respiratory disease, and to provide information on the role of mucociliary clearance in maintaining the integrity of the lung. The mucociliary dysfunction in Kartagener's syndrome is directly responsible for the development of recurrent respiratory tract infections, and eventually chronic bronchitis and bronchiectasis (Wanner, 1977).

Thus, mucociliary clearance plays an important role as a defense mechanism, and dysfunction may predispose to obstructive airway disease and respiratory infection. In this regard, it is interesting that the exposure-response trends observed for H_2SO_4 effects on bronchial clearance parallel those of previous studies conducted in our laboratory with cigarette smoke, a known causal factor in the complex etiology of chronic bronchitis. Bronchial clearance was markedly accelerated in both humans and donkeys exposed to the fresh smoke from several cigarettes (Albert et al., 1974, 1975). However, the fresh smoke from 10 or more cigarettes produced substantial retardation of bronchial clearance in donkeys (Albert et al., 1974). As indicated previously, studies in this laboratory have also known that repeated exposures to H_2SO_4 at 100 $\mu g/m^3$ produced a marked and persistent depression in bronchial clearance in the donkey comparable to that produced by chronic cigarette smoking, and did so in animals that demonstrated no transient effects following single exposures at 100 $\mu g\ H_2SO_4/m^3$.

Furthermore, the 4-wk series of 1-h daily exposures of rabbits at 250 and 500 $\mu g/m^3$ showed that modest changes in mucociliary clearance rates are associated with epithelial changes in small bronchi and bronchioles. The increases in thickness and secretory cell number are the kinds of responses that, if continued, could lead to clinical manifestations of bronchitis. Thus, although direct evidence of an association between chronic low level exposures to H_2SO_4 and chronic bronchitis is lacking, the similarity in responses between H_2SO_4 and cigarette-smoke exposures in terms of mucociliary clearance rates and epithelial morphology suggest that such an associated is possible.

The rate of mucociliary clearance may be a factor in the development of infectious diseases of the lung. The pharynx is continually colonized by potentially pathogenic organisms. For those reaching the lower conducting airways, these pathogens are confronted with the mucociliary barrier. In respiratory infection, especially in the pathogenesis of those due to viruses, the extent of penetration of disease vectors through the mucus to the underlying cells may be important in disease development. This rate, relative to the rate of mucociliary transport out of the lungs, could determine the capacity of, for example, inhaled viruses or bacteria to initiate disease (Proctor, 1979; Niedermann et al., 1983). It is known that in diseases characterized by retarded clearance from conducting airways—e.g., chronic bronchitis—there is a predisposition to respiratory infection (Cumming and Semple, 1980).

REFERENCES

Albert, R. E., Spiegelman, J. R., Lippmann, M., and Bennett, R. 1968. The characteristics of bronchial clearance in the miniature donkey. *Arch. Environ. Health* 17:50–58.

Albert, R. E., Lippmann, M., and Peterson, H. T. Jr. 1971a. The effects of cigarette smoking on the kinetics of bronchial clearance in humans and donkeys. In *Inhaled Particles* III, ed. W. H. Walton, pp. 165–168. Old Woking, Surrey: Unwin Bros.

Albert, R. E., Alessandro, D., Lippmann, M., and Berger, J. 1971b. Long-term smoking in the donkey: Effect on tracheobronchial particle clearance. *Arch. Environ. Health* 22:12–19.

Albert, R. E., Lippmann, M., Peterson, H. T. Jr., Berger, J., Sanborn, K., and Bohning, D. 1973. Bronchial deposition and clearance of aerosols. *Arch. Intern. Med.* 131:115–127.

Albert, R. E., Berger, J., Sanborn, K., and Lippmann, M. 1974. Effects of cigarette smoke components on bronchial clearance in the donkey. *Arch. Environ. Health* 29:99–106.

Albert, R. E., Peterson, H. T. Jr., Bohning, D. E., and Lippmann, M. 1975. Short-term effects of cigarette smoking on bronchial clearance in humans. *Arch. Environ. Health* 30:361-367.

Barclay, A. E., and Franklin, K. J. 1937. The rate of excretion of Indian ink injected into the lungs. *J. Physiol.* 90:482-484.

Breeze, R. G., and Wheeldon, E. B. 1975. The cells of the pulmonary airways. *Am. Rev. Respir. Dis.* 116:705-777.

Breuninger, H. 1964. Uber das physikalisch-chemische Verhalten des Nasenschleims. *Arch. Ohren-Nassen- und Kehlkopf.* 184:133-138.

Camner, P., and Philipson, K. 1972. Tracheobronchial clearance in smoking—discordant twins. *Arch. Environ. Health* 25:60-63.

Camner, P., Philipson, K., and Arvidsson, T. 1973. Withdrawal of cigarette smoking. A study of tracheobronchial clearance. *Arch. Environ. Health* 16:90-92.

Camner, P., Strandberg, K., and Philipson, K. 1974. Increased mucociliary transport by cholinergic stimulation. *Arch. Environ. Health* 29:220-224.

Camner, P., Strandberg, K., and Philipson, K. 1976. Increased mucociliary transport by adrenergic stimulation. *Arch. Environ. Health* 31:79-82.

Carson, S., Goldhamer, R., and Carpenter, R. 1966. Mucus transport in the respiratory tract. *Am. Rev. Respir. Dis.* 93(Suppl.):86-92.

Chan, T. L., and Lippmann, M. 1980, Experimental measurements and empirical modelling of the regional deposition of inhaled particles in humans. *Am. Ind. Hyg. Assoc. J.* 41:399-409.

Chan, T. L., Schreck, R. M., and Lippmann, M. 1980. Effects of the laryngeal jet on particle deposition in the human trachea and upper bronchial airways. *J. Aerosol Sci.* 11:447-459.

Chen, L. C., and Schlesinger, R. B. 1983. Response of the bronchial mucociliary clearance system in rabbits to inhaled sulfite and sulfuric acid aerosols. *Toxicol. Appl. Pharmacol.* 71:123-131.

Cumming, G., and Semple, S. J. 1980. *Disorders of the Respiratory System.* Oxford: Blackwell Scientific.

Cuddihy, R. G., Brownstein, D. G., Raabe, O. G., and Kanapilly, G. M. 1973. Respiratory tract deposition of inhaled polydispense aerosols in beagle dogs. *J. Aerosol Sci.* 4:35-45.

Dalhamn, T. 1956. Mucus flow and ciliary activity in the trachea of healthy cats and rats exposed to respiratory irritant gases. *Acta Physiol. Scand.* 123(Suppl.):9-161.

Davies, C. M. 1973. The deposition of aerosol in the human lung. *Aerosols in Physik, Medizine und Technik*, pp. 90-99. Bad Soden: Gesellschaft fur Aerosolforschung.

Ebert, R. R., and Terracio, M. J. 1975. Observation of the secretion on the surface of the bronchioles with the scanning electron microscope. *Am. Rev. Respir. Dis.* 112:491-496.

Emmett, P. C., Aitken, R. J., and Hannan, W. J. 1982. Measurements of the total and regional deposition of inhaled particles in the human respiratory tract. *J. Aerosol Sci.* 13:549-560.

Fairchild, G. A., Kane, P., Adams, B., and Coffin, D. 1975. Sulfuric acid and streptococci clearance from respiratory tracts of mice. *Arch. Environ. Health* 39:538-545.

Foord, N., Black, A., and Walsh, M. 1978. Regional deposition of 2.5-7.5 μm diameter particles in healthy male nonsmokers. *J. Aerosol Sci.* 9:343-357.

Forbes, A. R. 1976. Halothane depresses mucociliary flow in the trachea. *Anesthesiology* 45(1):59-63.

Foster, W. M., Bergofsky, E. H., Bohning, D. E., Lippmann, M., and Albert, R. E. 1976. Effect of adrenergic agents and their mode of action on mucociliary clearance in man. *J. Appl. Physiol.* 41:146-152.

Foster, W. M., Langenback, E., and Bergofsky, E. H. 1980. Measurement of tracheal and bronchial mucus velocities in man: Relation to lung clearance. *J. Appl. Physiol. Respirat. Environ. Exercise Physiol.* 48:965-971.

Frager, N. B., Phalen, R. F., and Kenoyer, J. L. 1979. Adaptation to ozone in reference to mucociliary clearance. *Arch. Environ. Health* 34:51-57.

Gastineau, R. M., Walsh, P. J., and Underwood, M. 1972. Thickness of bronchial epithelium with relation to exposure to radon. *Health Phys.* 23:857-860.

Gatto, L. A. 1981. pH of mucus in rat trachea. *J. Appl. Physiol.: Respirat. Environ. Exercise Physiol.* 50:1224-1226.

Gongora, G., Roy, M., Gongora, R., Drutel, P., Gaillard, J. P., and Jammet, H. 1981. Broncho-pulmonary clearance of radioactive labelled particles among 90 normal and pathological subjects. Preprint: *International Symposium on Deposition and Clearance of Aerosols in the Human Respiratory Tract,* May 22, 23. Bad Gleichenberg, Austria.

Grose, E. S., Gardner, D. E., and Miller, F. J. 1980. Response of ciliated epithelium to ozone and sulfuric acid. *Environ. Res.* 22:377–385.

Guerrin, F., Voisin, C., Macquet, V., Robin, H., and Lequien, P. 1971. Apport de la pH metric bronchique. *Prog. Respir. Res.* 6:372–383.

Gurman, J. L., Schlesinger, R. B., and Lippmann, M. 1980. A variable-opening mechanical larynx for use in aerosol deposition studies. *Am. Ind. Hyg. Assoc. J.* 41:678–680.

Holma, B. 1967. Short-term lung clearance in rabbits exposed to a radioactive bi-disperse (6 and 3 μm) polystyrene aerosol. In *Inhaled Particles and Vapours,* II, ed. C. N. Davies, pp. 189–195. Oxford: Pergamon.

Holma, B., Lindegren, J., and Andersen, J. M. 1977. pH effects on ciliomotility and morphology of respiratory mucosa. *Arch. Environ. Health* 32:216–226.

Iravani, J. 1971. Clearance function of the respiratory ciliated epithelium in normal and bronchitic rates. In *Inhaled Particles* III, ed. W. H. Walton, vol. I, pp. 143–146. Old Woking, Surrey, England: Unwin Bros.

Jeffery, P. K. 1982. The normal structure of bronchial epithelium. In *The Lung in Its Environment,* eds. G. Bonsignore and G. Cumming, pp. 57–72. New York: Plenum.

Kleinman, M. T., Bailey, R. M., Chang, Y-T.C., Clark, K. W., Jones, M. P., Linn, W. S., and Hackney, J. D. 1981. Exposures of human volunteers to a controlled atmospheric mixture of ozone, sulfur dioxide and sulfuric acid. *Am. Ind. Hyg. Assoc. J.* 42:61–69.

Kwart, H., Moseley, W. W., and Katz, M. 1963. The chemical characterization of human tracheobronchial secretion: A possible clue to the origin of fibrocystic mucus. *Ann. N.Y. Acad. Sci.* 106:709–721.

Landa, J., Hirsch, J., and Lebeaux, M. I. 1975. Effects of topical and general anesthetic agents on tracheal mucus velocity of sheep. *J. Appl. Physiol.* 38:946–948.

Lapp, N. L., Hankinson, J. L., Amandus, H., and Palmes, E. D. 1975. Variability in the size of airspaces in normal human lungs as estimated by aerosols. *Thorax* 30:293–299.

Last, J. A., and Cross, C. E. 1978. A new model for health effects of air pollutants: Evidence for synergistic effects of mixtures of ozone and sulfuric acid aerosols on rat lungs. *J. Lab. Clin. Med.* 91:328–339.

Leikauf, G. D. 1978. Particle deposition and clearance in the rabbit respiratory tract. M.A. thesis, New York University.

Leikauf, G. 1981. The effects of sulfuric acid aerosol on respiratory mechanics and mucociliary particle clearance in healthy and asthmatic adults. Ph.D. thesis, Graduate School of Arts and Sciences, New York University.

Leikauf, G., Yeates, D. B., Wales, K. A., Spektor, D., Albert, R. E., and Lippmann, M. 1981. Effects of sulfuric acid aerosol on respiratory mechanics and mucociliary particle clearance in healthy nonsmoking adults. *Am. Ind. Hyg. Assoc. J.* 42:273–282.

Leikauf, G. D., Spektor, D. M., Albert, R. E., and Lippmann, M. 1984. Dose-dependent effects of submicrometer sulfuric acid aerosol on particle clearance from ciliated human lung airways. *Am. Ind. Hyg. Assoc. J.,* in press.

Lippmann, M. 1977. Regional deposition of particles in the human respiratory tract. In *Handbook of Physiology Section IX—Reactions to Environmental Agents,* eds. D. H. K. Lee, H. L. Falk, and S. D. Murphy, pp. 213–232. American Physiological Society, Bethesda, Md.

Lippmann, M., and Albert, R. E. 1967. A compact electric-motor driven spinning disc aerosol generator. *Am. Ind. Hyg. Assoc. J.* 28:501–506.

Lippmann, M., and Albert, R. E. 1969. The effect of particle size on the regional deposition of inhaled aerosols in the human respiratory tract. *Am. Ind. Hyg. Assoc. J.* 30:257–275.

Lippmann, M., Schlesinger, R. B., Leikauf, G., Spektor, D., and Albert, R. E. 1982. Effects of sulphuric acid aerosols on the respiratory tract airways. *Ann. Occup. Hyg.* 26:677–690.

Lippmann, M., Gurman, J., and Schlesinger, R. 1983. Role of particle deposition in occupational lung disease. In *Aerosols in the Mining and Industrial Work Environments,* Vol. I, eds. V. A. Marple and B. Y. H. Liu, pp. 119-138. Ann Arbor, Mich.: Ann Arbor Science.

Lopez-Vidriero, M., and Ried, L. 1978. Bronchial mucus in health and disease. *Br. Med. Bull.* 34:63-74.

Martin, M. G., and Morrow, P. E. 1969. Effect of changing O_2 and CO_2 levels on tracheal mucociliary transport rate. *J. Appl. Physiol.* 27:385-388.

Mariassy, A. T., and Plopper, C. G. 1983. Tracheobronchial epithelium of the sheep: I. Quantitative light-microscopic study of epithelial abundance and distribution. *Anat. Rec.* 205:263-275.

Matsuba, K., and Thurlbeck, W. M. 1971. The number and dimensions of small airways in non-emphysematous lungs. *Am. Rev. Respir. Dis.* 104:516-524.

Newhouse, M. T., Dolovich, M., Obminski, G., and Wolff, R. K. 1978. Effect of TLV levels of SO_2 and H_2SO_4 on bronchial clearance in exercising man. *Arch. Environ. Health* 33:24-31.

Niederman, M. S., Rafferty, T. D., Sasaki, C. T., Merrill, W. W., Matthay, R. A., and Reynolds, H. Y. 1983. Comparison of bacterial adherence to ciliated and squamous epithelial cells obtained from the human respiratory tract. *Am. Rev. Respir. Dis.* 127:85-90.

Pavia, D., and Thomson, M. L. 1971. Inhibition of mucociliary clearance from the human lung by hyoscine. *Lancet* 1:449-450.

Pavia, D., Thomson, M. D., and Pocock, S. J. 1971. Evidence for temporary slowing of mucociliary clearance in the lung caused by tobacco smoking. *Nature* 231:325-326.

Phalen, R. F., Kenoyer, J. L., Crocker, T. T., and McClure, T. R. 1980. Effects of sulfate aerosols in combination with ozone on elimination of tracter particles inhaled by rats. *J. Toxicol. Environ. Health* 6:797-810.

Proctor, D. R. 1979. Tests of airway defense mechanisms. In *The Lungs in the Transition between Health and Disease,* eds. P. E. Macklem and S. Permutt, pp. 227-241. New York: Marcel Dekker.

Raabe, O. G., Yeh, H-C., Newton, G. J., Phalen, R. F., and Velasquez, D. J. 1977. Deposition of inhalable monodisperse aerosols in small rodents. In *Inhaled Particles IV,* ed. W. H. Walton, pp. 3-21. Oxford: Pergamon.

Reid, L. 1970. Evaluation of model systems for study of airway epithelium, cilia, and mucus. *Arch. Intern. Med.* 126:428-434.

Reid, L. 1973. The landing site in airborne transmission and airborne infection. IV. In *International Symposium on Aerobiology,* eds. J. F. Pk.Vlers and K. C. Winkler. Utrecht: Oostheoh.

Ross, S. M., and Corrsin, S. 1974. Results of an analytical model of mucociliary pumping. *J. Appl. Physiol.* 37:333-340.

Sackner, M., Ford, D., Fernandez, R., Cipley, J., Perez, D., Kwoka, M., Reinhart, M., Michaelson, E. D., Schreck, R., and Wanner, A. 1978. Effects of sulfuric acid aerosol on cardiopulmonary function of dogs, sheep and humans. *Am. Rev. Respir. Dis.* 118:497-510.

Sanchis, J., Dolovich, M., Chalmers, R., and Newhouse, M. T. 1971. Regional distribution and lung clearance mechanisms in smokers and non-smokers. In *Inhaled Particles* III, ed. W. H. Walton, vol. I, pp. 183-191. Old Woking, Surrey, England: Unwin Bros.

Schiff, L. J., Bryne, N. M., Fenters, J. D., Graham, J. A., and Gardner, D. E. 1979. Cytotoxic effects of sulfuric acid mist, carbon particulates and their mixtures on hamster tracheal epithelium. *Environ. Res.* 19:339-354.

Schlesinger, R. B. 1975. The intrabronchial pattern of particle deposition. Ph.D. Dissertation, New York University.

Schlesinger, R. B. 1980. Particle deposition in model systems of human and experimental animal airways. In *Generation of Aerosols,* ed. K. Willeke, pp. 553-575. Ann Arbor, Mich.: Ann Arbor Science.

Schlesinger, R. B. 1984. Comparative irritant potency of inhaled sulfate aerosols. Effects on bronchial mucociliary clearance. *Environ. Res.,* in press.

Schlesinger, R. B., and Lippmann, M. 1972. Particle deposition in casts of the human upper tracheobronchial tree. *Am. Ind. Hyg. Assoc. J.* 33:237-251.

Schlesinger, R. B., and Lippmann, M. 1976. Particle deposition in the trachea: *In vivo* and in hollow casts. *Thorax* 31:678-684.

Schlesinger, R. B., and Lippmann, M. 1978. Selective particle deposition and bronchogenic carcinoma. *Environ. Res.* 15:424–431.

Schlesinger, R. B., Bohning, D. E., Chan, T. L., and Lippmann, M. 1977. Particle deposition in a hollow cast of the human tracheobronchial tree. *J. Aerosol Sci.* 8:429–445.

Schlesinger, R. B., Lippmann, M., and Albert, R. E. 1978. Effects of short-term exposure to sulfuric acid and ammonium sulfate aerosols upon bronchial airway function in the donkey. *Am. Ind. Hyg. Assoc. J.* 39:275–286.

Schlesinger, R. B., Halpern, M., Albert, R. E., and Lippmann, M. 1979. Effect of chronic inhalation of sulfuric acid mist upon mucociliary clearance from the lungs of donkeys. *J. Environ. Pathol. Toxicol.* 2:1351–1367.

Schlesinger, R. B., Gurman, J. L., and Lippmann, M. 1982a. Particle deposition within bronchial airways: Comparisons using constant and cyclic inspiratory flows. *Ann. Occup. Hyg.* 26:47–64.

Schlesinger, R. B., Concato, J. B., and Lippmann, M. 1982b. Particle deposition during exhalation: A study of replicate casts of the human upper tracheobronchial tree. In *Aerosols in the Mining and Industrial Work Environment*, eds. V. A. Marple and B. Y. H. Liu, pp. 165–176. Ann Arbor, Mich.: Ann Arbor Science.

Schlesinger, R. B., Naumann, B. D., and Chen, L. C. 1983. Physiological and histological alterations in the bronchial mucociliary clearance system of rabbits following intermittent oral or nasal inhalation of sulfuric acid mist. *J. Toxicol. Environ. Health* 12:441–465.

Spiegelman, J. R., Hanson, G. D., Lazarus, A., Bennett, R. J., Lippmann, M., and Albert, R. E. 1968. Effect of acute sulfur dioxide exposure on bronchial clearance in the donkey. *Arch. Environ. Health* 17:321–326.

Stahlhofen, W., Gebhart, J., and Heyder, J. 1980. Experimental determination of the regional deposition of aerosol particles in the human respiratory tract. *Am. Ind. Hyg. Assoc. J.* 41:385–398.

Sturgess, J. M. 1977. Mucous lining of major bronchi in the rabbit lung. *Am. Rev. Respir. Dis.* 115:819–827.

Van As, A. 1977. Pulmonary airway clearance mechanisms: A reappraisal. *Am. Rev. Respir. Dis.* 115:721–725.

Van As, A., and Williams, E. 1983. The distribution of tracheal mucus in four species of mammal. *Am. Rev. Respir. Dis.* 127(4:2):295 (abstr.).

Wanner, A. 1977. Clinical aspects of mucociliary transport. *Am. Rev. Respir. Dis.* 116:73–125.

Wolff, R. K., Dolovich, M., Rossman, C., and Newhoust, M. T. 1975. Sulfur dioxide and tracheobronchial clearance in man. *Arch. Environ. Health* 32:521–527.

Wolff, R. K., Dolovich, M., Obminski, G., and Newhouse, M. T. 1977. Effect of sulfur dioxide on tracheobronchial clearance at rest and during exercise. In *Inhaled Particles* IV, ed. W. H. Walton, pp. 321–330. Oxford: Pergamon.

Wolff, R. K., Muggenburg, B. A., and Silbaugh, S. A. 1981. Effect of 0.3 and 0.9 μm sulfuric acid aerosols on tracheal mucus clearance in beagle dogs. *Am. Rev. Respir. Dis.* 123:291–294.

Yeates, D. B., Aspin, N., Bryan, A. C., and Levison, H. 1973. Regional clearance of ions from the airways of the lung. *Am. Rev. Respir. Dis.* 107:602–608.

Yeates, D. B., Aspin, N., Levison, H., Jones, M. T., and Bryan, A. C. 1975. Mucociliary tracheal transport rates in man. *J. Appl. Physiol.* 39:487–495.

Yeates, D. B., Pitt, B. R., Spektor, D., Karron, G. A., and Albert, R. E. 1981. Coordination of mucociliary transport in the human trachea and intrapulmonary airways. *J. Appl. Physiol.: Resp. Environ. Exercise Physiol.* 51:1050–1064.

SIMILARITIES OF HOST DEFENSE MECHANISMS AGAINST PULMONARY INFECTIOUS DISEASES IN ANIMALS AND MAN

Gareth M. Green

Department of Environmental Health Sciences,
The Johns Hopkins University, School of Hygiene
and Public Health, Baltimore, Maryland

Evidence linking exposure to air pollutants with increased susceptibility to infectious diseases in humans comes from epidemiological, clinical, and experimental laboratory studies. The data suggest that the most common, and perhaps the most sensitive, index of the pulmonary effect of air pollutant exposure is on post upper respiratory infection, prolonged cough, phlegm, and purulent sputum. Experimental models of these relationships for extrapolation to humans should be able to measure such minor changes in symptomatology and physiology rather than require major lethal events. The bacterial aerosol model for quantifying nonspecific defense mechanisms of the bronchopulmonary tree utilizing nonpathogenic organisms fulfills this criterion. The function of the six major components of pulmonary antimicrobial defense mechanisms—including aerodynamic filtration, secretory respiratory tract fluid, fluid transport at the alveolar and bronchial levels, the phagocytic function of alveolar macrophages, the augmenting mechanisms of blood-derived inflammatory cells, and the secretory and cellular-specific immune mechanisms and their mediator products—can all be quantified by this experimental animal model system. The defensive functions are remarkably similar across animal species, and available human data suggest that findings obtained using the model may be extrapolated to humans.

INTRODUCTION

Evidence linking exposure to air pollutants with increased susceptibility to infectious disease in humans comes from three lines of epidemiological investigation: (1) increased morbidity and hospitalization of patients with chronic pulmonary disease associated with air pollution episodes (Lawther, 1958); (2) increased respiratory symptomatology of cough, phlegm, and "chest infection" in children correlated with air pollution indices (Speizer et al., 1980); and (3) long-term clinical studies of chronic bronchitis (Fletcher, 1968). Although the association of air pollution level with excessive hospitalization in chronic lung disease patients and respiratory symptomatology in children seems valid, the interpretation of this association as mediated through host defense mechanisms and pulmonary infection is conjectural. The increased cough and phlegm that often lead to hospitalization of patients with chronic pulmonary disease may be induced as an irritant response to air pollutant exposure rather than an infective response. It is not clear that the increased symptomatology in children associated with pollutant exposure is attributable to pulmonary infection. Although antibiotic treatment helps to control infective complications of these acute exacerbations, the inflammatory and secretory derangements responsible for symptoms are not clearly a complication of respiratory-tract infection. Furthermore, long-term clinical studies have demonstrated that control of infection through antibiotic chemoprophylaxis and therapy of acute infective complications of exacerbations have no effect on the progression of the underlying process of progressive airways obstruction (Tager and Speizer, 1975). In fact, this extensive review of epidemiologic studies performed over two decades comes to the conclusion that although acute exacerbations and pulmonary infection appear to be related, infection cannot be affirmed as causative.

RELEVANT MODELING

The significance of these observations and conclusions for the question at hand is that animal modeling of the human problem should model the relevant infective relationships with air pollution found in humans. Since, after 25 years of research, this relationship is still incompletely characterized, conclusions drawn from existing animal models must be accepted as applying to humans with considerable caution. In humans the association is with increased symptoms of cough, phlegm, and colds going into the chest. These are nonspecific symptomatic changes that could be attributable to irritant, allergic, or climatic as well as infectious etiologies. When infective episodes are apparent, they more often have the character of purulent bronchitis than pneumonia, and are related to autochthonous flora rather than newly acquired external pathogens. Such infections are nonlethal by replication of bacteria and invasion of tissues and the bloodstream. The underlying health effect, except where severe chronic disease is present, is characterized as a minor morbidity rather than a mortality incident.

The significance of these considerations for relevant animal models is clear. Experimental models of extrapolation should measure minor changes in symptomatology and physiology rather than major lethal events, should involve autochthonous rather than exogenous microbial flora, should measure physiologic derangement or minor morbidity rather than mortality, should not depend on the virulence of pathogenic microbial species, should involve infection as an outcome and not a cause of the physiologic derangement, and should concentrate on physiologic rather than pathologic measures of host defense (Colucci et al., 1973).

The bacterial aerosol model for quantifying nonspecific defense mechanisms of the bronchopulmonary tree utilizing nonpathogenic organisms fulfills all of these criteria. With the use of nonpathogenic organisms, minor changes in physiologic derangement of phagocytic or transport mechanisms can be quantified without coincident morbidity, pathologic inflammatory changes, or mortality, and the physiologic derangement precedes the infection rather than the other way around. Furthermore, the relevance of an exposure to invasive infection and pneumonia can be proven by the appropriate choice of the agent of exposure or the species of the test organism. The use of Gram-positive bacterial species will sense alterations of pulmonary macrophage function, whereas many Gram-negative organisms test the effectiveness of secondary polymorphonuclear leukocyte chemotaxis and phagocytosis (Pierce et al., 1977). The role of transport systems can be simultaneously quantified with suitable radiotracers, and the effectiveness of specific immune augmentive mechanisms may be examined as an additional variable. The model fulfills "the guidelines for experiments to predict the degree of safety of a material for man" as laid down by Weil (1972), in that (1) a variety of mammalian species handle the material biologically in an identical or very similar manner to humans, (2) dose-response testing is possible, (3) dose levels exist for humans and animals below which the adverse effect does not occur, (4) measurements can be made in groups or individuals and analyzed accordingly, and (5) effects are specific and different for different routes of administration.

DEFENSE MECHANISMS

With minor variations, a wide variety of mammalian species, including humans, share in common an array of defensive mechanisms anatomically and physiologically integrated in the respiratory tract to prevent and control most exposures to infective organisms. The integrated systems have the following components (Green, 1970).

First, an aerodynamic filtration mechanism dependent on the inertial impaction of suspended particulate and the absorption of gaseous material onto the respiratory mucosal surface as the inhaled airstream encounters turbulence in the nasal air passages and repeatedly changes direction as it courses through the pharynx and along the bifurcations of the tracheobronchial tree. This mechanism limits the penetration of particulate material to the so-called respirable size range, between about 0.5 and 3 μm in aerodynamic diameter. This mechanism is exceedingly effective in limiting total dosage to the respiratory membrane, since only finely divided particulates of small mass penetrate to that depth. It is therefore very difficult to exceed certain dose levels of particulate material to the respiratory membrane surfaces. Dosage to the respiratory tract is determined both by particle concentration and size, and also by the volume of air inhaled into the lungs in the unit time. Since tidal volume bears a predictable relationship to body size and weight, it tends to proportionalize the dose delivered to target tissue to the size of the animal across a wide range of species (Krasovskii, 1976). A similar relationship proportionalizes the dose of pollutant gaseous material according to body size of animal and strengthens the similarities of dose to target tissue across a wide range of animal species (Book, 1982). The anatomical structure of the airways and lungs is, however, not identical among mammalian species. Smaller species, for example, may have fewer generations of bronchi and a greater respiratory function in the distal bronchioles than larger species such as primates and man. Interlobular septae in primates may influence the pathways of fluid transport in the lung.

The secretory fluid layer of the respiratory tract is the second defensive system and is dependent primarily on the metabolic, synthetic, and secretory functions of the type II cells in the alveoli, the Clara cells, and the goblet-, serous-, and mucous-gland epithelium of the bronchial tree. This system is remarkably similar in a variety of animal species and serves to bind inhaled irritants, absorb and neutralize acidic or alkaline substances, and provide a vehicle for transport through and out of the lung. It also carries specific immunoglobulins, secreted locally by submucosal plasma cells, that immunologically bind foreign antigens to prevent their penetration and injury to underlying tissues. The secretions on the respiratory membrane may be augmented under conditions of irritant or other stresses by tissue- and plasma-based edema fluids, which can increase the volume of fluid available for the above functions and accelerate the transport process of substances out of the lung. Unfortunately, little is known in different animal species about factors governing the rate and volume of secretion, so that this is a difficult parameter with which to establish interspecies comparisons.

The third defensive system of the lung is the collection of fluid transport mechanisms present in the interstitium, on the alveolar surfaces, and on the surfaces of the tracheobronchial tree. Interstitial fluids are derived by the action of oncotic forces on the capillary circulation of the respiratory membrane. These fluids percolate through the interstitial space toward peribronchial, perivenous, paraseptal, and subpleural lymphatics, transporting with them particulate or other material that has penetrated through the epithelial boundary of the type I cells. The movement of the fluid layer along the alveolar membrane is not well understood, but is thought to involve alveolus-to-alveolus transport through pores of Kohn toward peribronchial, paraseptal, and subpleural positions. The energy for this fluid transport is not well understood, but may derive from pressure and tension gradients related to the elastic forces developed by ventilation and intrinsic in the gradients of surface activity dependent on the lipid surfactant secretions in the alveolus. These fluid transport pathways converge at the terminal bronchioles, where cells and deposited debris along with secretions are moved onto the mucociliary escalator and thence to the trachea, glottis, and to the gastrointestinal tract (Green, 1973). Transport rates, as measured by radioactive tracer clearance, are much faster on the mucociliary stream ($T_{\frac{1}{2}}$ = 30–90 min) than from the alveolar to the bronchiolar level of the lung ($T_{\frac{1}{2}}$ = 24 h). Late alveolar clearance is even slower. In addition, a small fraction of inhaled particles that fails to gain access to exit transport pathways is retained in the lung for years and perhaps for a lifetime. Although there are large individual variations, these kinetics appear to be quite comparable in a wide variety of animal species and in humans.

The fourth principal defense mechanism that operates primarily at the respiratory membrane level of the lung depends on the phagocytic properties of alveolar macrophages, which are the first cells to interact with inhaled bacteria (Green and Kass, 1964). Alveolar macrophages are large histiocytic mononuclear cells found in all species of air-breathing animals and comprise more than 95% of the free cells obtainable by bronchopulmonary lavage from healthy lungs. A greater admixture of polymorphonuclear leukocytes, lymphocytes, monocytes, eosinophils, and plasma cells may be obtained in a variety of different disease states. These findings are surprisingly consistent in all animal species tested to date, including humans. In addition to their phagocytic capacity, alveolar macrophages contain a wealth of proteolytic, enzymatic, and biochemical synthetic capabilities that enhance the power and diversity of their bactericidal mechanisms and elaborate chemotactic and other messenger substances that augment the defensive armamentarium of the alveolus or institute pathologic tissue reactions. Much of the genetic variation that occurs within and between species is expressed through defects or deficiencies in these biochemical cellular mechanisms. Furthermore, chemical and other environmental toxicants appear to exert their primary focus of impairment on these cellular systems.

The fifth set of defense mechanisms that protect the lung against infection are the inflammatory cells called in from the tissues and the bloodstream by chemotaxins inherent in the bacteria themselves or elaborated by interaction of the bacteria with the alveolar macrophages. These cells are the polymorphonuclear leukocytes and the mononuclear phagocytes from the bloodstream and are called in early against Gram-negative organisms and as a secondary response when alveolar macrophages fail to gain control of other bacterial species (Pierce et al., 1977). The participation of these blood-borne phagocytes can be recognized by the changing cellular constituents in bronchoalveolar lavage samples. These cells are thought to contribute to inflammatory damage caused in the lung by the release of proteolytic enzymes in the alveolus and in connective tissue spaces. They may be in part responsible for the changing symptomatology and respiratory dysfunction so frequently found in individuals with air-pollutant exposures. The inflammatory processes follow a variety of pathologic sequences well described in studies of pneumonia, bronchopneumonia, granulomatous disease, bronchiolitis, and other similar inflammatory pulmonary infections.

A sixth and most essential mechanism of defense of the bronchopulmonary tree is the augmentation of the natural defenses provided by secretory and cellular-specific immune cells and their mediator products (Kaltreider, 1976). The lymphocytic system accompanies the development of the bronchial tree with bronchial-associated lymphoid tissue occurring at bifurcations from the major bronchi to the terminal bronchiole. These nodules serve both as foci for the proliferation of lymphocytes and as nodal points between afferent and efferent lymphatics from the lung parenchyma. It is at this point that antigenic material transported to the bronchi from the lung parenchyma may first interact with the immunologic apparatus and generate an immune response to accelerate the resistance mechanisms of phagocytosis and bactericidal activity. As in other locations, the lymphocyte immune responses may generate immunopathology (Daniele et al., 1981), not only to damage tissues but to interfere with the normal defensive responses of the phagocytic mechanisms of the lung parenchyma (Jakab, 1982).

In healthy animals, bronchial-associated lymphoid tissue may be barely visible, particularly at the periphery of the bronchopulmonary tree. But in a number of chronic diseases that frequently infest these animals, the bronchial-associated lymphoid tissue becomes greatly hypertrophied and can be readily visualized at the bronchial bifurcations and in the submucosa along the bronchial epitheliaum. The precise function of the bronchial lymphoid tissue is not well understood, but it may be the site for local immunologic response to low levels of inhaled antigen and for transmitting the immunologic messages to the systemic immunological system.

An additional immunological system along the bronchial tree is the secretory immunoglobulin mechanism that depends on submucosal plasma cells, which secrete the dimeric immunoglobulin A, and the epithelial tissues, which add the bridging secretory piece as the immunoglobulin passes to the bronchial surface. Secretory immunoglobulin A acts to prevent penetration and injury by foreign antigens deposited on the respiratory epithelium, but its role in protection against infection is not really understood since it does not appear to function as an opsonin.

The immunologic responses facilitate phagocytosis and bactericidal activity by the elaboration of antigen-specific opsonins, which bind to inhaled microorganisms and acclerate the process of particle adhesion, ingestion, and intracellular digestion. A significant part of the effect of the immunologic response is the activation of alveolar macrophages, which, in response to a recognized antigen, elaborate intracellular metabolic and biochemical systems that lead to a hyperactive phagocytic response to specific antigen and enhance nonspecific activity against a variety of unrelated antigens (Kaltreider, 1976). It may be this macrophage activity that accounts for the more effective bactericidal response after aerosol than parenteral immunization. Antibodies are particularly important in response to viral infection, not only to accomplish the eradication of the viral agent but because the immunocytological response to cell-associated viral antigen can temporarily impair the effectiveness of the primary cellular responses to secondary bacterial invaders. This appears to be the mechanism whereby viral infection alters susceptibility to secondary bacterial infection in influenza and other species of viral pneumonias (Jakab, 1982).

TECHNIQUES FOR EXTRAPOLATION

The defense mechanisms of the bronchopulmonary tree show functional as well as anatomical similarities across a broad variety of mammalian species. While transport kinetics can be assessed *in vivo* by radiotracer-tagged particle techniques and by direct observation at bronchoscopy, and while phagocytic and immunologic responses can be determined on specimens isolated by bronchopulmonary lavage techniques, the most efficient means of evaluating the integrated systems in experimental animals is the use of the radiotracer-tagged fine-particle bacterial aerosol. Bacterial species are labeled during growth with a radioactive tracer in a defined ratio of tracer label activity to viable bacterial units. After aerosolization, the rate of clearance of the radiotracer measures the mucociliary and deep-lung clearance times, while the change in the ratio between the tracer and the viable bacterial count reflects the course of bactericidal activity or bacterial replication (Green and Goldstein, 1966). Clearance pathways can be quantified by measuring tracer label activity in lungs, bronchial lymph nodes, spleen, liver, or other systemic sites, and gastrointestinal contents. In small animal species such as rats and mice, this is most effectively accomplished by postmortem analyses of organs, while in larger species with suitable labels the clearance kinetics can be measured *in vivo*. Utilizing different tracer labels and selective bacterial growth media, mixed bacterial challenges can be given as is common in natural exposures to bacterial infection (Green and Green, 1968). The most notable advantage of this method is that it allows physiological assessment of bactericidal mechanisms in individual animals and thus permits the detection of individual variations in host responses to pollutant exposures.

Utilizing these techniques, a variety of animal species have been studied and show remarkably similar responses to bacterial challenge aerosols, although with significant variations between species in quantitative effectiveness of eradicating specific bacterial organisms (Jakab and Green, 1975). However, highly variable responses are found with different kinds and doses of toxic exposures (Goldstein et al., 1976). The major drawback of this method is that it has not been tested in humans, although its components of tracer transport and *in vitro* phagocytosis have been studied extensively. All the information available to date would indicate that extrapolation of data observed in animals to humans is valid qualitatively, and in many instances may be valid quantitatively. However, more careful clinical studies in humans are required before this statement can be made with certitude or generality.

REFERENCES

Book, S. A. 1982. Scaling toxicity from laboratory animals to people: An example with nitrogen dioxide. *J. Toxicol. Environ. Health* 9:719.

Colucci, A. V., Hammer, D. I., Williams, M. E., Hinners, T. A., Pinkerton, C., Kent, J. L., and Love, G. J. 1973. Pollutant burdens and biological response. *Arch. Environ. Health* 27:151.

Daniele, R. P., Henson, P. M., Fantone, J. C. III, Ward, B. A., and Dreisin, R. B. 1981. State of the art. Immune complex injury of the lung. *Am. Rev. Respir. Dis.* 124:738.

Fletcher, C. M. 1968. Bronchial infection and reactivity in chronic bronchitis. *J. R. Coll. Phys. (Lond.)* 2:183.

Goldstein, E., Jordan, G. W., MacKenzie, M. R., and Osebold, J. W. 1976. Methods for evaluating the toxicological effects of gaseous and particulate contaminants on pulmonary microbial defense systems. *Annu. Rev. Pharmacol. Toxicol.* 16:447.

Green, G. M. 1970. The J. Burns Amberson Lecture. In defense of the lung. *Am. Rev. Respir. Dis.* 102:691.

Green, G. M. 1973. Alveolobronchiolar transport mechanisms. *Arch. Intern. Med.* 131:109.

Green, G. M., and Goldstein, E. 1966. A method for quantitating intrapulmonary bacterial inactivation in individual animals. *J. Lab. Clin. Med.* 68:669.

Green, G. M., and Kass, E. H. 1964. The role of the alveolar macrophage in the clearance of bacteria from the lung. *J. Exp. Med.* 119:167.

Green, L. H., and Green, G. M. 1967. Selective multiplication of a single bacterial species in a mixed infection of the lung. *Antimicrob. Agents Chemother.* 7:67–69.

Jakab, G. J., and Green, G. M. 1975. Variations in pulmonary antibacterial defenses among experimental animals. *Infect. Immun.* 11:601.

Jakab, G. J. 1982. Immune impairment of alveolar macrophage phagocytosis during influenza virus pneumonia. *Am. Rev. Respir. Dis.* 126:778.

Kaltreider, B. H. 1976. State of the art. Expression of immune mechanisms in the lung. *Am. Rev. Respir. Dis.* 113:347.

Krasovskii, G. N. 1976. Extrapolation of experimental data from animals to man. *Environ. Health Perspect.* 13:51.

Lawther, P. J. 1958. Climate, air pollution and chronic bronchitis. *Proc. R. Soc. Med.* 51:262.

Pierce, A. K., Reynolds, R. C., and Harris, G. D. 1977. Leukocytic response to inhaled bacteria. *Am. Rev. Respir. Dis.* 116:679.

Speizer, F. E., Ferris, B. Jr., Bishop, Y. M. M., and Spengler, J. 1980. Respiratory disease rates and pulmonary function in children associated with NO_2 exposure. *Am. Rev. Respir. Dis.* 121:3.

Tager, J., and Speizer, F. E. 1975. Role of infection in chronic bronchitis. *N. Engl. J. Med.* 292:563.

Weil, C. S. 1972. Guidelines for experiments to predict the degree of safety of a material for man. *Toxicol. Appl. Pharmacol.* 21:194.

INDEX